AF708996

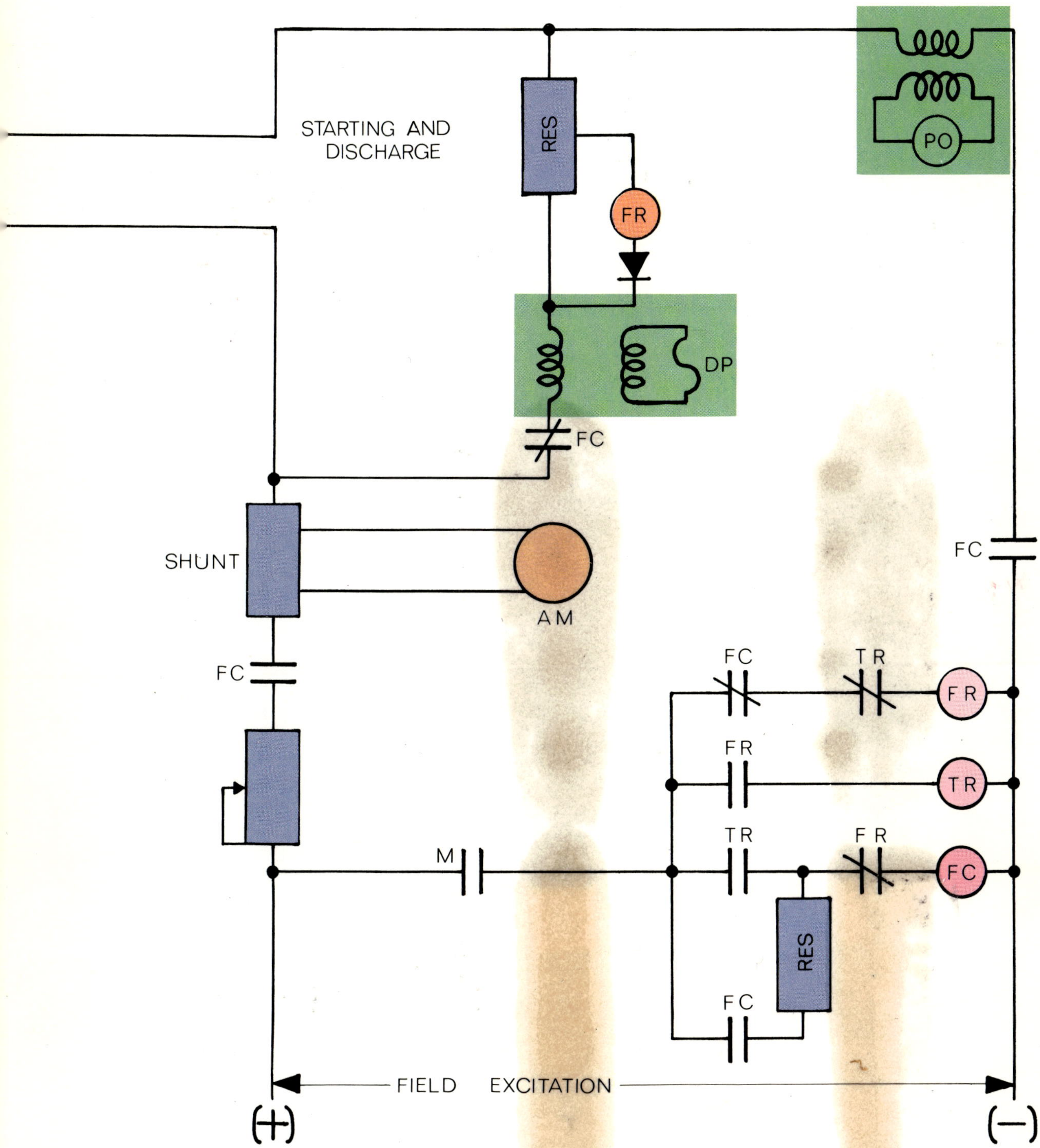
STARTING AND DISCHARGE
RES
FR
DP
PO
FC
SHUNT
AM
FC
M
FC
TR
FR
TR
TR
FR
FC
RES
FC
FIELD EXCITATION
(+)
(−)

A10202 629673

ELECTRICAL-ELECTRONIC DRAFTING AND DESIGN

WITHDRAWN

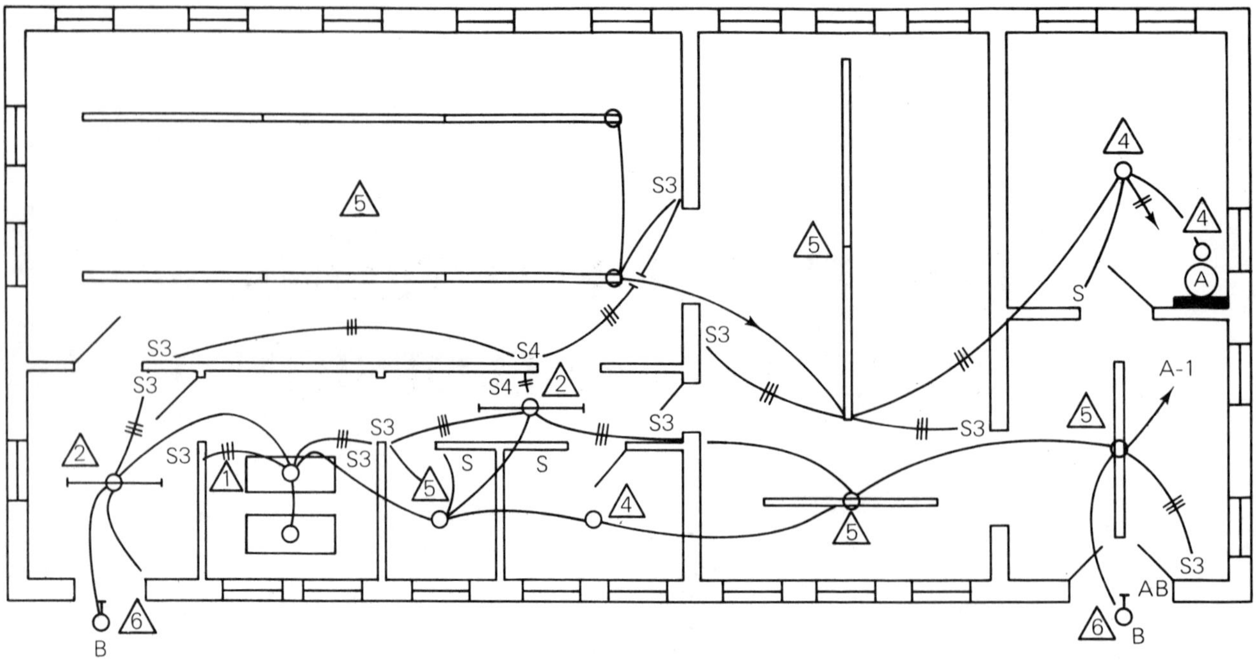

RESTON PUBLISHING COMPANY, INC.
Reston, Virginia
A PRENTICE-HALL COMPANY

ELECTRICAL-ELECTRONIC DRAFTING AND DESIGN

John E. Traister

TK
431
.T73

Library of Congress Cataloging in Publication Data

Traister, John E.
Electrical-electronic drafting and design.

Includes index.
1. Electric drafting. 2. Electronic drafting.
I. Title.
TK431.T73 621'.2'6213 78-7432
ISBN 0-87909-212-2

© 1979 by
Reston Publishing Company, Inc.
A Prentice-Hall Company
Reston, Virginia 22090

All rights reserved. No part of this book may be reproduced in any way, or by any means, without permission in writing from the publisher.

10 9 8 7 6 5 4 3 2 1

Printed in the United States of America.

CONTENTS

PREFACE

This book is designed as a practical guide to the drafting of electrical and electronic systems. It contains much of the knowlege and experience gained over a period of many years by the author and by draftsmen, technicians, consulting engineers, and many manufacturers in the fields. It is tailored to the specific needs of the person responsible for transmitting technical information though the preparation and interpretation of electrical and electronic drawings; therefore, the first consideration has been the requirements of the draftsmen employed by consulting engineers, contractors, and manufacturers.

If any one word best describes this work, it is hoped to be the word *practical*. It is arranged in a logical order of learning difficulty; it is designed so that each student can work on his own job and may proceed with a minimum of assistance according to his or her own ability; it provides a well-rounded course in basic drafting with emphasis on the electrical and electronics fields and is ready for use without further preparation on the part of the instructor; it is usable at many educational levels; it bridges the gap between theory and problems encountered in actual employment at consulting engineering firms, electrical contracting firms, and manufacturers.

In summary, this book is a quick reference for those actively engaged in the drafting of electrical and electronic systems, a learning method for those studying electrical drafting and design, and a refresher for those with wide experience in the field.

A deep and grateful bow must be made in the direction of the draftspersons, designers, and engineers who helped supply material for this book. I am especially indebted to the many manufacturers who supplied several of the illustrations and reference data.

If this book can take a few of the stumbling blocks out of the student's path and light the way somewhat, the purpose of the book will have been fulfilled.

JOHN E. TRAISTER

ELECTRICAL-ELECTRONIC DRAFTING AND DESIGN

Chapter 1

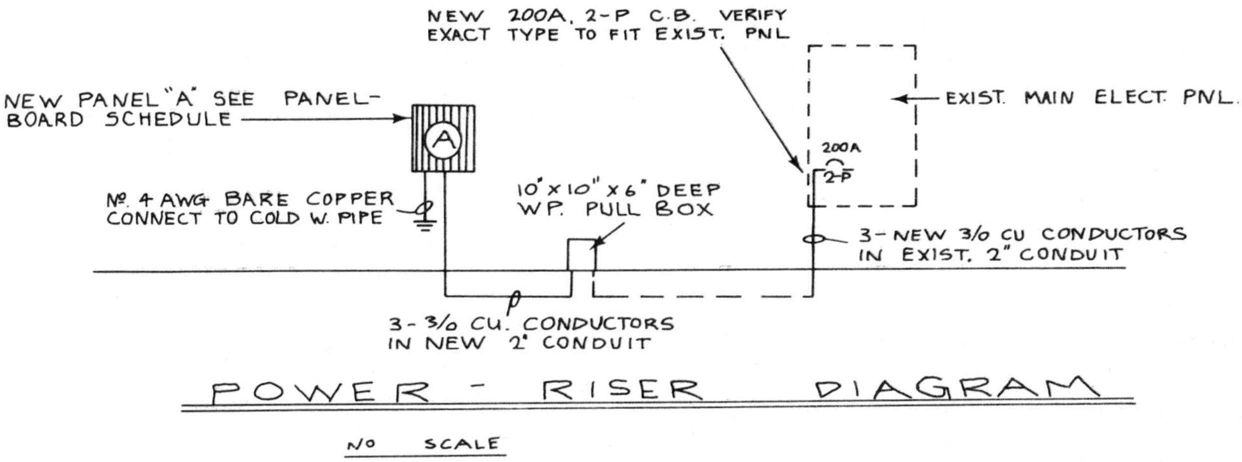

INTRODUCTION TO ELECTRICAL AND ELECTRONIC DRAFTING

The primary objective of this chapter is to give the reader an overall picture of the electrical and electronic industry and the draftsperson's relationship to it. Familiarity with these relationships is considered necessary to give draftspersons a proper background for approaching their work more intelligently.

This chapter will also introduce fundamental drafting procedure, types of electrical and electronic drawings, a survey of electrical and electronic drafting, and other basic essentials necessary to the electrical and electronic drafting profession.

1.1 Fundamentals of Electrical and Electronic Drafting

During the early days of the electrical construction industry, the electrical installations were usually laid out by the electrician on the job—often only as the work progressed. As these systems became more extensive and complex, contractors began hiring mechanical draftsmen to prepare working drawings to supplement the sketchy layouts on the architect's drawings. These electrical working drawings provided a basis for preparing the cost estimates and for giving instructions to the electricians installing the systems. In fact, these are still the primary purposes of most electrical drawings today.

As the electrical systems in buildings continued to become a more important part of the general construction and the systems became still more extensive and complex, persons with the proper knowledge and training began to devote their time exclusively to designing and laying out electrical systems as consulting engineers, selling their services to architects. These firms probably accounted for the hiring of more electrical draftsmen than any other industry at the time.

Engineers, technicians, designers, and draftspersons are still constantly working on drawings and construction documents for electrical systems in new buildings as well as plans for the renovation of existing buildings. Much of the design for such systems is accomplished by the use of mathematics, good judgment, and drafting. Mathematical equations are used in calculating the amount of illumination for a given area, circuit and feeder wire sizes, the service-entrance size, and the like. Good judgment is gained from experience and is used to determine such items as the type of lighting fixtures to be used in a given area, etc. Drafting is the means of using lines, symbols, dimensions, and notations to convey the engineer's calculations and design to the workmen on the job. Therefore, drafting is really a language—a means of communication—and any potential draftsman must master this language so that his drawings will show a complete description of the finished product and so that the workmen can understand what is expected of them.

Unlike freehand artists, who require a certain amount of natural talent, draftsmen are made—by acquiring a good knowledge of the profession and then applying this knowledge to practical applications. However, there are a few personal characteristics that, by the very nature of the work, a person should have to be a successful draftsman regardless of his previous training or background. These characteristics include:

1. A reasonable amount of mathematical ability
2. Patience
3. An orderly mind
4. Imagination and ability to visualize
5. Ability to concentrate
6. Neatness

Any potential draftsman who lacks the preceding background should take steps to improve himself.

1.2 Types of Electrical and Electronic Drawings

Those involved in electrical and electronic drafting will encounter many types of drawings during the preparation of even a single project. A brief sampling of the various types that are likely to be encountered in this field of the drafting profession is in order. A more detailed study—including step-by-step procedures for drawing them—will be covered in the chapters to follow.

Pictorial drawings are those drawings in which the objects are drawn in one view only; that is, three-dimensional effects are simulated on the flat plane of drawing paper by drawing several faces of an object in a single view. This type of drawing is very useful to describe objects and convey information to those who are not trained in blueprint reading or else to supplement the conventional orthographic (multiview) drawings in the more complex systems.

Examples of the use of pictorial drawings would be the instructions for the many do-it-yourself kits now on the market for beginners; drawings used in training manuals in various industries and the armed forces; catalog illustrations, and those drawings used by assembly-line workers for assembling electronic equipment.

The types of pictorial drawings most often found in the electrical and electronic fields include:

1. Isometric drawing
2. Oblique drawing
3. Perspective drawing

All of these types are more difficult to draw than conventional multiview drawings. Further disadvantages of pictorial drawings are that intricate parts cannot be pictured clearly and that they are difficult to dimension. Still, as mentioned previously, they are being used more and more, and every electrical and electronic draftsman must be able to draw such views professionally.

A view projected onto a vertical plane in which all of the edges are foreshortened equally is called an *isometric projection*. Figure 1-1 shows an isometric drawing of a cube. In this view, the edges are 120 degrees apart and are called the isometric axes, while the three surfaces shown are called the isometric planes. The lines parallel to the isometric axes are called the isometric lines.

This type of pictorial drawing is usually preferred over the other two types mentioned for certain electrical details and wiring diagrams on working drawings, because it is possible to draw isometric lines to scale in the same manner as floor plans or multiview plans are drawn in orthographic views. Isometric drawings may be made with the 30–60-degree triangle.

The *oblique drawing* is similar to the isometric drawing in that one face of the object is drawn in its true shape and the other visible faces are shown by parallel lines drawn at the same angle (usually 45–30 degrees) with the horizontal. However, unlike an isometric drawing, the lines drawn at a 30-degree angle are shortened to preserve the appearance of the object and are therefore not drawn to scale. The drawing in Fig. 1-2 is an oblique drawing of a cube.

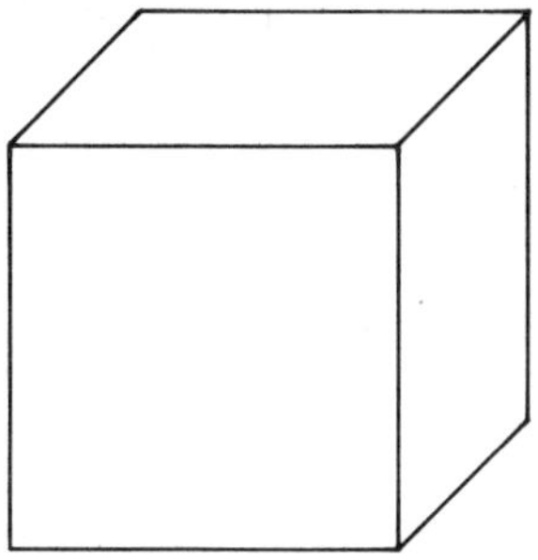

Figure 1-2. The oblique drawing shown here is another form of pictorial drawing.

The two methods of pictorial drawing described so far produce only approximate representations of objects as they appear to the eye, as each type produces some degree of distortion of any object so drawn. However, because of certain advantages (they can be drawn to scale, they are easy to draw, etc.) they are the types of pictorial drawings most often used in the electrical/electronic industry.

Sometimes—as for a catalog illustration—it is desired to draw an exact pictorial representation of an object as it actually appears to the eye. A drawing of this type is called a *perspective drawing*; one such drawing appears in Fig. 1-3.

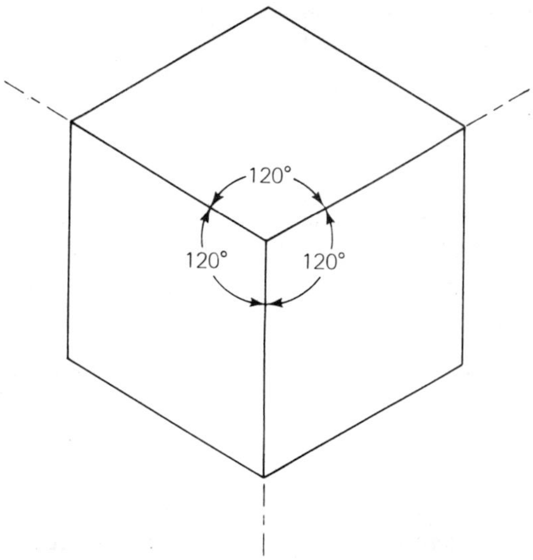

Figure 1-1. An isometric drawing of a cube showing the isometric axis 120 degrees apart.

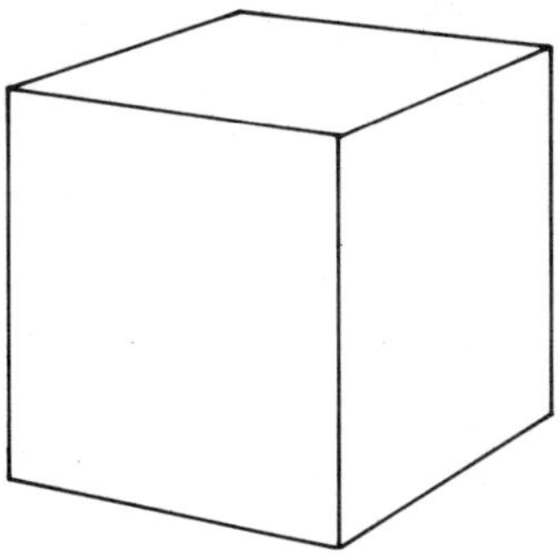

Figure 1-3. A perspective drawing is used when an exact pictorial representation of an object is required.

An *orthographic-projection drawing* is one that represents the physical arrangment and views of specific objects. These drawings give all plan views, elevation views, dimensions, and other details necessary to construct the project or object. Although the pictorial drawing in Fig. 1-4 suggests the form of a block, it does not show the actual shape of the surfaces, nor does it show the dimensions of the object so that it may be constructed.

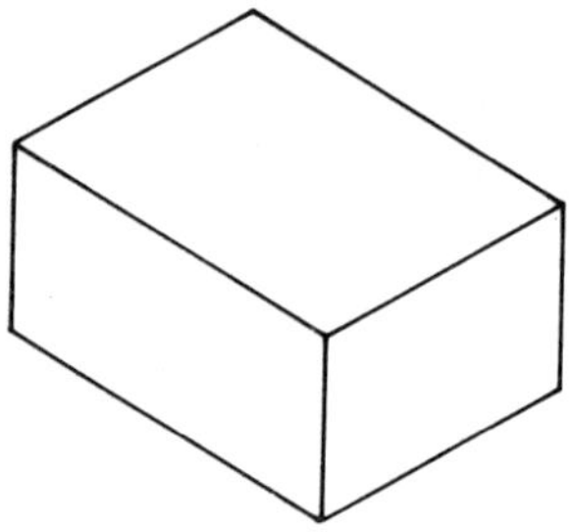

Figure 1-4. A pictorial drawing suggesting the form of a block.

An orthographic projection of the block in Fig. 1-4 is shown in Fig. 1-5. One of the drawings in this figure shows the block as though the observer were looking straight at the front; one, as though the observer were looking straight at the left side; one, as though the observer were looking straight at the right side; and one, as though the observer were looking at the rear of the block. The remaining view is as if the observer were looking straight down on the block. These views, combined with the dimensions, will allow the object to be constructed properly from materials such as metal, wood, plastic, etc.

Electrical diagrams are drawings that are intended to show, in diagrammatic form, electrical components and their related connections. Such drawings are seldom drawn to scale, and show only the electrical association of the different components. In diagram drawings, symbols are used extensively to represent the various pieces of electrical equipment or components, and lines are used to connect these symbols—indicating the size, type, and number of wires.

In general, the types of diagrams that will be encountered by the electrical/electronic draftsman will include: flow diagrams (Fig. 1-6), single-line block diagrams (Fig. 1-7), and schematic wiring diagrams (Fig. 1-8). The chapters to follow will show the reader exactly how to plan and draw all of these types of diagrams.

1.3 Survey of Electrical and Electronic Drawings

You have just reviewed the various types of drawings most often encountered in electrical and electronic drafting, but most of the examples given were of a cube or block—hardly the typical drawings you will be making as a professional draftsperson. The examples to follow are intended to show how the previous types of drawings relate to actual applications in the electrical and electronic drafting fields.

Electrical and electronic drawings that are prepared by consulting engineers, contractors, manufacturers, and others involved in the industry are very unique drawings. Most of these drawings will encompass all of the previously described drawings regardless of the firm involved. For example, drawings produced by manufacturers of electronic equipment (Radio Shack, Heath Company, etc.) usually consist of the following:

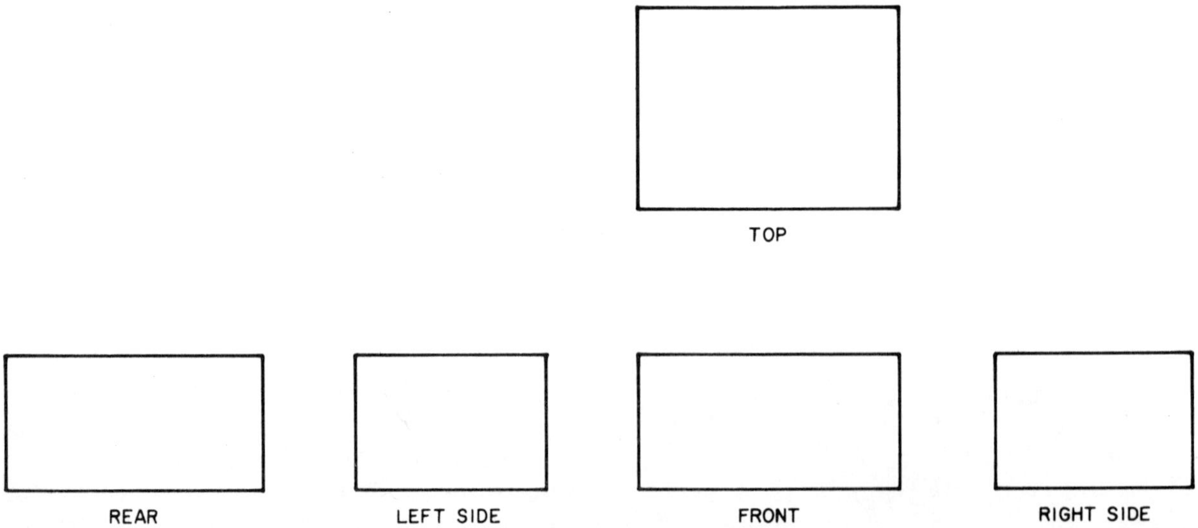

Figure 1-5. An orthographic projection of the block shown in Fig. 1-4.

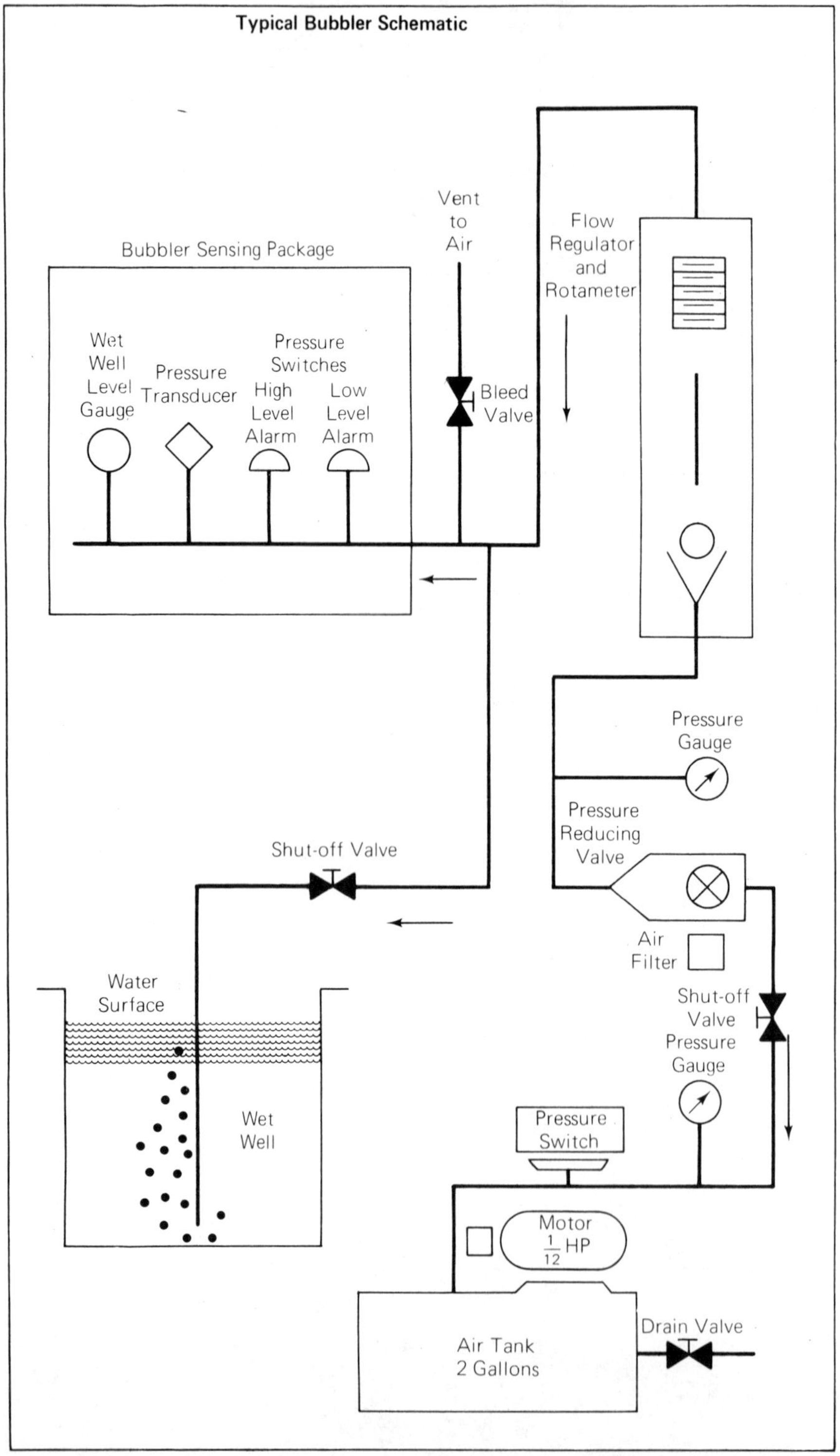

Figure 1-6. A typical flow diagram of the type that will be encountered by the electrical draftsman.

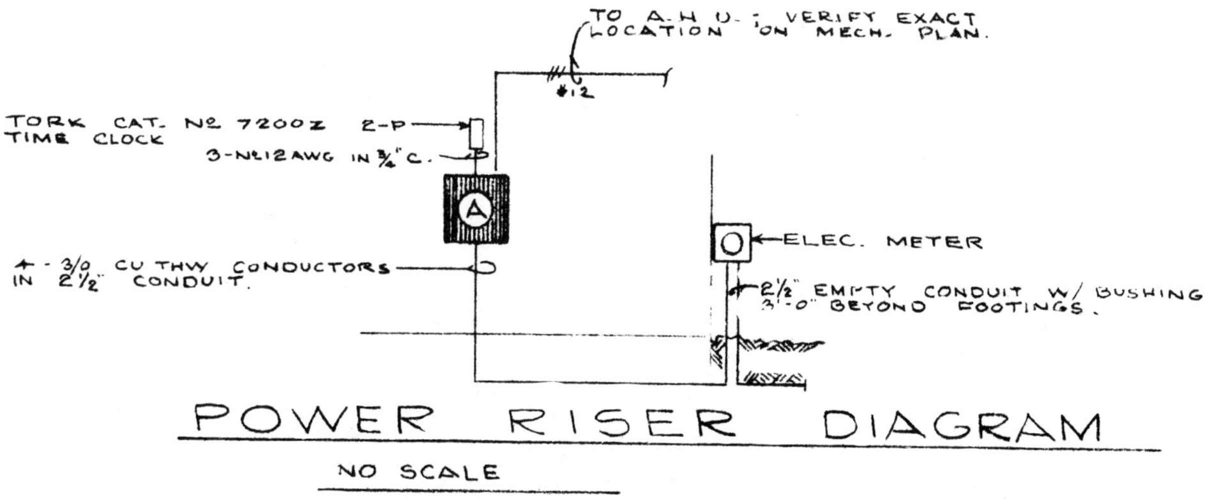

Figure 1-7. A power riser diagram using the single-line block method of representing the components and their related wiring.

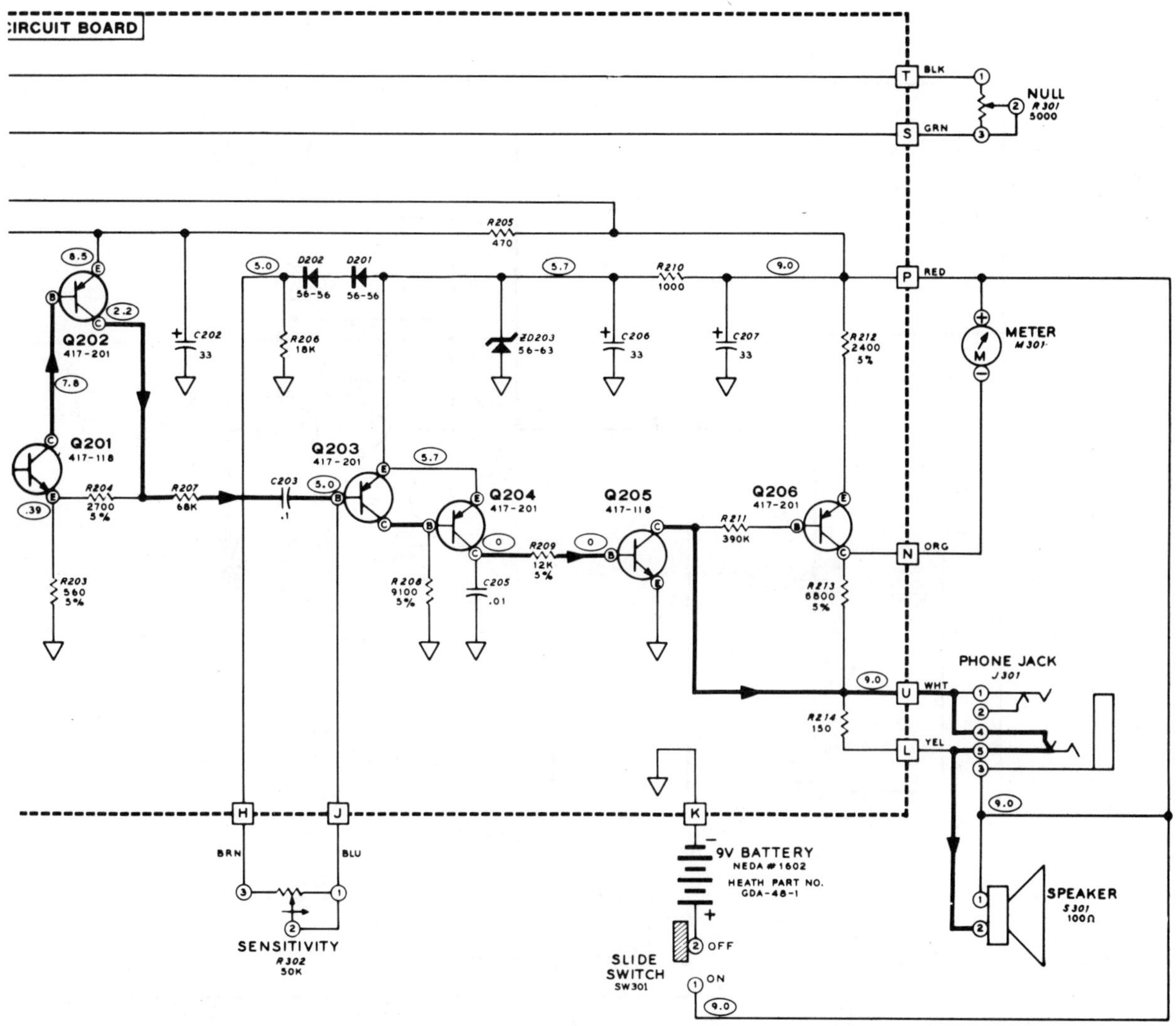

Figure 1-8. The schematic wiring diagram shown here typifies the type that electronic draftsmen will be required to draw. (*Courtesy of Heath Co.*)

Figure 1-9. Pictorial drawings are required to illustrate various components in electrical and electronic systems.

1. A schematic diagram (Fig. 1-8) to show—in diagrammatic form—the connections of the various components for the project.
2. Pictorial drawings to illustrate the various components (Fig. 1-9) or to show more clearly how the mechanical and electrical work is to be done on a project (Fig. 1-10). Perspective drawings will also be required to illustrate the piece of equipment in catalogs.
3. Orthographic views to show how such items as, say, an equipment chassis are to be constructed.

Electric lighting manufacturers need orthographic views as a guide for manufacturing the various lighting fixtures and for use as reference by consulting engineers and electrical contractors. See Fig. 1-11. They may also require oblique or isometric drawings similar to the one in Fig. 1-12.

Draftsmen working for architects, consulting engineers, electrical contractors, and similar firms will be required to make the following drawings:

1. A plot plan (Fig. 1-13) showing the location of the building on the property and all outside electrical wiring.
2. Floor plans (Fig. 1-14) showing the walls and partitions for each floor level with the physical location of all outlets and related wiring.
3. Single-line block diagrams such as the power-riser diagram in Fig. 1-15 showing the service entrance and the panelboard components.
4. Schematic wiring diagrams (Fig. 1-16) showing details of control wiring for various systems.

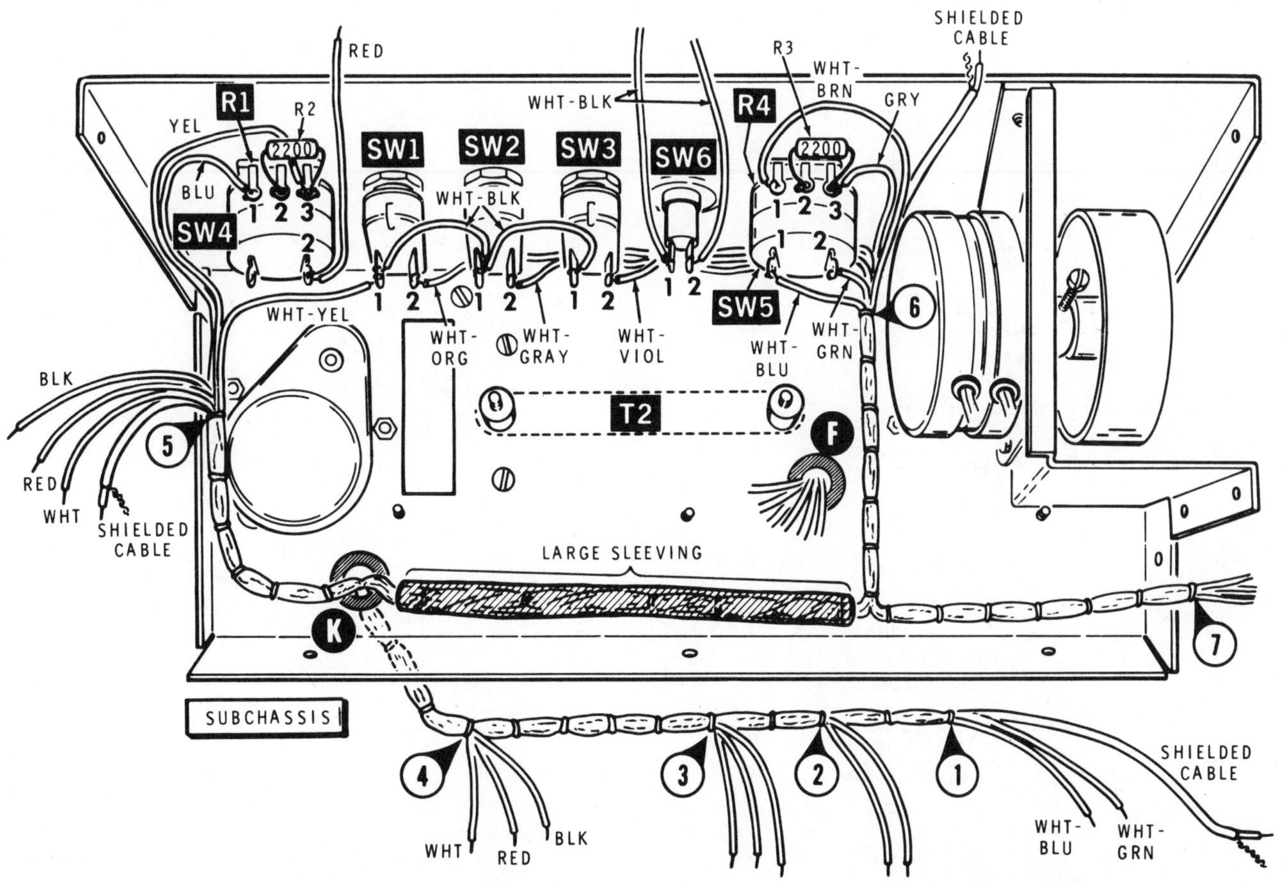

Figure 1-10. Perspective drawings, like the one shown here, are sometimes required to illustrate pieces of equipment in catalogs and instruction manuals.

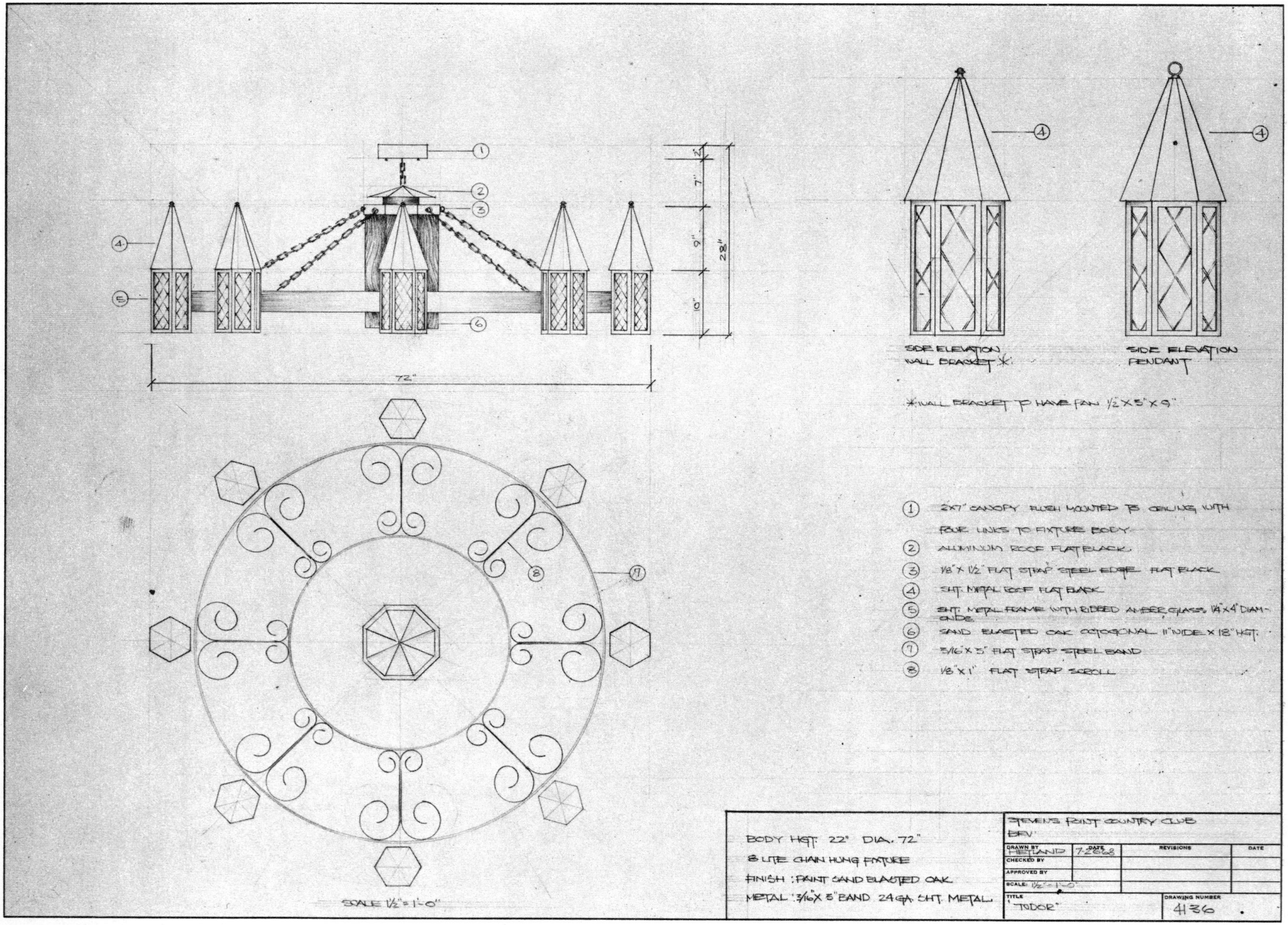

Figure 1-11. Orthographic views of a lighting fixture as used by an electric lighting manufacturer. (*Courtesy of Winona Studio of Lighting.*)

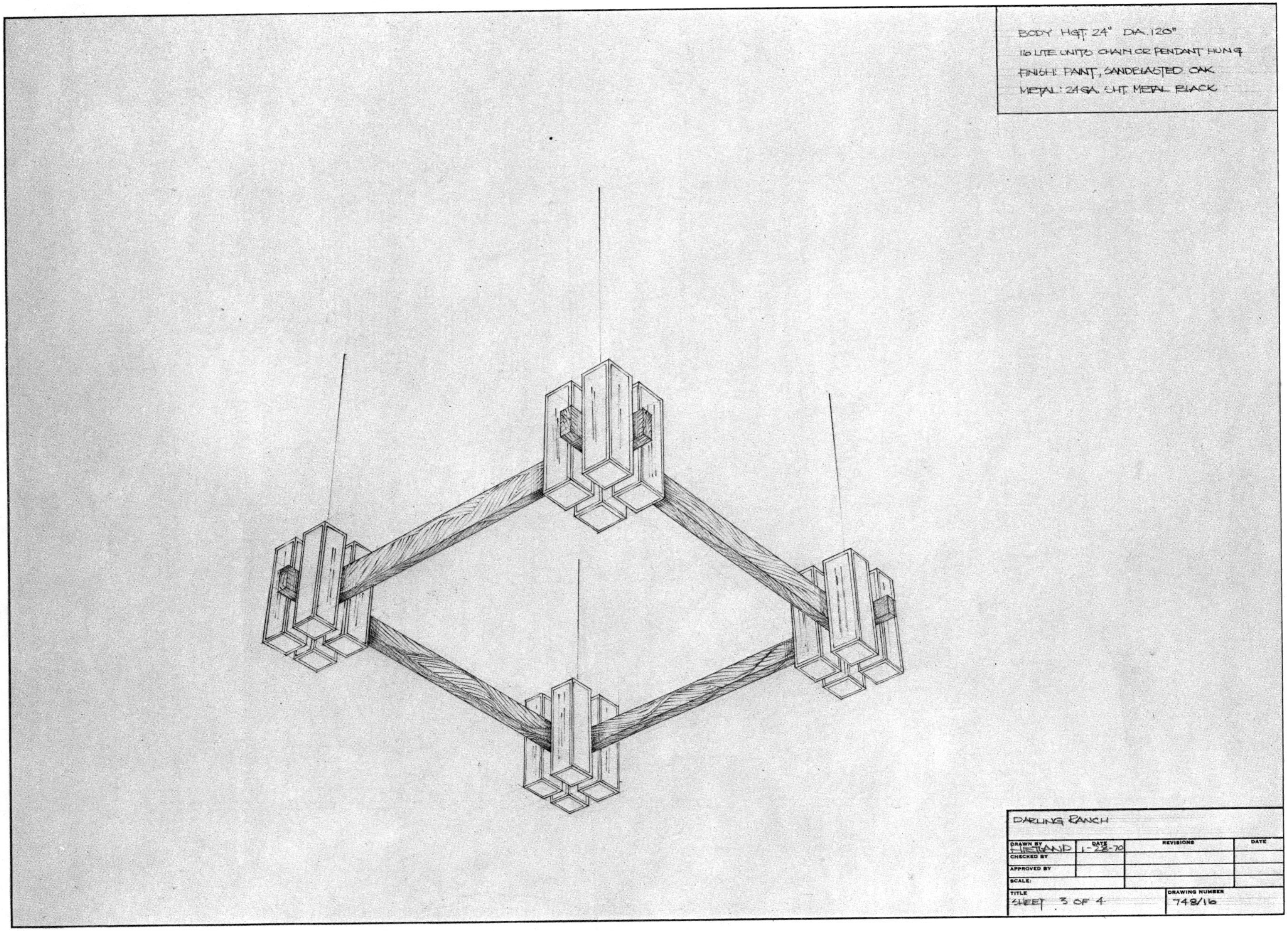

Figure 1-12. Lighting manufacturers may also require oblique or isometric drawings similar to the one shown here. (*Courtesy of Winona Studio of Lighting.*)

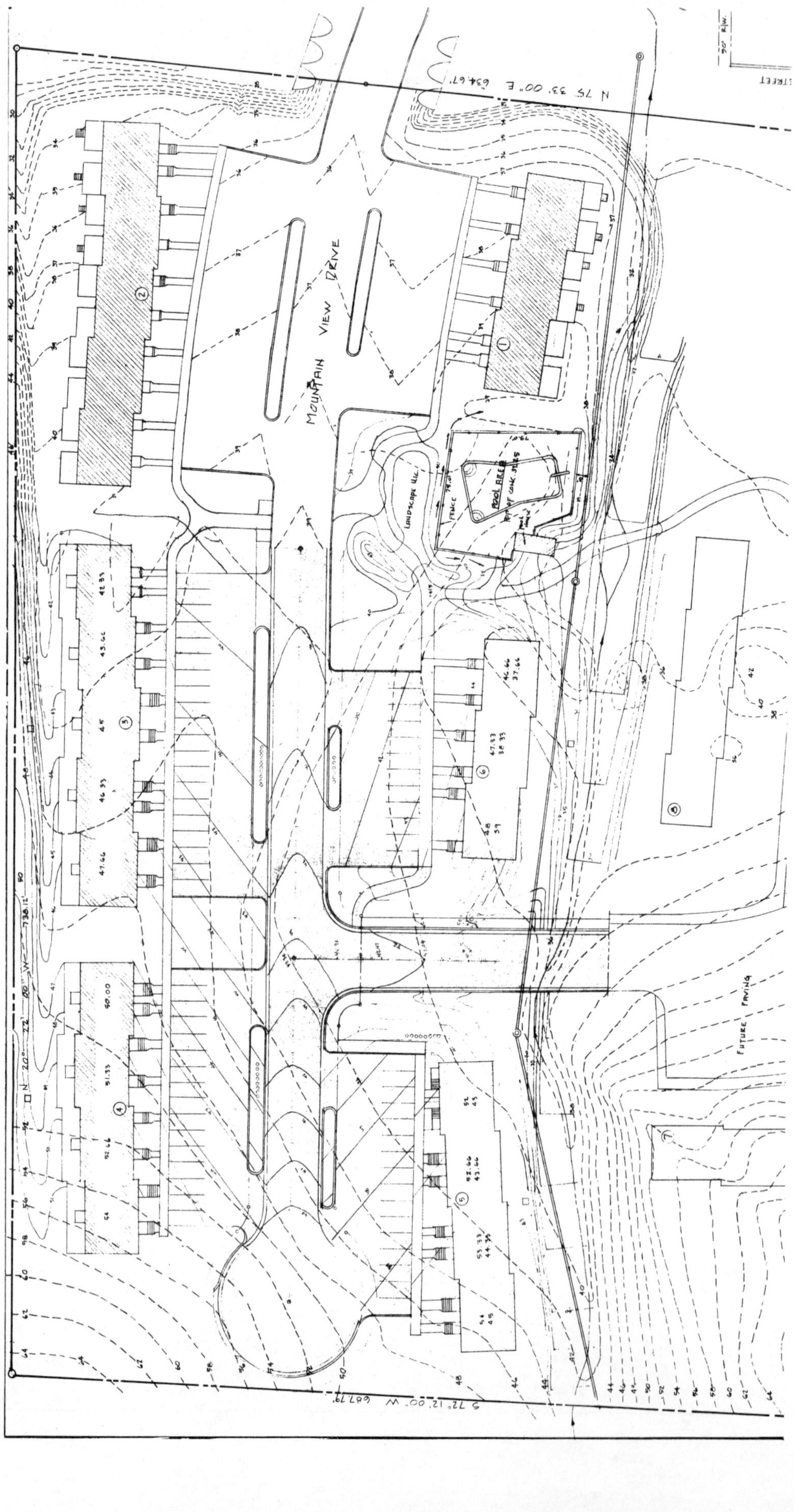

Figure 1-13. Site plan showing the location of buildings on the property and all outside utilities.

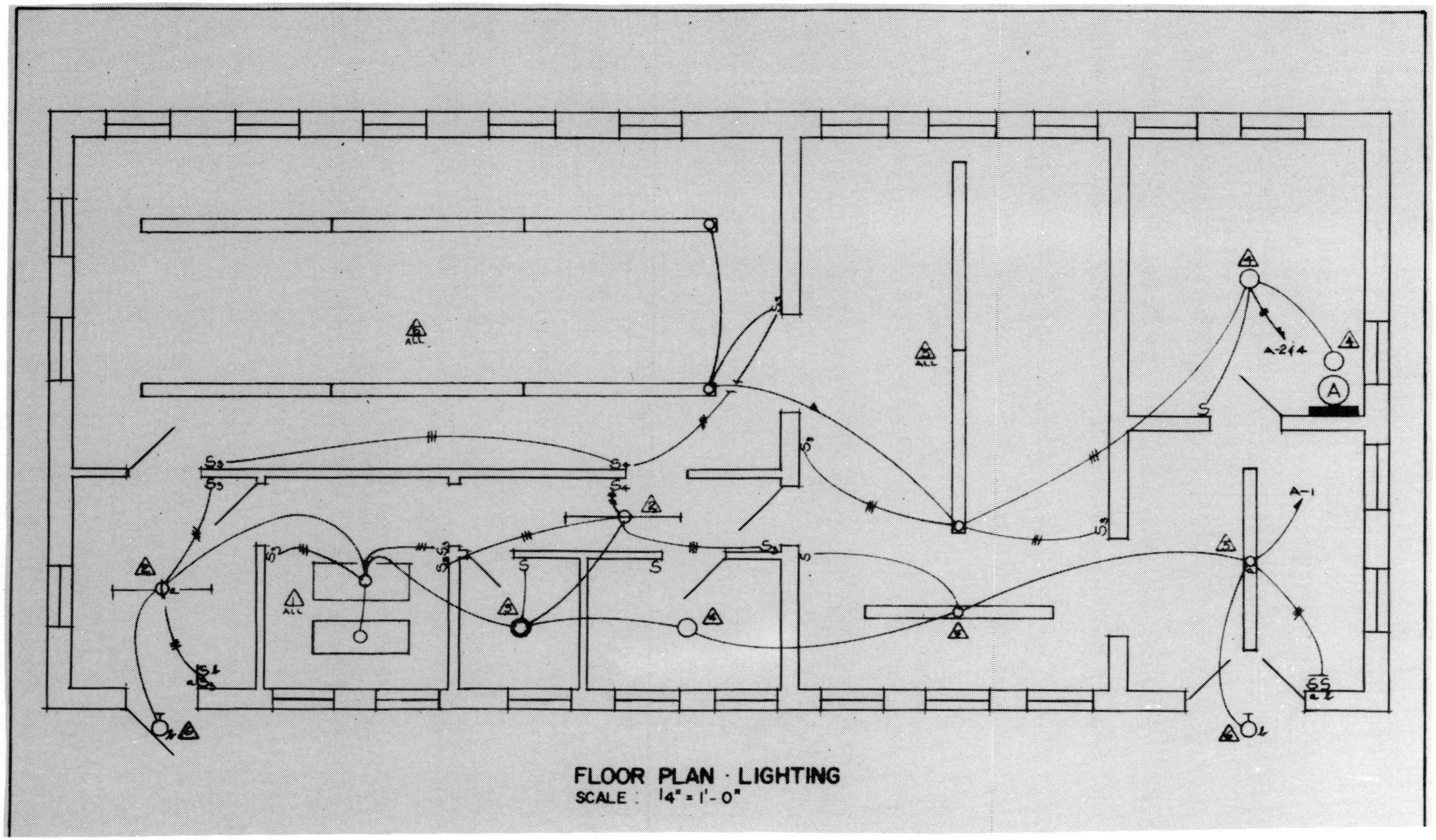

Figure 1-14. Floor plan showing the walls, partitions, windows, doors, lighting outlets, and related wiring.

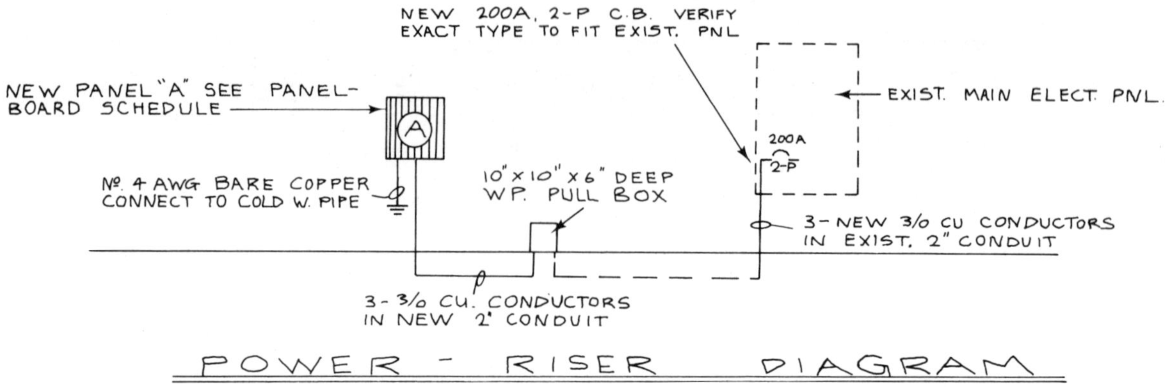

Figure 1-15. Single-line block diagram showing the service entrance and related wiring for a building project.

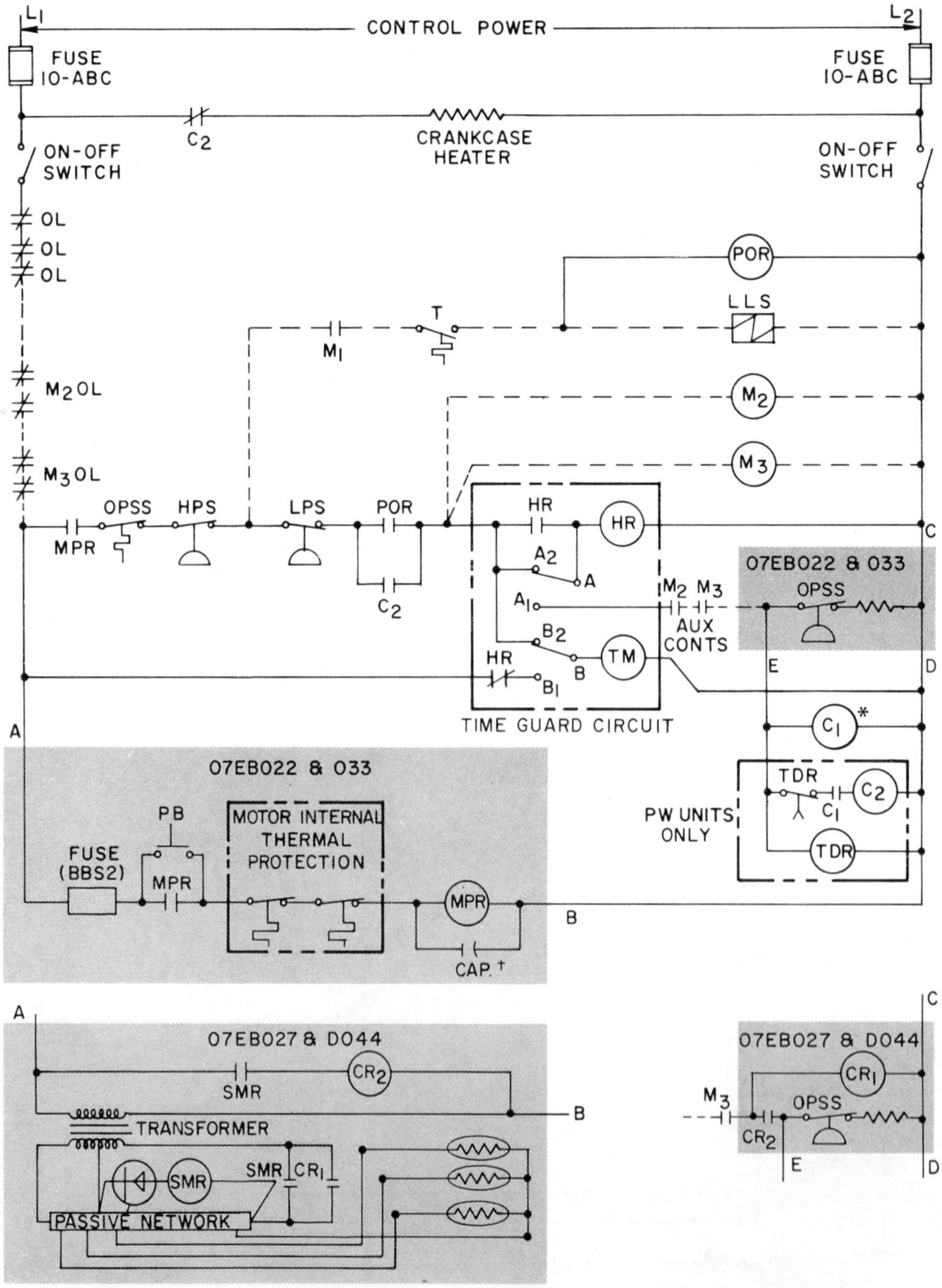

Figure 1-16. Schematic wiring diagram showing details of the control wiring for a heating and cooling system.

SUMMARY

An electrical/electronic drawing is an abbreviated language for conveying exact, detailed information in the form of lines, symbols, abbreviations, and notes, which would otherwise take many pages of manuscript or hours of verbal instruction to convey. It is also a means of keeping a permanent record of projects already constructed.

The types of drawings most often encountered by the electrical/electronic draftsman are

1. Pictorial drawings
   a. Isometric
   b. Oblique
   c. Perspective
2. Orthographic projections
   a. Plan views
   b. Elevations
   c. Sections
3. Single-line block diagrams
4. Schematic wiring diagrams

QUESTIONS

Answer these questions by filling in the blanks.

1. Pictorial drawings that are made with a 30–60-degree triangle and with all lines drawn to scale are called ______drawings.
2. A pictorial representation of an object as it actually appears to the eye is called a _____drawing.
3. The drawing in Fig. 1-17 is a pictorial view of a housing for electronic components. The _____is what you see when you look directly down at the housing. In doing so, the sides, the bottom, the front, and the _____are cut from your view.
4. The _____of the housing is what you see when the housing is directly in front of you. In this case, you cannot see the top, bottom, _____, or the two sides.
5. A _____ side view is what would be seen if the right side of the housing was turned towards you.
6. A drawing that is meant to show, in diagrammatic form, electrical components and their related connections is called a _____.
7. The three categories in which most pictorial drawings fall are
   a. _____
   b. _____
   c. _____

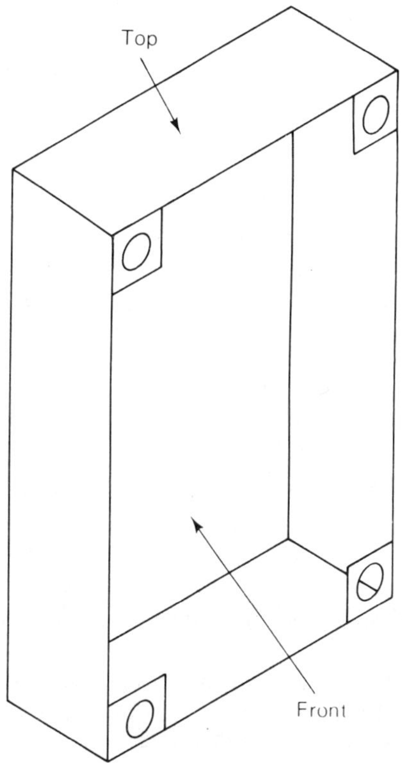

Figure 1-17. A pictorial view of a housing for electrical components.

8. During the early days of electrical construction, the electrical work in buildings was usually laid out by the _____ on the job.
9. List three characteristics that a good draftsman should have. 1._____2._____3._____
10. Using Fig. 1-4 and Fig. 1-5 as guides, roughly sketch an orthographic projection (multiview drawing) of the housing in Fig. 1-17.

1.4 Electronic Components

The parts used in electronics are often simply called parts, but are more often referred to on electronic drawings as components. The electronic draftsman must become familiar with these components, and this portion of the chapter is designed to introduce the reader to a few of them. A more detailed study will take place in Chapters 6 and 7.

Electronic symbols are used to represent the various components on electronic drawings. The drawings—usually in schematic form—show how the various components are connected. It is, therefore, of great importance that the electronic draftsman learn to read schematic diagrams at a very early date. While it may be difficult to learn all of the electronic symbols all at once, they are quite easy to learn if you take a few of them at a time, learning to read the simpler diagrams before going to the more difficult ones.

When parts are connected in an electronic circuit, there are certain paths through which the electrons must flow and other paths where the flow of current must be avoided. *Conductors* are used to provide paths for the current flow, and *insulators* are used to prevent the current flow.

Conductors consist of such materials as copper, silver, aluminum, iron, and similar metals. Insulators include rubber, plastic, wood, paper, or a combination of these materials. Wires, for example, are usually covered with an insulating material such as rubber or plastic. Its main purpose is to keep the current flowing through the wire to the component it is supposed to reach and to prevent the current from becoming sidetracked should the wire touch another conductor.

Batteries are used as a power supply in electronic circuits and consist of one or more cells, each cell having two terminals—a positive terminal and a negative terminal. When a battery is connected to a circuit, the electrons flow from the negative terminal through the circuit and back to the battery through the positive terminal.

Capacitors, consisting basically of two metal plates separated by air or some other material, are very important electronic components. They are used in nearly every electronic circuit.

Resistors are used in electronic circuits to offer an opposition or resistance to the flow of electric current. They are made of a carbon-type composition in most cases. They are rated in ohms—from only a few ohms to well over one million ohms.

Transistors are made of materials called semiconductors. As the name implies, a semiconductor is a device that falls halfway between an insulator and a conductor; that is, it is neither a good conductor nor a good insulator. The transitor is a comparatively new device, but much progress has been made with it in a relatively short period of time. Almost all radios, TV sets, and other electronic equipment use transistors rather than the vacuum tubes of a few years ago. For this reason, the electronic draftsman should learn all he can about transistors and their use in electronic circuits.

The symbols for the components just described are shown in Fig. 1-18. The reader should learn these, as they will be seen quite often throughout this book.

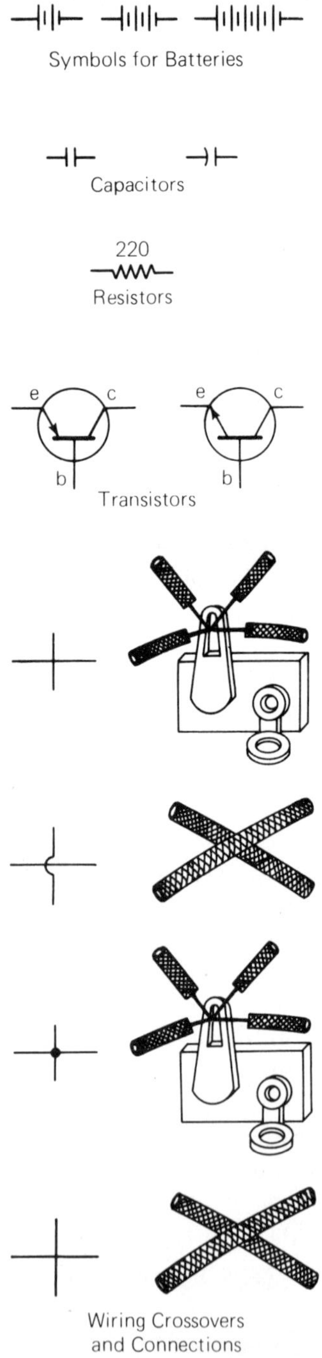

Figure 1-18. Typical electronic symbols.

Chapter 2

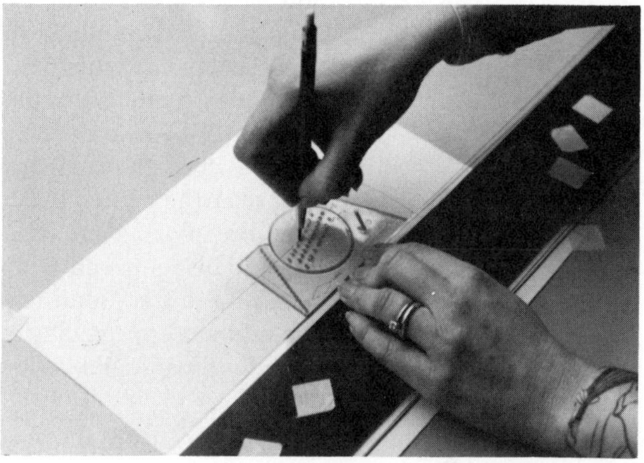

SELECTION AND USE OF DRAFTING EQUIPMENT

The main objective of this chapter is to learn what tools are required for high-quality electrical and electronic drafting. Further objectives include the proper use of these tools and how to select the best ones available for electrical drafting in general.

2.1 Basic Drafting Tools

The equipment commonly used by the electrical and electronic draftsman is listed and described in the paragraphs to follow. This equipment list covers the minimum essential tools needed to do good work on the usual kind of electrical drawing. They should be of the best quality that the draftsman can afford and include:

- Drawing board
- T-square
- Architect's scale
- Engineer's scale
- Assorted leads
- Lead pointer
- Eraser
- Erasing shield
- 30–60-degree triangle
- 45-degree triangle
- Irregular curves
- Various templates
- Lead holders
- Set of instruments
- Drafting tape
- Lettering guide
- Tracing paper
- Protractor
- Dry-clean pad
- Dust brush

Drawing board. Drawing boards vary greatly in size, from very small ones to large, vertical boards that are 6 or 7 feet high by several feet long. Regardless of the size, the drawing board should have a smooth flat surface on which the drawing paper is placed. It should be made of soft pine or bass wood and be at least 18 inches wide by 24 inches long for the student draftsman. For professional application, a board 31 inches wide by 42 inches long should be considered as minimum, since most electrical drawings used for building construction are made on 24- by 36-inch tracing paper. If a T-square is to be used in conjunction with the board, the left and right edges of the board must be perfectly straight so that they may accurately guide the T-square.

Most drafting rooms utilize drawing tables similar to that shown in Fig. 2-1. The less expensive ones are made to a standard height; others can be adjusted to any desired working height. Levers or knobs on the tables also permit the top to be regulated to various angles. The table tops are usually covered with a special cover, which not only minimizes glare but provides a smooth, firm working surface for the drawing sheets.

Although many draftsmen prefer to work standing over the board, they have to sit occasionally during the work day. Therefore, the drawing board should be equipped with a posture chair. This type of chair has a free-floating back rest with a seat that can be adjusted to the desired height, and also an independently adjustable floor ring.

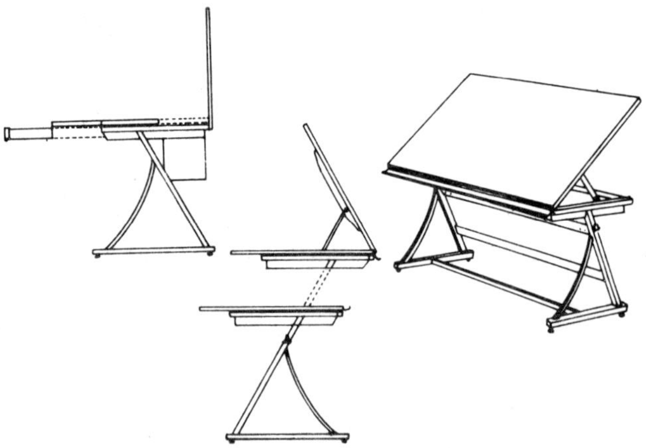

Figure 2-1. Most drafting rooms utilize drafting tables similar to the one shown here.

T-square. The common T-square is shaped like the letter "T" and consists of two parts, called the blade and the head. The head of the T-square is always held against the left working edge of the board by right-handed draftsmen, and against the right edge of the board by left-handed draftsmen. Horizontal lines are drawn along the upper edge of the blade from left to right (right to left by left-handed draftsmen). The length of the T-square should be long enough to cover the entire length of the drawing board. The most popular T-squares have plastic or celluloid edges, which permit lines to be visible underneath the edge of the blade. Figure 2-2 shows a typical T-square.

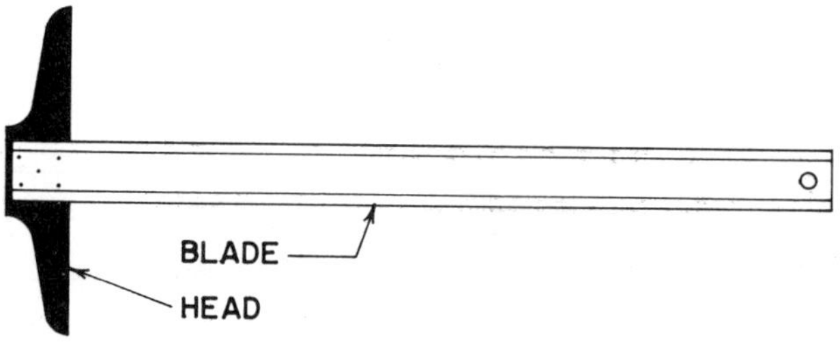

Figure 2-2. A T-square showing the head and blade.

Parallel straightedge. In most professional drafting rooms, a parallel edge (Fig. 2-3) is used instead of a T-square. The parallel edge is used just like the T-square, except that the ends are permanently attached to the board by a wire-cable arrangement. These wires, passing through pulleys within the parallel edge, keep the edge parallel to the board at all times, as it is moved up and down the board.

Parallel edges can be mounted on either drawing tables, as shown in Fig. 2-1, or on student drafting boards. Some makes of parallel edges have spring-mounted roller bearings installed on the underside of the straightedge in order to reduce smudging of drawing lines. This type of straightedge can be recommended as the best all-around type for most phases of electrical drafting.

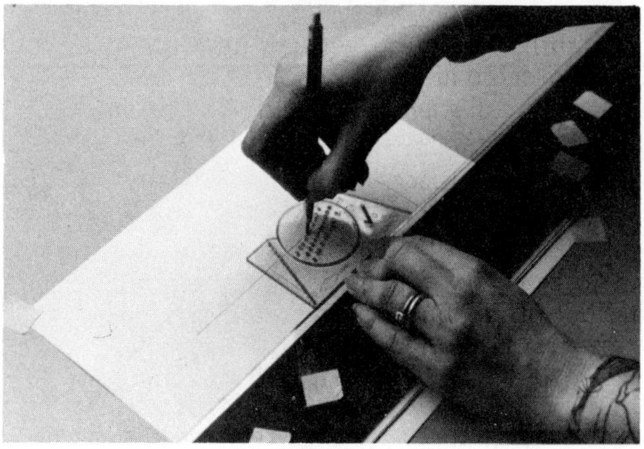

Figure 2-3. Parallel straightedge is used instead of a T-square in most professional drafting rooms.

Drafting machine. The drafting machine (Fig. 2-4) is used in some drafting rooms in place of the T-square/parallel edge, triangles, and scales. In some types of drafting, this machine can save much time, but the student draftsman must learn to use the basic tools first and then decide for himself if he desires to use the drafting machine—many do not.

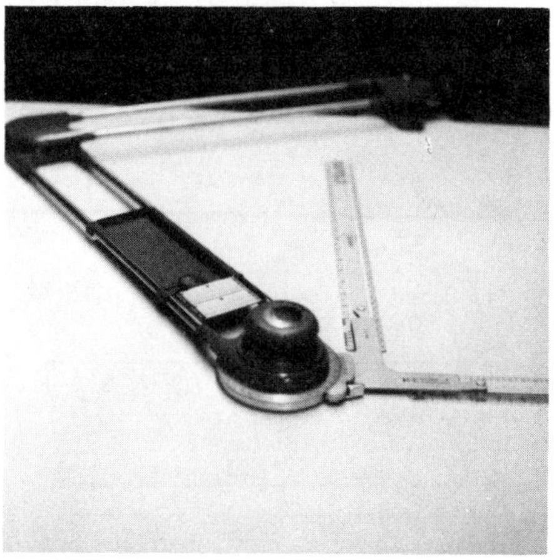

Figure 2-4. A drafting machine takes the place of the T-square, triangles, and scales.

The drafting machine's major time-saving value lies in the fact that many drawing operations can be combined, such as laying out horizontal and vertical building lines, scaling, and laying out angles. The centralized control knob allows the draftsman to accomplish these operations with his left hand, leaving his right hand free for drawing.

Architect's scale. The architect's scale is used for measuring. The basic unit at the end of the scale represents one foot and is subdivided into twelve parts per inch. On the larger scales, such as 1″ = 1′ 0″, the inch division is further divided so that the smallest subdivision may represent ¼ in. (Fig. 2-5). The smallest division on smaller scales does not have the basic unit divided into so many divisions. For example, the smallest division on the ⅛-in. scale represents 2 inches (Fig. 2-6). The use of the architect's scale will be further explained in Chapter 4, Dimensioning.

Engineer's scale. The engineer's scale is similar to the architect's scale, except that an inch is divided into 10, 20, 30, 40, 50, or 60 parts (Fig. 2-7). This scale is useful when one is working with map drawings or building site plans. The engineer's scale marked 10 is read 1″ = 10′ or some other integral power of 10; for example, 1′ = 100′, 1′ = 1000′, etc. The use of the engineer's scale will be fully explained in Chapter 12, Site Plans.

Triangles. Triangles are the instruments used by the draftsman, in combination with a straightedge, to

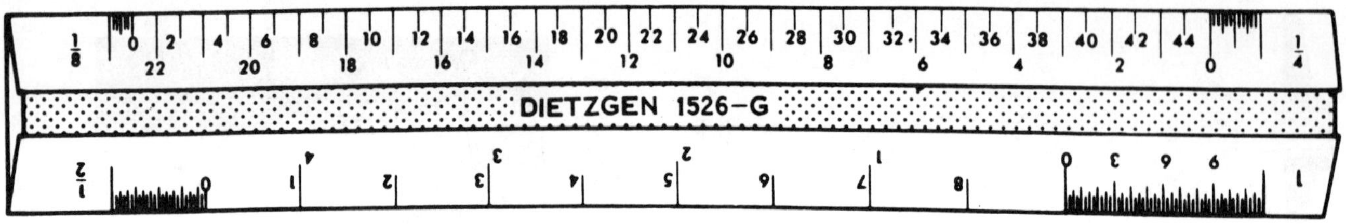

Figure 2-5. Typical graduations on an architect's scale.

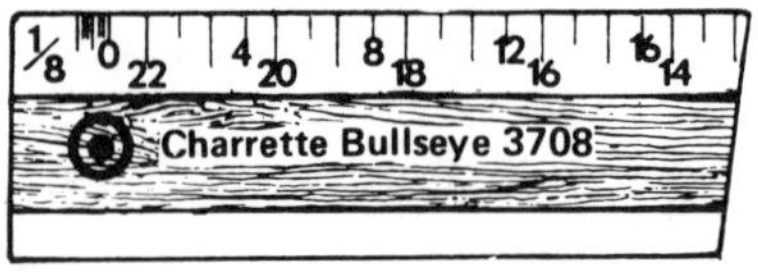

Figure 2-6. Architect's scale showing the ⅛-in. graduations.

1" = 1'-0".

Figure 2-7. Engineer's scale with ten graduations per inch.

draw both vertical lines and sloping lines. One type has angles of 30, 60, and 90 degrees and is called the 30–60-deg triangle (Fig. 2-8). The minimum size of this triangle should be 10 inches. The 45-deg triangle has one 90-deg angle and two 45-deg angles (Fig. 2-9). The 45-deg triangle should be at least six inches long. Both types of triangles are held against the upper edge of the T-square or parallel edge, and lines are drawn from the bottom of the triangle toward the top.

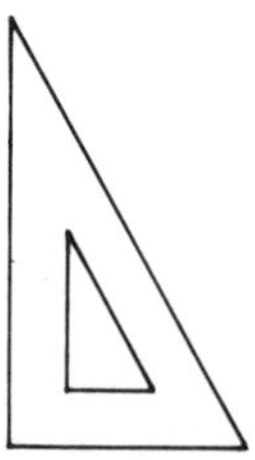

Figure 2-8. 30–60-deg triangle.

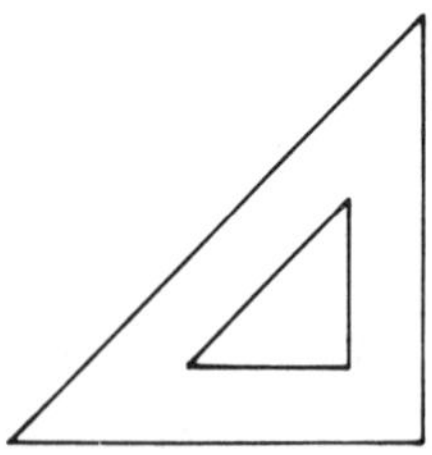

Figure 2-9.

If both triangles are used in combination, angles of other degrees can be drawn. Two such combinations are illustrated in Fig. 2-10.

Adjustable triangles are becoming more popular

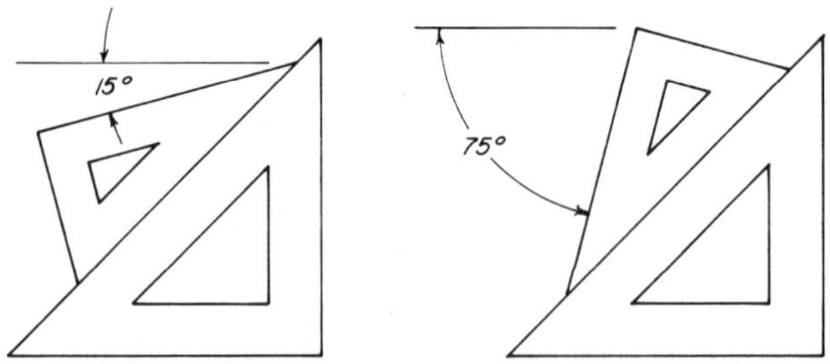

Figure 2-10. Combining two triangles to obtain other angles.

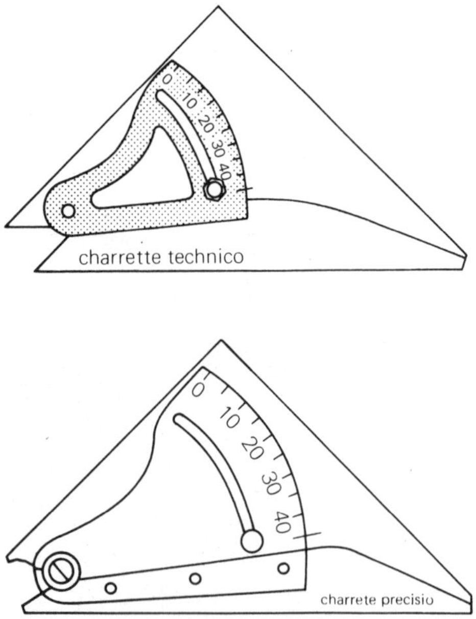

Figure 2-11. An adjustable triangle capable of angles from zero to 90 deg.

and may save the draftsman much time. However, as pointed out previously, the student should become familiar with the use of conventional drafting tools before attempting the use of the various "short-cut" devices. An adjustable triangle is shown in Fig. 2-11.

Irregular curves. Irregular curves are used in electrical drafting mainly to draw circuit lines, and to

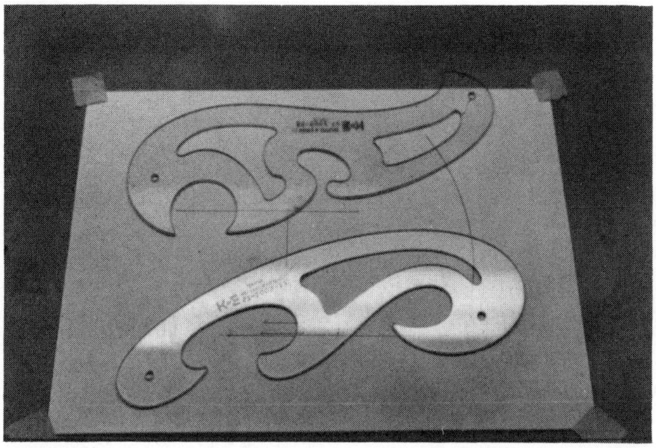

Figure 2-12. Irregular French curves are available in many different shapes and sizes and can produce practically any desired curvature.

draw curves other than standard circles or arcs. Irregular curves (two of which are shown in Fig. 2-12) are available in many different shapes and sizes and can produce practically any desired curvature. They are also the most difficult drafting instruments to use skillfully.

Lead holder. Mechanical lead holders have all but replaced wooden pencils in most drafting rooms (Fig. 2-13). They are supplied in various shapes and sizes and are made from metal, plastic, or various combinations of the two. Most lead holders operate by pressing a button at the top of the holder which instantly projects the lead, permitting a quick adjustment to any desired length. Upon release of the button, the holder automatically grips the lead firmly and securely. The lead is then pointed by using a mechanical lead pointer or a sandpaper board.

The electrical draftsman should begin with a minimum of two lead holders. However, most professional draftsmen like to have at least five or six different colors; each color represents a different degree of lead. After some time, the draftsman automatically knows that a certain color lead holder contains a certain degree of lead and therefore saves much time when different degrees of lead are frequently used on the drawings. Other lead holders have adjustable degree markings on their barrels and can be adjusted for the size of lead used at any time. However, we have found that the color-coded system is still the best way to distinguish between different degrees of lead.

Lead. Drawing lead is graded according to its degree of hardness. This ranges from 7B, the softest and blackest, to 9H—the hardest. Although practically all manufacturers use the same system of marking their lead, wide variations in hardness will be found among different manufacturers. It is for this reason that most draftsmen usually prefer to use only one brand of lead on each drawing. The following hardnesses are recommended for electrical drafting:

- 9H Used for guidelines.
- 6H Used for light sketch lines.
- 4H Used for medium-weight lines, such as building outlines for electrical drawings.
- 2H Used for drawing electrical symbols, circuits, and equipment on working drawings.
- H Used for sharp, dark lines, such as border lines, and for lettering.
- F Used for lettering in some cases.

A conical-shaped point on the lead is preferred for all electrical drawing and lettering (Fig. 2-14).

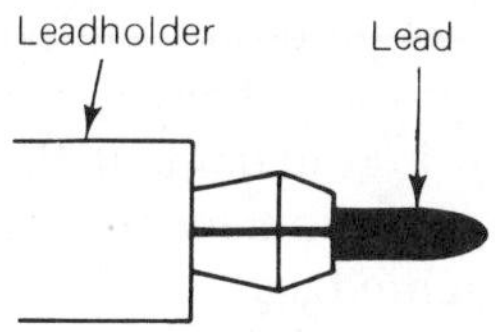

Figure 2-14. A conical-shaped point on the lead is preferred for all electrical and electronic drawing and lettering.

Eraser. Since we all make mistakes, the eraser needs little explanation. Use soft or medium-soft pencil erasers with beveled edges.

Erasing shield. When there are many lines close together and only one line needs to be erased, the others may be protected by using an erasing shield such as the one shown in Fig. 2-15. Erasing shields are thin metal plates with openings of various shapes

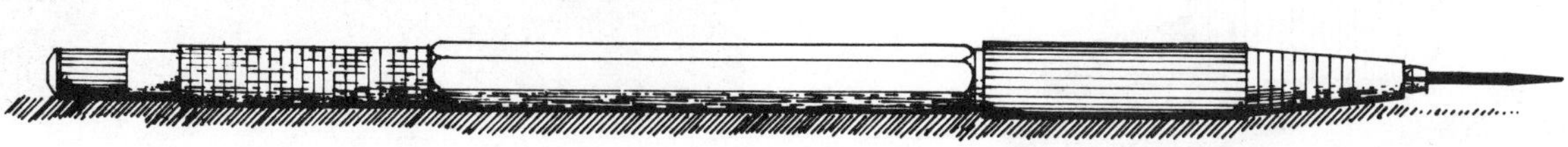

Figure 2-13. A typical lead holder.

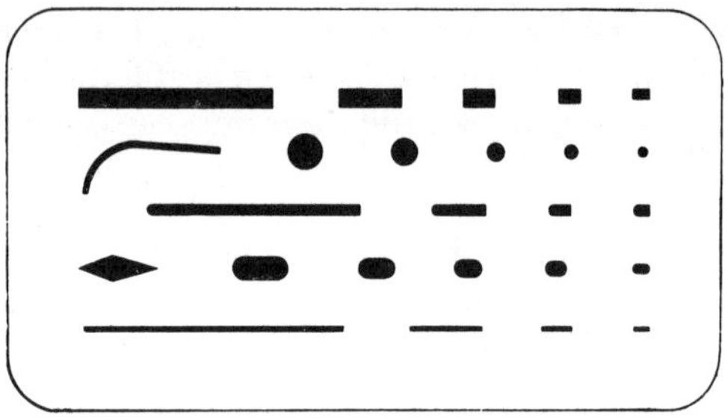

Figure 2-15. Erasing shield used to protect lines near the one being erased.

in them. The shield is placed on the drawing so that the line to be erased appears in one of the openings. Only the line in this opening will be erased, as the thin metal shield protects other lines near the one being removed.

Lead pointer. The lead pointer is used to point or "sharpen" the refill drawing lead used in mechanical lead holders. Three types are in general use: the portable type (illustrated in Fig. 2-16), the desktop type, and the clamp-on type shown in Fig. 2-17. The last is ideal for the electrical draftsman.

Drafting tape. Thumbtacks and staples have been used for many years to hold the drawing paper in place on the board. Both damage the board as well as the drawing paper, and they also obstruct the movement of the straightedges.

Masking or drafting tape is the preferred method of securing drawing paper to the board. Merely stick small pieces of tape across each corner of the paper. Drafting tape will do no damage to either the board or the paper and will offer little or no obstruction to the T-square or parallel edge. Drafting tape ¾ in. wide is the preferred width in most drafting rooms.

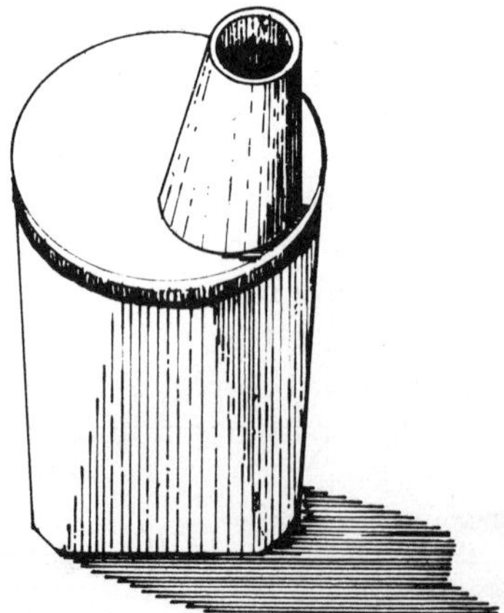

Figure 2-16. Portable type lead pointer.

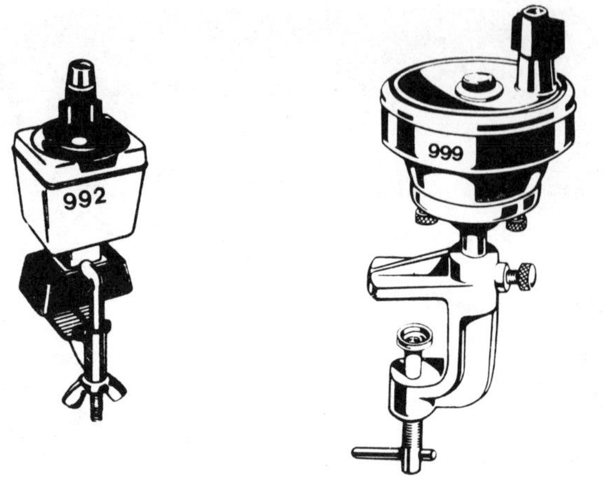

Figure 2-17. Clamp-on type lead pointer.

Dust brush. The dust brush shown in Fig. 2-18 (a) is a very necessary item for the electrical draftsman, since it is extremely valuable for removing eraser particles and dust from a drawing.

Figure 2-18. Typical dust brush.

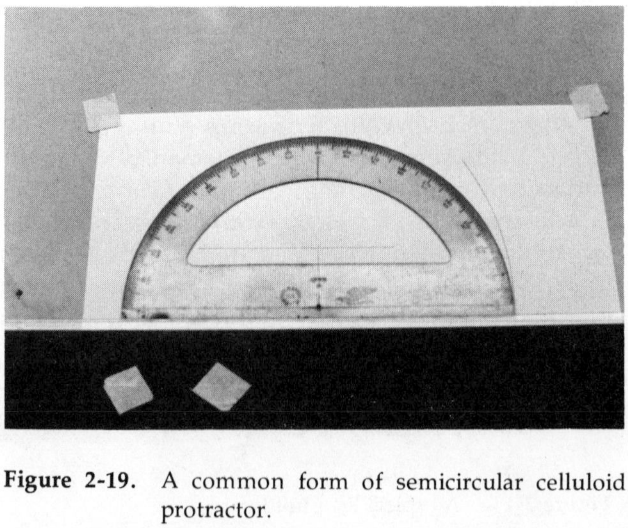

Figure 2-19. A common form of semicircular celluloid protractor.

Protractors. The protractor is a device for laying off and measuring angles on drawings. The best form for electrical drafting consists of a full circle graduated in degrees or fractions of a degree.

Figure 2-19 shows the most common form of circular celluloid protractor. To lay off an angle, the center "0" of the protractor is placed at the vertex of the angle with the edge of the bar coinciding with the "1" line to which the angle is referred. A light mark is then made on the drawing at the proper graduation of the protractor arc. After the protractor is removed, a line is drawn joining this mark with the vortex. Angles can be measured in a similar manner.

Set of instruments. A set of drawing instruments is usually found in the possession of many draftsmen, although plastic templates are rapidly taking their place. A typical set consists of a compass used for drawing arcs and circles, a divider used for laying out or measuring distances, and a beam compass for the precise drawing of arcs or circles having radii greater than 6 inches. It is recommended that the draftsman not purchase a set of drawing instruments until the situation warrants.

Figure 2-20 shows a set of good drawing instruments which are inexpensive. This set will handle most of the requirements met in electrical/electronic drafting.

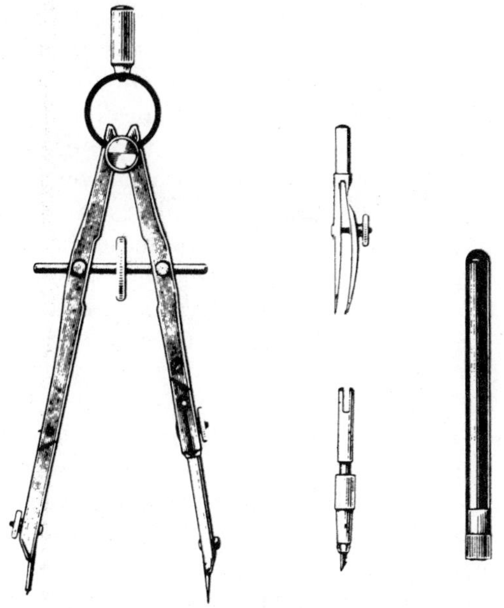

Figure 2-20. A set of good drawing instruments that will handle most of the requirements met in electrical and electronic drafting.

Miscellaneous instruments. Other special drafting instruments will be covered in Section 2.3. Electrical templates (useful time-saving devices) will be fully covered in Chapter 6.

From time to time, new drafting tools appear on the market. Many are useful time-saving devices, while others are gimmicks. Always ask for a demonstration of such items before they are purchased for use in the drafting room; better yet, ask another draftsman who has used the particular tool to see how he or she likes it.

2.2 Care and Use of Drafting Instruments

Neatness and accuracy are the two most important qualities of any drawing, and these qualities can be obtained only by the use of good-quality drafting instruments that are in first-class condition. We shall define first-class condition as excellent physical and mechanical shape, and clean—the last is extremely important.

The prerequisite for clean drawings is a clean draftsman. The draftsman's hands should be washed frequently in order to avoid smearing dirt onto the drawing. After handling coffee or food, the hands should be washed before one returns to the drawing board. In other words, avoid—as much as possible—introducing any foreign matter into the drafting area.

The drawing board and drawing paper should be brushed every few minutes of drawing to clean away eraser particles and graphite. A dry-clean pad used on the drawing prior to the brushing will eliminate smudge marks on the drawings and will result in the neatest of drawings (all other things being equal).

When drawing instruments are not being used, they should be kept off the drawing board, because the edges of such instruments pick up graphite; if they are then laid on the drawing board, this graphite will come loose, rest on the drawing, and cause smudging if not cleaned at once.

The drawing board should be cleaned once a week (more often if necessary)—at a slack time, if at all possible. It stands to reason that if the drafting room is pushing to turn out a rush job, or perhaps you are working on a project that is already late, nothing will irritate the job captain more than for one of the draftsmen to choose this time to clean his equipment. Late Friday afternoon is a good time for cleaning the drawing board—just before closing time. Then you will have a clean fresh board awaiting you on Monday morning. The most popular cleaning products include: a cleaning powder, Windex®, or several types of commercially manufactured drafting equipment cleaner.

Two types of board covers are presently in use. One type is applied to the drawing board by dampening the under side of the cover with a wet cloth or sponge, and then stapling it to the board at two- or three-inch intervals. The best method, however, is to use a double-faced tape along the top edge of the board, and then press the board cover on the board. If the cover is left for 24 hours, the surface will be perfectly flat and will hug the drawing board surface without the need of fasteners. After the cover is installed in this manner, the edges of the cover should be trimmed with a razor blade or sharp knife—flush with the board edges.

Other drafting instruments may be cleaned with dry-clean pads, cleaning powder, etc., except ink pens. The latter may be cleaned with warm water immediately after use (be sure not to lose any of the points down the drain), or one of the commercial pen cleaners like the ultrasonic cleaner or liquid pen cleaner may be used. Both types remove dried waterproof India ink from nonabsorbent surfaces such as pens, lettering instruments, brushes, etc.

2.3 Time-saving Drafting Techniques

Many contractors, consulting engineers, manufacturers, and others involved in electrical and electronic drafting have recently whacked their draftsmen's board time—per project—to more than 55 percent while increasing comprehension of the drawings by field personnel. All this is made possible by incorporating several new drafting techniques within the department.

Basic modifications over earlier techniques include the use of improved ghost-free stick-ons printed with symbol lists, equipment schedules, details, and other frequently used data; new and easy-to-draw electrical and electronic symbols and construction details to better convey the information to the workmen on the job; mechanical lettering machines for notes on drawings rather than hand lettering; and special time-saving drafting instruments and supplies.

Most of these drafting practices will find immediate use in all electrical drafting departments—even those turning out the minimal amount of drawings. On the other hand, a few of the devices—such as the mechanical lettering machine—can pay for itself only when used by firms turning out drawings in quantity. Many of the smaller firms merely have their secretaries type the required notes on 8½" × 11" stick-on paper for reproduction. You will seldom see many of these devices in drafting classes; you are supposed to learn the use of basic tools first, and then progress to the less conventional ones.

The *Vari Typer* is one type of mechanical lettering device that reduces the time involved in lettering drawings. This machine is particularly valuable for filling in bills of materials, specifications, notes, as well as for inserting dimensional figures. It is operated much like a conventional electric typewriter, except that the carriage will accept drawing paper of practically any size and the type compares to regular drafting lettering styles in a variety of different sizes. Since this machine produces uniform, clear, and sharp composition of type directly on the tracing paper, it is especially useful for lettering drawings that are to be microfilmed.

A typical application of this machine would be as follows: The draftsman, using lines and symbols, draws all necessary data on the tracing paper in the usual way, except that all notes, dimensions, and similar data requiring lettering are omitted. He then makes a print of the tracing (for reference) and writes all notes, dimensions, etc., in longhand on the print in the correct location and of the approximate size of the lettering. This is normally done with a red pencil. The typist then attaches the print on the wall directly in view (behind the lettering machine), inserts the original tracing in the carriage of the Vari Typer, and types data supplied by the draftsman on the tracing paper. The draftsman also uses some code or other means to indicate the size type of each note or dimension.

For firms requiring this type of lettering machine, they can be leased for approximately $65–$70 per month. Many suppliers also have a lease/purchase plan; that is, you lease the machine, say, for three years, and at the end of that time, you own the machine outright. Free maintenance of the machine usually is included during the leasing period.

As mentioned previously, many drafting firms may not turn out enough drawings to warrant leasing or purchasing a Vari Typer. However, much board time can be saved by grouping as many notes as possible in one location on the drawing and then by having the firm's secretary type these notes on stick-on paper, which is then attached to the original tracing. A typical example of such notes on drawings appears in Fig. 2-21. The method of attaching the stick-on paper to the tracing paper is shown in Fig. 2-22.

UNDERGROUND DISTRIBUTION

1. Concrete pad may be poured in place or may be precast.
2. All conduits shall be installed before placing pad.
3. Conduits shall not be placed under the section of the pad supporting the transformer so that the original ground is not disturbed.
4. Conduit shall be rigid polyvinyl chloride.

Figure 2-21. Example of drawing notes, which may be typed on stick-on paper prior to attaching to the drawing sheet.

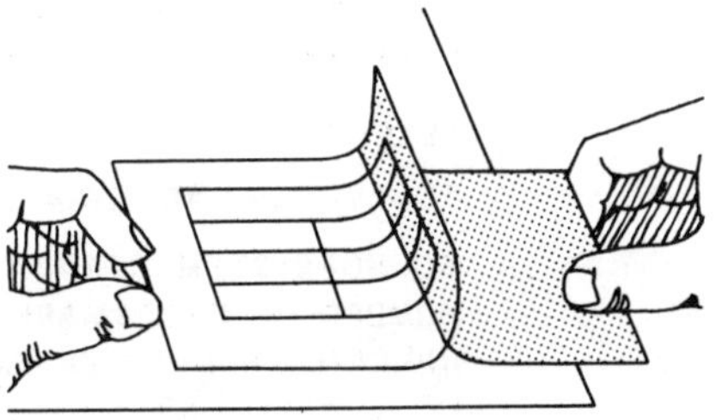

Figure 2-22. Method of attaching the stick-on paper.

The typewriter/stick-on method is utilized by having the draftsman complete all lines, symbols, dimensions, and short notes in the usual manner. However, space is left on the drawing to attach a sheet of typewritten stick-on notes. Then rather than the usual hand lettering of long notes at various locations on the drawing, the draftsman letters, "See Note No. 1 This Sheet," and draws an arrowhead to the item or location to which the note applies; the secretary types the note on stick-on paper for attachment to the tracing at a later date.

When one is using this method, it is best to have the sheet of typewritten notes appear on the sheet to which they apply. Sometimes, however, this is not possible because of either the length of the notes or the amount of drawings on the particular sheet of tracing paper. In such a case, all notes for the project can appear on an extra sheet of tracing paper for the necessary data—usually at the end of the set. For example, if the given project requires three sheets of tracing paper for the working drawings, a fourth sheet may be utilized for all the notes. Then, where long notes are necessary on the other sheets, the draftsman letters (for example), "See Note No. 5 on Sheet E-4."

When it is desired to have the written specifications incorporated into the set of drawings rather than in a separate booklet, this same method may be used. The specifications are typed—as usual—on 8½" × 11" stick-on paper and then attached (in proper order) to the tracing paper of the same size as the working drawings; that is, if the working drawings are drawn on 24" × 36" tracing paper, the 8½" × 11" stick-ons are attached to a sheet of this size. They may then be secured with the set of working drawings—all construction documents being kept together.

If a larger size lettering is necessary than can be produced with the conventional typewriter type, lettering templates such as those shown in Fig. 2-23 may be used without necessitating the purchasing of the more expensive mechanical lettering devices.

Figure 2-23. Typical lettering templates.

Symbols—like the ones described in Chapter 6—have been used for a long time to simplify the draftsman's work. Although engineers and designers attempt to use the symbols adopted by the United States Standards, most consulting engineering firms, as well as contractors and manufacturers, frequently modify these standard symbols to fit their own particular needs. When standard symbols are modified, it becomes necessary to provide a symbol list, or legend, on the drawings for the workmen's interpretation. Drawing and hand lettering a symbol list like the one in Fig. 2-24 can take from two to four hours of the draftsman's time, but if the drafting department would have 25 or so such lists offset printed on stick-on paper, the entire list could be attached to any set of drawings in only a matter of minutes. When additional data are needed, notes may be used to cover the situation at hand.

Electrical detail drawings of a separate item or portion of a system giving a complete and exact description to show the workmen exactly what is required have long been used on electrical and electronic working drawings. But many firms have cut these details to a minimum because of the amount of board time required to produce most of them.

Now, much time can be saved in reproducing necessary electrical and electronic details by having the ones most frequently used printed on stick-on

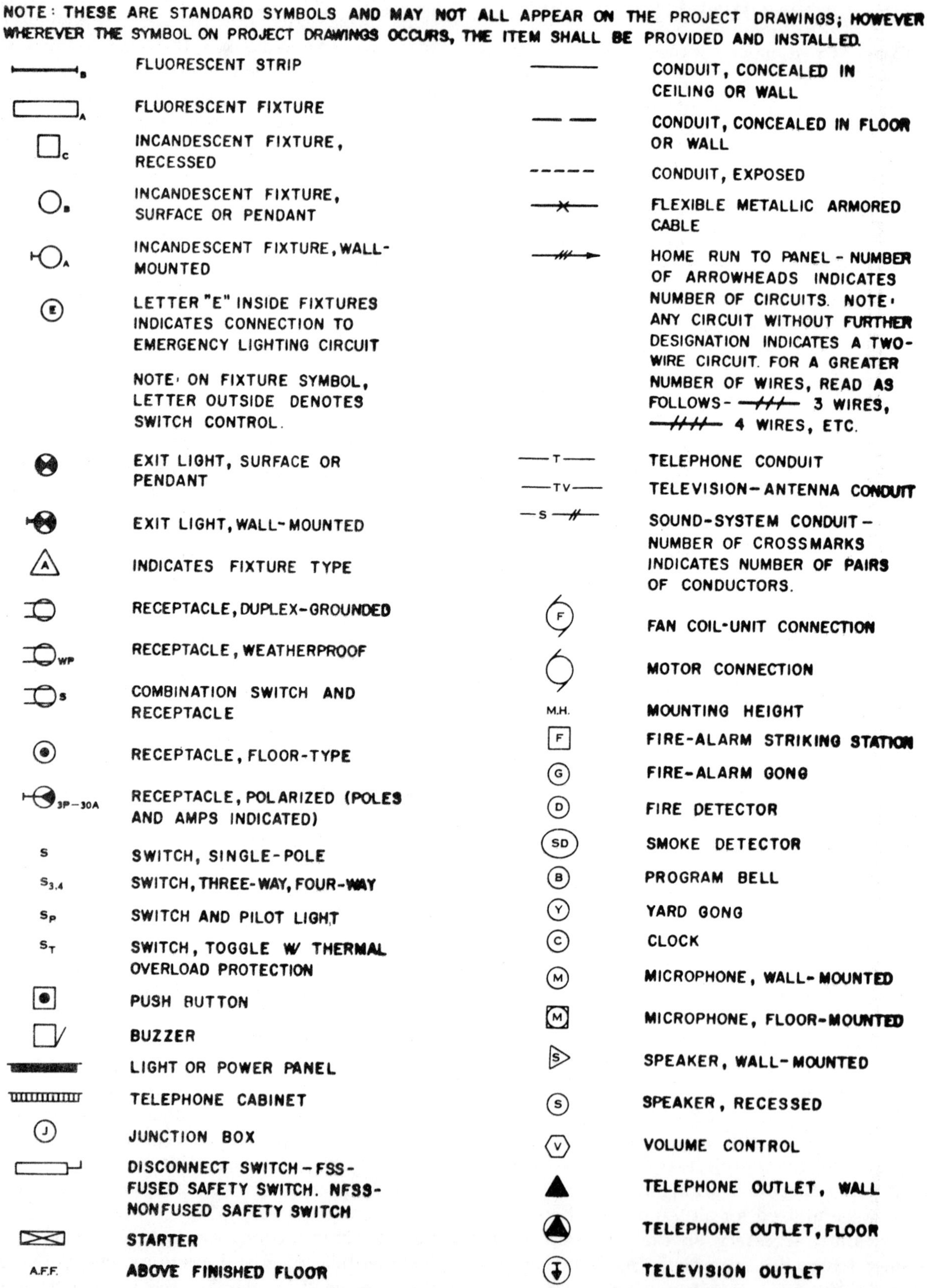

Figure 2-24. (a) A symbol list like this one can take from two to four hours of hand lettering, but only a minute or so if the same list were printed on stick-on paper and attached to the drawing.

PANELBOARD SCHEDULE

PANEL Nº.	TYPE CABINET	PANEL MAIN AMPS	VOLTS	PHASE	BRANCHES 1P	2P	3P	PROT	FRAME	ITEM FED OR REMARKS
MDP	SURFACE	1200	120/208V	3φ,4W	–	–	1	200	200	PANEL A
	SQ."D" TYPE "I"-LINE				–	–	1	200	200	PANEL B
	W/ M.L.O.				–	–	1	200	200	PANEL C
					–	–	1	800	800	PANEL P
					–	–	1	——	200	PROVISIONS ONLY
							1	100	100	PANEL D
"A"	FLUSH	200	120/208V	3φ,4W	39	–	–	20	70	LIGHTING
	SQ.D" TYPE NQO				3	–	–	20	70	SPARES
	W/ M.L.O.									
"B"	FLUSH	200	120/208V	3φ 4W	39	–	–	20	70	RECEPTACLES
	SQ."D" TYPE NQO				1	–	–	20	70	SPARES
	W/ M.L.O.				–	1	–	30	70	WATER HEATER
"P"	SURFACE	800	120/208V	3φ 4W	–	–	5	125	200	ROOF TOP UNITS
	SQ."D" TYPE "I" LINE				2	–	—	20	100	.7 KW DUCT HTR. & [illegible] KW
	W/M.L.O.				1	–	—	20	100	1 KW DUCT HTR.
					–	–	1	50	100	15 KW DUCT HTR.
					–	–	1	90	100	24 KW DUCT HTR.
					–	–	2	—	—	PROVISIONS ONLY
					–	1	–	20	100	1.5 KW DUCT HEATER
					–	–	1	100	100	ROOF TOP UNIT
"C"	SURFACE	200	120/208V	3φ 4W	18	–	–	20	100	LIGHTS, RECEPTACLES
	SQ."D" TYPE NQO				–	1	–	50	100	220 V RECEPT.
	W/ M.L.O.				10	–	–			PROVISIONS
					–	1	–	30	100	WATER HTR.
					–	–	1	20	100	CONVEYOR 1½ HP.
					1	–	–	20	100	EXH. FANS
					4	—	–	20	100	SPARES
					2	--	—	20	100	REAR SIGN
"D"	FLUSH	100	120/240V	3φ 4W	8	–	–	20	100	SIGN & SHOW WINDOW OUTLET & LIGHTS
					8	–	–	20	100	SPARES
						FURNISH HANDLE LOCKS FOR DESIGNATED BRANCH CIRCUITS - SEE DRAWINGS				

Figure 2-24(b). A well-organized schedule that would otherwise take several paragraphs of notes to convey the same information.

paper and filed. Whenever the detail is needed for a particular project, the backing paper is removed, and the detail is then applied to the tracing paper in only a few minutes.

Those details that are not used as frequently may be drawn freehand to a large scale on scaled ruled paper. At full size these freehand drawings do not have a neat appearance, but when they are reduced to, say, one-fourth their original size on a Xerox reduction copier, they will appear to have been drawn with a straightedge and templates if the draftsman uses reasonable care on the enlarged version. This method has cut the drawing time for certain details in half for many firms employing draftsmen.

The reduction in board time required to turn out these details means that the firm can show more of them on the project drawings, which in turn will help reduce labor on the project itself; complex systems will be better understood by the workmen.

The electrical/electronic draftsman should also make use of schedules on the working drawings whenever practical. When properly organized and presented, not only are schedules powerful time-saving devices for the draftsman and engineer, but they also save the specification writer and the workmen on the job much valuable time. Figure 2-24b shows one type of well-organized schedule that would otherwise take several paragraphs of notes to convey the same information.

Those in charge of drafting departments should always be on the lookout for new drafting tools and supplies that will lessen board time while turning out neater and more accurate drawings. The draftsman himself should also be on the lookout; if he comes up with a new device or method that will save his boss time, you can be certain that it will not go unnoticed. Many such tools have been on the market for years, but for some reason many draftsmen have never used them. Some of these tools include a lead pointer instead of the old sandpaper board; proportional dividers (Fig. 2-25) or a panograph (Fig. 2-26) for making copies of drawings to a different scale; crosshatch and lettering templates, and the like.

Figure 2-25. Proportional dividers used to make copies of drawings to a different scale.

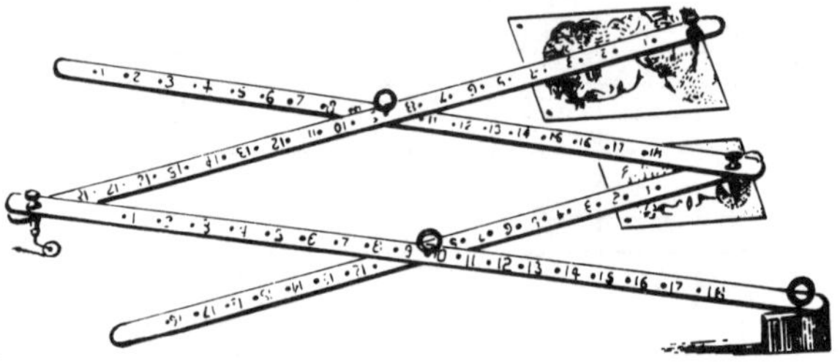

Figure 2-26. Another device for making copies of drawings to a different scale is a panograph.

The possibilities are almost endless, and we have only touched upon the subject, but the ones described should be sufficient to start draftsmen and those in charge of drafting departments thinking in the right direction.

The drafting facilities also play an important role in productivity. The draftsmen should have a place to work in the firm's establishment where they will not be disturbed by the usual office routine—the coming and going of salesmen and general shop activities.

A separate drafting room should be provided and, if possible, separate rooms for each designer. The room should be provided with good illumination and heating and cooling facilities, and the surfaces should be painted a pleasing color. A dingy, poorly lighted and ventilated drafting room is certainly not conducive to the preparation of good drawings.

SUMMARY

High-quality tools and the ability to use them correctly are necessary for any type of high-quality work. The draftsman should also know how to care for his tools and most important, to keep them clean at all times.

There are many time-saving tools and methods used in many drafting rooms to save board time, but the student draftsman should master the use of conventional drafting tools before attempting the various "short-cut" devices.

QUESTIONS

Check the correct answer for the following true or false questions. If any part of the statement is false, the statement should be marked "false."

		T	F
1.	A 45-deg triangle and a 30–60-deg triangle may be used in combination to form a 15-deg triangle.	___	___
2.	The architect's scale is the type most often used on maps and building site plans.	___	___
3.	The compass found in most sets of drawing instruments is seldom used today, because templates are easier to use when one is drawing arcs or circles.	___	___
4.	Irregular curves are used in electrical drafting for drawing some circuit lines and in electronic drafting for ruling curved lines other than circle arcs.	___	___
5.	Sandpaper boards are more useful to the modern draftsman than mechanical lead pointers.	___	___
6.	Erasing shields are used only when the entire drawing must be erased.	___	___
7.	Masking or drafting tape is the preferred method of securing drawing paper to the board.	___	___
8.	A lead pointer is used to point out the various grades of lead used in electrical/electronic drafting.	___	___
9.	All drafting areas should be provided with good illumination, but ventilation is never worth bothering with.	___	___
10.	A mechanical lettering machine can pay for itself only in firms turning out large quantities of drawings.	___	___

Chapter 3

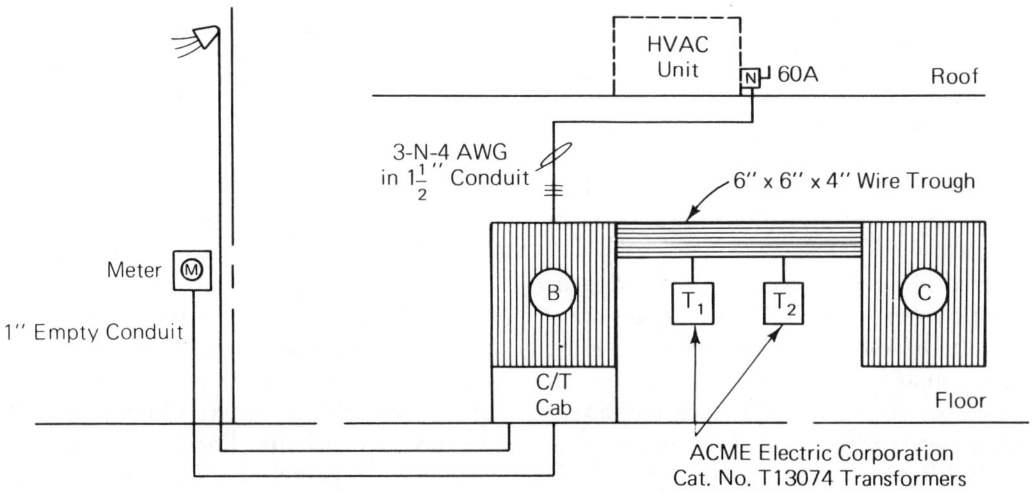

BASIC DRAFTING

Previous chapters have dealt with preliminary requirements of electrical and electronic drafting in order to give the reader a solid foundation for approaching the drafting profession in an intelligent manner. This chapter will begin with actual drawing. The techniques described will be used throughout your drafting career.

3.1 Line Work

All drafting lines have one thing in common—they are all the same color. However, good easy-to-read contrasting lines can be made by varying the width of the lines as well as by selecting various hardnesses of lead.

Drawing lead is graded according to its degree of hardness, which ranges from 7B (the softest and blackest) to 9H, the hardest. The degree of lead to be used depends on such factors as grade of paper and the kinds of lines to be drawn. The lines illustrated full size in Fig. 3-1 are recommended for all electrical and electronic drafting and should be used for all assignments in this book.

The *extraheavy full line* [Fig. 3-1(a)] is used for border lines on drawing sheets and also around schedules or any other place where emphasis is required.

The *heavy full line* [Fig. 3-1(b)] is used for all circuit lines and symbols and for other electrical/electronic components or equipment supplied by the circuits.

The *light full line* [Fig. 3-1(c)] is used for section lines, background outlines of circuit boards, building floor plans, and similar uses where the object to be drawn must be seen but is secondary to the electrical/electronic system. This type of line is also used for dimensioning.

The *extralight full line* is used for guidelines for lettering, layout lines, and other uses where the object drawn should not show up on the final drawing or blueprint.

The *broken line* [Fig 3-1(d)] is used for showing hidden objects or to indicate certain types of electrical circuits, such as existing or future circuits, shielding, and mechanical connections.

The *centerline* [Fig. 3-1(e)] is used to indicate the centerline of an object or parts of an object.

The *cutting plane line* [Fig. 3-1(g)] is used to show the position of the cutting plane.

The *light break line* [Fig. 3-1(i)], which is drawn partially with a straightedge and partly with freehand zigzags, is used for long breaks.

The *heavy break line* [Fig. 3-1(h)]—drawn freehand—is used for short breaks.

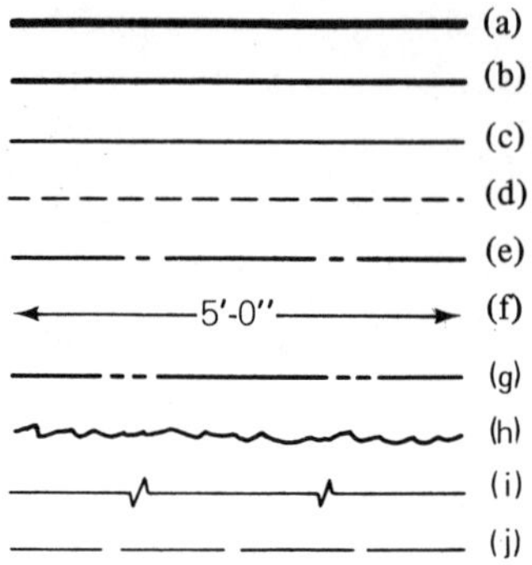

Figure 3-1. Typical drafting lines.

The drawing in Fig. 3-2 shows the practical application of some of the lines previously described.

In order to produce clear sharp lines, your drawing lead must be kept sharp at all times with a conical point—the best type of point for most work in electrical and electronic drafting. The lead holder should be tilted about 60 degrees from the paper in the same direction in which the line is being drawn (Fig. 3-3). Also tilt the pencil about 15 degrees away from the edge of the straightedge, as shown in Fig. 3-4. The lead holder or pencil should be rotated slowly in your hand as the line is being drawn; doing so will keep the lines uniform in width and will prevent your point from becoming flat on one side—resulting in a "fuzzy" line. Several strokes should be used for each visible line.

To draw horizontal lines, place the head of the T-square firmly against the left edge of the board for right-handed draftsmen, and against the right edge of the board for left-handed draftsmen. While holding the T-square against the appropriate edge, slide it near the point where the line is to be drawn (Fig. 3-5). Then position the hand so that fingers are on the blade of the T-square and the thumb is on the board, as shown in Fig. 3-6. Then move the blade to the exact position where the line is to be made. Then make horizontal lines from left to right (right to left for left-handed draftsmen), following the directions given earlier. It is also recommended that the little finger on the hand holding the pencil glide lightly on the blade of the T-square as the line is drawn.

Triangles are used in combination with the T-square for drawing vertical lines. In drawing such lines, the triangle is held against the upper edge of the T-square (Fig. 3-7) or parallel edge, and lines are drawn from the bottom toward the top. Again, hold the pencil so that it slants at an angle of about 60 deg to the paper in the direction the line is drawn and slightly (15 deg) away from the triangle.

Although circles and arcs can be drawn with a

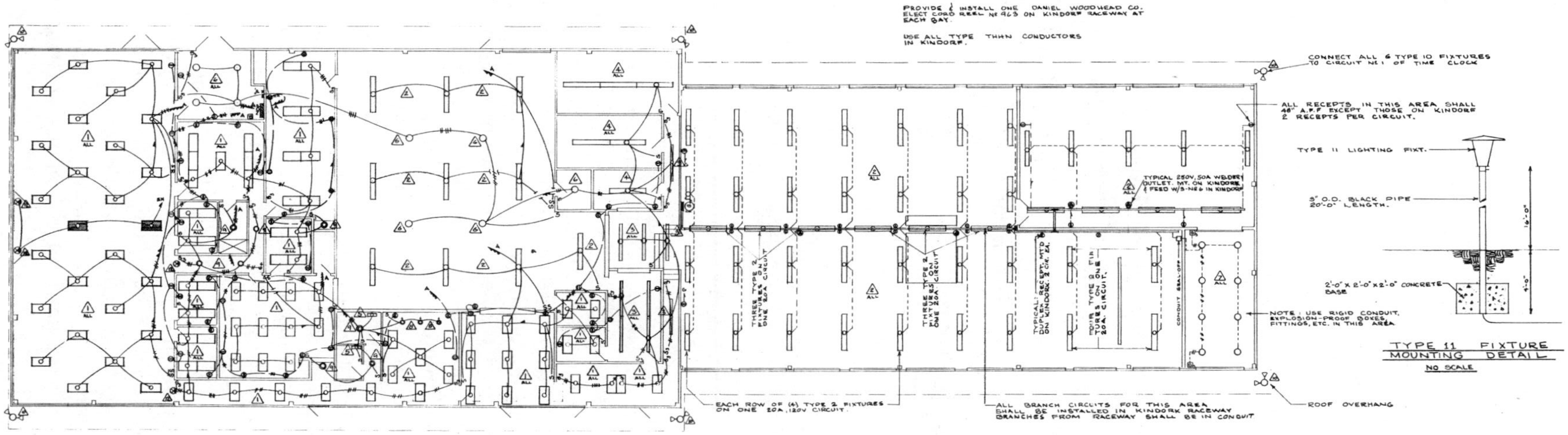

Figure 3-2. Practical application of the lines shown in Fig. 3-1.

Figure 3-3. Tilt lead holder about 60 degrees from paper in the same direction the line is being drawn for horizontal lines.

Figure 3-4. Tilt lead holder about 15 degrees from triangle for vertical lines.

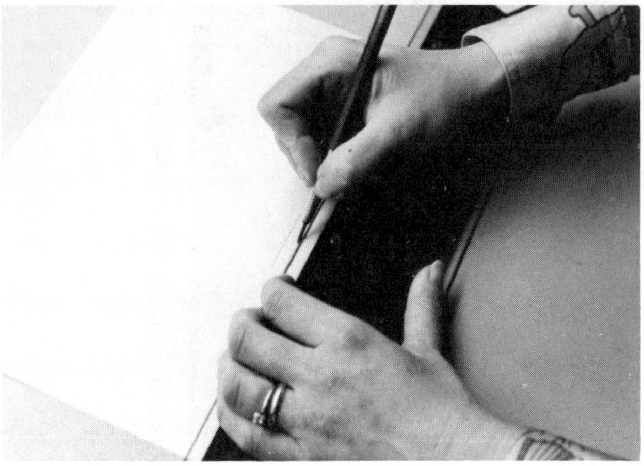

Figure 3-5. Sliding the edge of the T-square where the line is to be drawn.

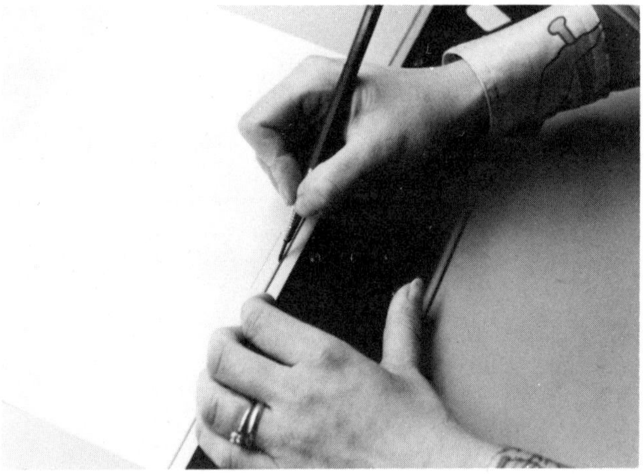

Figure 3-6. Position of the hand and fingers for drawing lines with the T-square.

Figure 3-7. Method of positioning triangle on T-square for drawing vertical lines.

bow compass, it is recommended that circle templates, as will be described later, be used whenever possible. A more uniform line with better contrasts can be made with the template, and the use of the bow compass necessitates piercing a hole in the drawing paper when the point is placed in position on the drawing. Sometimes, however, when it becomes necessary to use very large circles on the drawing, the beam compass must be used.

To use the beam compass, set the points to the required radius. Then hold the beam compass with two hands, as shown in Fig. 3-8, and draw the circle or arc.

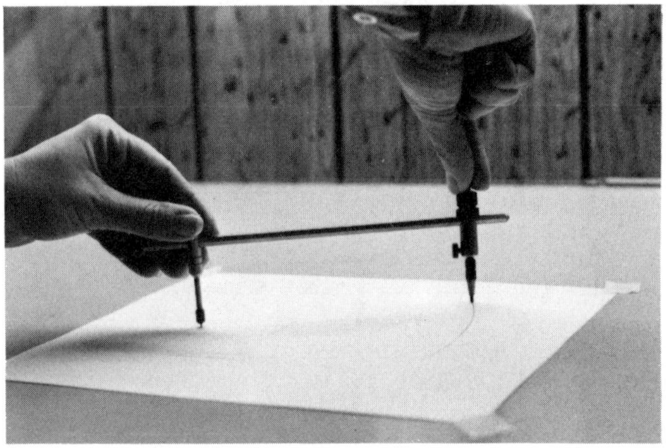

Figure 3-8. Method of holding beam compass for drawing large circles or arcs.

As mentioned previously, line contrast is obtained by either varying the widths of lines or by using different hardnesses (degrees) of lead or a combination of the two. To illustrate, the *extralight full line*—used for lettering guidelines or layout work—should be made with lead of between 6H and 9H hardness; the point should be very sharp and drawn with the minimum of pressure. In doing so, be extremely careful not to rip the drawing paper, as these hard leads will easily do so. The *light full line* can be made with lead hardness of between 6H and 4H.

For heavier lines, such as the types used for border lines and emphasis, use lead of either 2H or H and draw a sharp glossy jet-black line by making at least three strokes back and forth along the straightedge or triangle. Remember to slowly rotate the pencil at all times to prevent the lead from becoming flat on one side. Once the first line is drawn in the desired location, move the straightedge slightly away from the line that was just drawn—not enough to make two separate lines, but just enough to make the first line wider. If done properly, the result will be a neat, extraheavy, solid line.

SUMMARY OF 3.1

All drafting lines are the same color—black. Therefore, contrast can be obtained only by varying the width of the line and by using lead of different degrees of hardness.

A conical-shaped point on the drawing lead is preferred for all electrical and electronic drawing and lettering.

Drawing lead is rated in hardness, which ranges from 7B (the softest and blackest) to 9H, the hardest.

The drawing pencil should always be rotated when drawing lines to insure sharpness in both the lead and the line being drawn.

ASSIGNMENT

Before we continue with this chapter, it is recommended that the reader perform the following exercises:

1. Clean the dust from your drawing board with your dust brush. Then place a sheet of 11″ × 17″ tracing paper on the board. Adjust the paper so that its top edge is parallel to the upper edge of the T-square. Holding the paper in this position, fasten the corners of the paper to the board with small strips of drafting tape. Your board should now look similar to the one in Fig. 3-9.

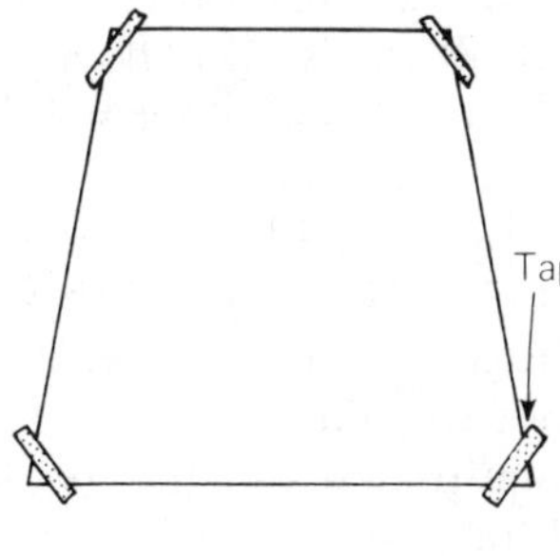

Figure 3-9. Appearance of drawing paper once it is fastened to the drawing board with drafting tape.

2. Measure ¼ inch in from all sides of the paper, making very light dots to indicate these measurements. Now draw a border line completely around the paper, as shown in Fig. 3-10. Use the extraheavy full line described in this chapter for the border.
3. Measure in from your left-hand border line and draw one vertical extralight line from the top border line to the bottom border line. Now measure in from the right-hand border line and draw another vertical line at this location.

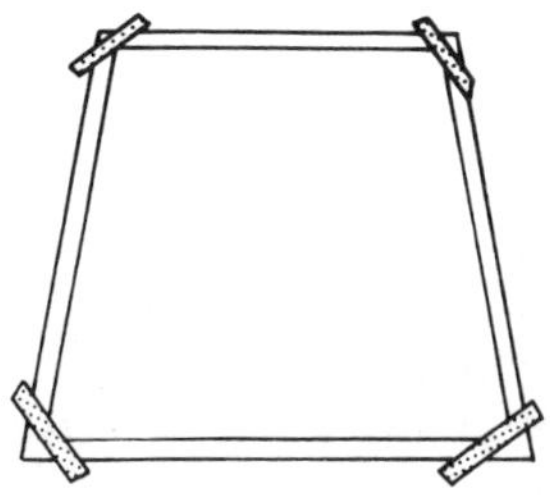

Figure 3-10. Drawing sheet with ¼ in. border lines drawn around the perimeter of the paper.

4. Starting approximately one inch below the top border line, practice drawing each type of line covered in this chapter from the left extralight line to the right extralight line. Each type of line should be drawn at least ten times. Use Fig. 3-1 as a guide.
5. Continue on down the sheet, but leave a blank space of at least 3 in. horizontally and 2 in. vertically in the lower right-hand corner (within the border line) of the drawing sheet. This will be used to insert other information at a later date. If your name needs to be placed on the drawing for identification purposes, do so in this blank space, using a very light line so that it may be erased later when you letter in the title block.

QUESTIONS

1. Why is it necessary to have different line widths and different types of lines in electrical and electronic drawings?
2. What kind of point is preferred for all electrical and electronic drafting?
3. List the various hardnesses of lead from the softest to the hardest.
4. Name two ways to draw circles on working drawings.
5. What drafting tools are used in combination with the T-square to draw vertical lines?
6. Why should the lead holder or pencil be rotated slightly when one is drawing lines?
7. What type of line should be used to show the position of a cutting plane?
8. For what purposes should the *extralight full line* be used on electrical and electronic drawings?
9. For what purposes are *extraheavy lines* used?
10. The *light break line* is drawn partially with what two methods?

3.2 Lettering

The appearance of any drawing depends to a great extent upon the appearance of the lettering on the drawing. Sufficient data, such as figures, dimensions, and explanatory notes, are used extensively on practically all electrical and electronic drawings, and it is essential that they be lettered on the drawing in a neat and legible manner.

There are several styles of lettering in use, but since lettering on electrical and electronic drawings must be of such a nature that it can be executed easily and rapidly by the draftsman, the single-stroke Gothic letters are the most popular. This type of lettering does not mean that an entire letter is formed with a single stroke; rather, it means that a line of the required thickness is made with one stroke. This type of lettering may be either vertical or inclined (Fig. 3-11). Both types are acceptable in electrical and electronic drawings, but some industries may prefer one over the other for their particular drawings. For this reason, the student should become proficient in using both the inclined and vertical letters. He will then have little trouble in adapting to the particular style of his employer when he obtains a job.

THIS IS VERTICAL LETTERING

THIS IS INCLINED LETTERING

Figure 3-11. Examples of vertical and inclined lettering.

In order that the lettering on a drawing may have a legible and pleasing appearance, the letters must be well formed, must have a good relative width, and must be spaced so that the distances between them appear to be uniform. Methods of forming all letters and numerals are listed in Fig. 3-12 for vertical letters and in Fig. 3-13 for inclined letters. These are recommended strokes for the beginner. However, after the draftsman gains skill in forming neat and well-proportioned letters with this method, he may find that a method of forming letters with slightly different strokes may be easier and faster.

Before you begin practicing the art of lettering, you should learn to make guidelines and use them whenever you are lettering a drawing. Such guidelines enable draftsmen to form neat and professional-looking letters and numbers at all times. As mentioned earlier in this chapter, all guidelines should be made with lead no harder than 6H; many prefer 9H.

The lettering guide is a drafting aid that enables the draftsman to draw evenly spaced parallel guidelines for lettering and numbering rquirements on working drawings. This type of guide is referred to as an Ames lettering guide. It is used by sliding it along the T-square with the point of a pencil placed in the proper hole.

With the Ames lettering guide, it is possible to draw guidelines and slope lines (used for inclined lettering) for letters and numbers from 1/16 in. to 2 in. in height. The numbers from 10 through 2 on the disc denote the height of letters in thirty-seconds of an inch. To illustrate, assume that you want letters ¼ in. high. Since the numbers are gauged in 1/32 of an inch, rotate the disc with the right hand until the number "8" is at the frame index, as shown in Fig. 3-14. Eight thirty-seconds is ¼ in.

To turn the disc clockwise, hold the frame at the upper left-hand corner with the left hand and rotate the disc with the right hand. To turn the disc counterclockwise, use the reverse procedure.

Again, always use 6H or harder drawing lead for all guidelines. Then, with the readable side up, position it on the straightedge as shown in Fig. 3-14. Insert a conical-shaped lead into the proper hole, and with the lead holder perpendicular to the drawing paper, slide the lettering guide across the drawing paper at the location desired for the guidelines. Make certain that the guide frame rests flat on the straightedge at all times. When one line is drawn, move the pencil lead to another hole within the column to produce another; then move from hole to hole to produce a series of guidelines for lettering.

If the number 10 is placed on the center mark, the holes in the disc form three vertical columns. The center column is the one used to make guidelines for all lettering requirements in this book. This column may be called the ½-ratio column, because all of the lines are equally spaced. When using this column for guidelines, start with the top hole and then use every second hole, as shown in Fig. 3-15. For multiple lines, this gives the same distance between lines as the height of the letter. However, in most cases, drafting departments prefer the spaces between the letters—when two or more lines occur—to be made half the height of the letter, as shown in Fig. 3-16.

Figures 3-17 through 3-20 show the recommended adjustments of the Ames lettering guide for various applications on electrical and electronic drawings.

The use of the triangle in lettering will help the beginning draftsman to form the vertical strokes uni-

Figure 3-12. Method of forming vertical letters and numerals.

Figure 3-13. Method of forming inclined letters and numerals.

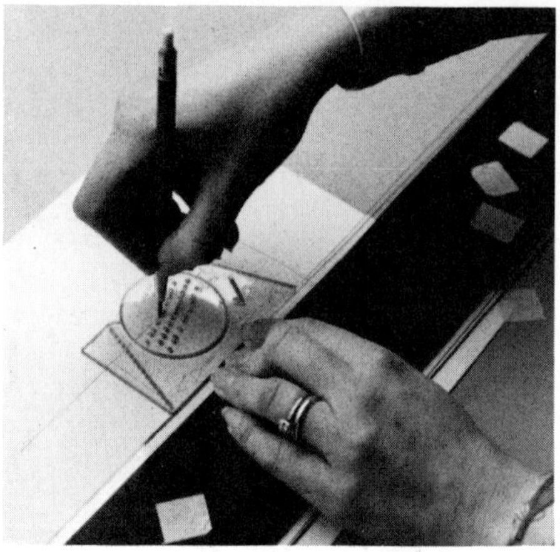

Figure 3-14. Method of positioning the Ames lettering guide on the straightedge for drawing guidelines.

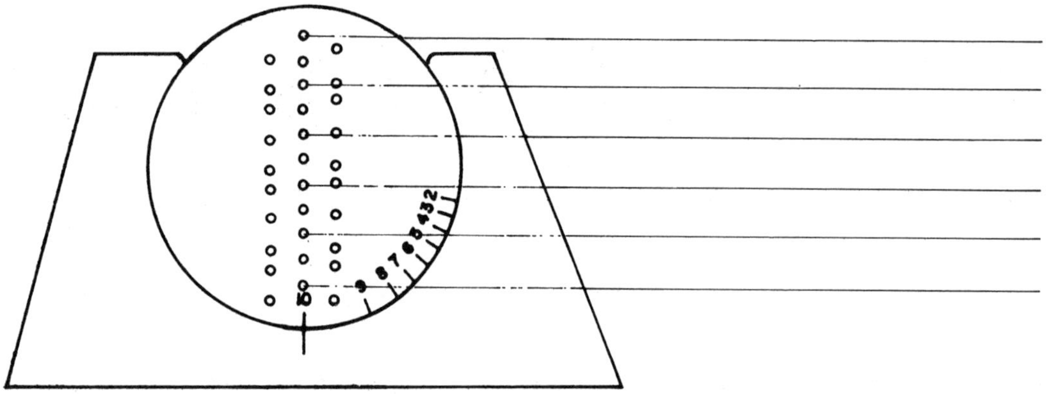

Figure 3-15. Lettering guide set at No. 10, showing method of drawing guidelines equally spaced.

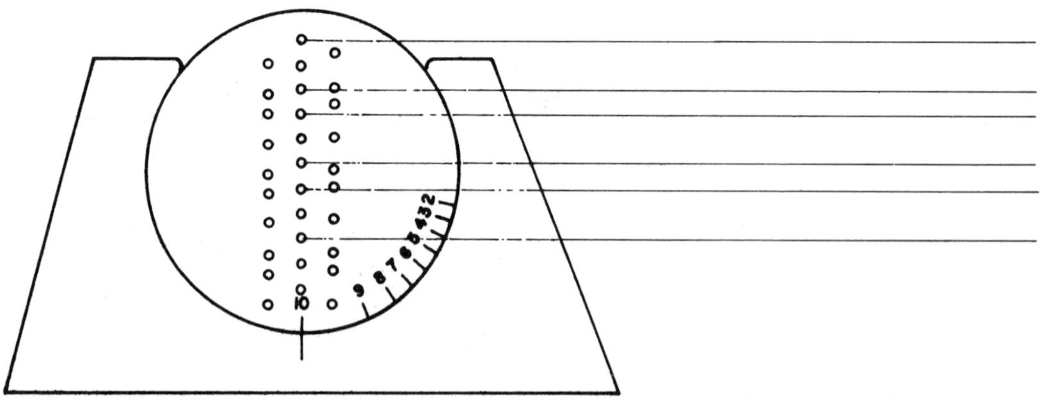

Figure 3-16. Lettering guide set at No. 10, showing method of drawing guidelines when spaces between lines are to be half the height of the letters.

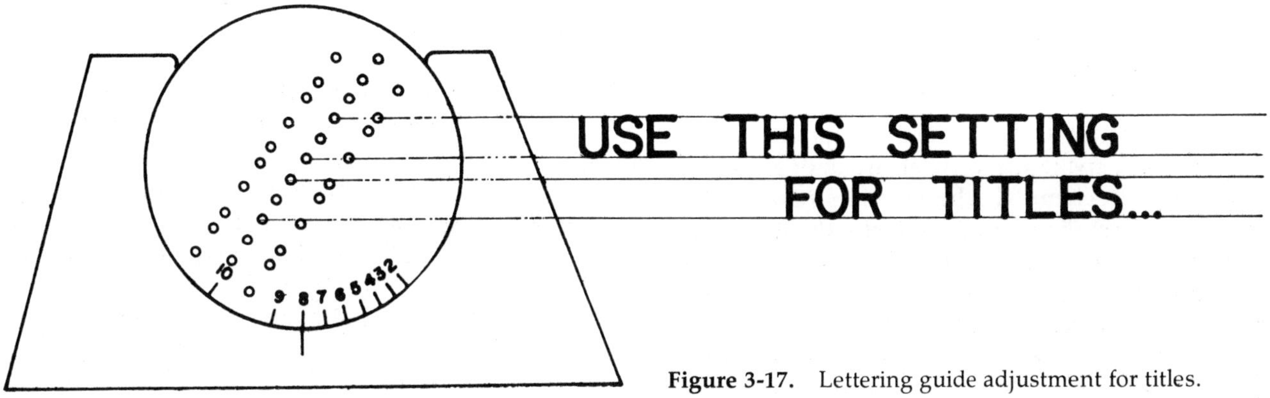

Figure 3-17. Lettering guide adjustment for titles.

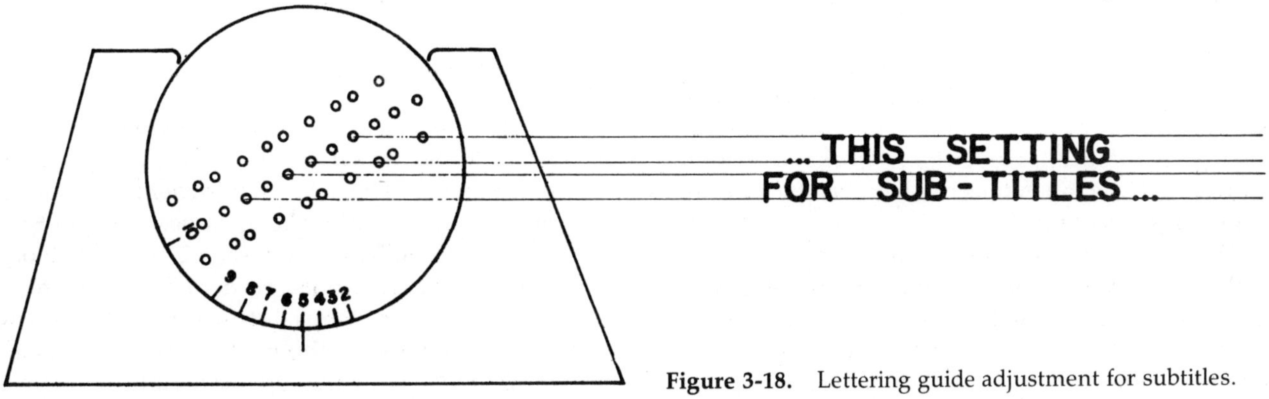

Figure 3-18. Lettering guide adjustment for subtitles.

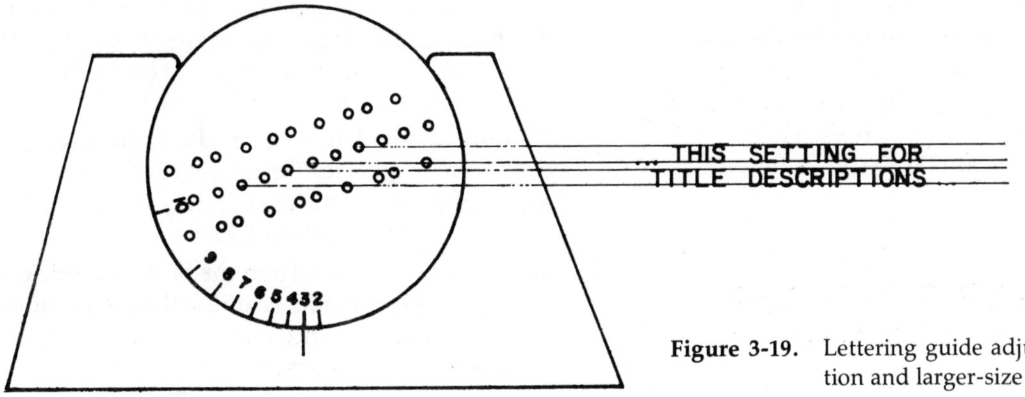

Figure 3-19. Lettering guide adjustment for title description and larger-size notes.

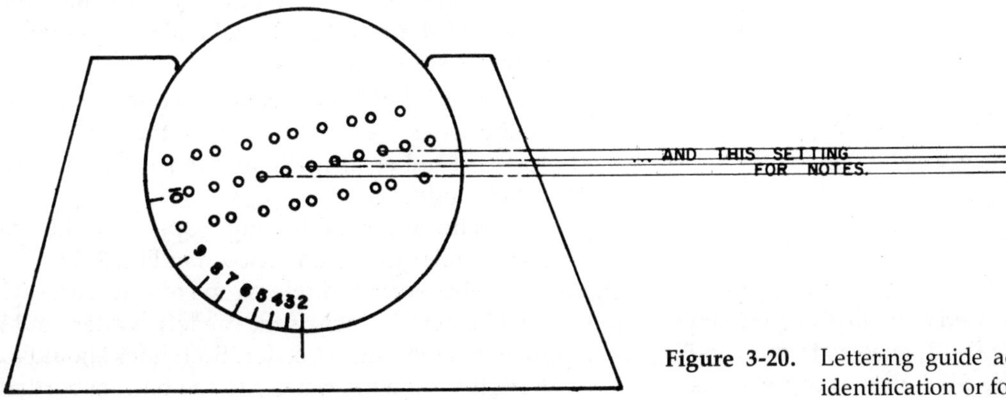

Figure 3-20. Lettering guide adjustment for component identification or for smaller-size notes.

formly when using the vertical style of single-stroke Gothic lettering. For example, if the word "letter" is to be lettered, first draw the guidelines with a 6H or harder lead. Then with the use of a straightedge and a 90-deg triangle, make the vertical stroke of the letter L; then draw freehand the horizontal stroke of this letter. Again form the vertical stroke for the first letter E with the straightedge and triangle; then form the three remaining horizontal strokes of the letter freehand. The two vertical strokes of the two T's in the word are also made with the 90-deg triangle, while the horizontal tops of the T's are made freehand. The next letter E is made as before. The letter R is formed by using the triangle to form the vertical stroke and then drawing freehand the remaining portion of the letter.

This type of lettering aid is a great help in making uniform vertical strokes which, in turn, help to insure a neat appearance. However, the draftsman should strive to master an entirely freehand method of lettering, since much time can be saved by lettering all drawings freehand.

The area between letters should appear to be equal, although some letters and numbers, because of their shape, must be wider, whereas others need to be narrower. Still, by varying the spaces correctly, the overall impression will be that the letters are spaced evenly and will give a pleasing appearance. Equally important is the spacing between the words, sentences, and lines. For example, the distance between words should be approximately equal to the height of the letters, while the spacing between sentences should be twice the height of the letters. Figure 3-21 illustrates these points. Figure 3-22 shows how fractions should be lettered.

THIS IS AN EXAMPLE OF PROPER
SPACING BETWEEN LETTERS, ETC.

Figure 3-21. Example of proper spacing between letters and sentences.

$1\frac{1}{2}$ $2\frac{3}{4}$ OR $1^1/_2$ $2^3/_4$

Figure 3-22. Method of lettering fractions.

While a hard lead is used to make the guidelines for lettering, a much softer lead, such as H or even F, should be used for making the letters and numerals.

SUMMARY OF 3.2

The appearance of any drawing depends a great deal upon the appearance of the lettering on the drawing sheet.

Single-stroke Gothic letters are the most popular style in use today for electrical and electronic drawings. Both vertical and inclined methods should be learned.

Neat lettering requires that the letters

1. Be well formed.
2. Have good relative widths.
3. Be spaced so that the distance between them appears to be uniform.

ASSIGNMENT FOR 3.2

1. Secure a sheet of 11" × 17" drawing or tracing paper to the drawing board with small pieces of drafting tape. Draw ½-inch border lines around the perimeter of the sheet, as described in Sec. 3.1. Be sure to leave room for a 2" × 3" title block in the lower right-hand corner of the drawing.

   Set your Ames lettering guide at the number 8 mark on the center mark and, beginning 1 inch below the top border line of the tracing paper, draw a series of three sets of guidelines as shown in Fig. 3-23. Use 6H or harder lead. Draw these guidelines from the left border to the right border, as shown in Fig. 3-17.
2. Measure down 1 inch from the bottom guideline; make a light dot for reference. Readjust the Ames guide to the number 5; then make another series of three sets of guidelines.
3. Measure down 1 inch from the bottom guideline in the series just made; readjust the Ames guide to the number 4, and make another series of three sets of guidelines.
4. Measure down 1 inch from the bottom guide line in the series just made; readjust the Ames guide to the number 3; make another series of three sets of guidelines.
5. Measure down 1 inch from the bottom guideline in the series just made; readjust the Ames guide to the number 2; make another series of three sets of guidelines.

   The sheet of tracing paper should now look similar to the one shown in Fig. 3-24.

   Use your triangle to make one very light vertical line ½ inch from the left border and ½ inch from the right border. Both lines should extend from the top border to the bottom border.

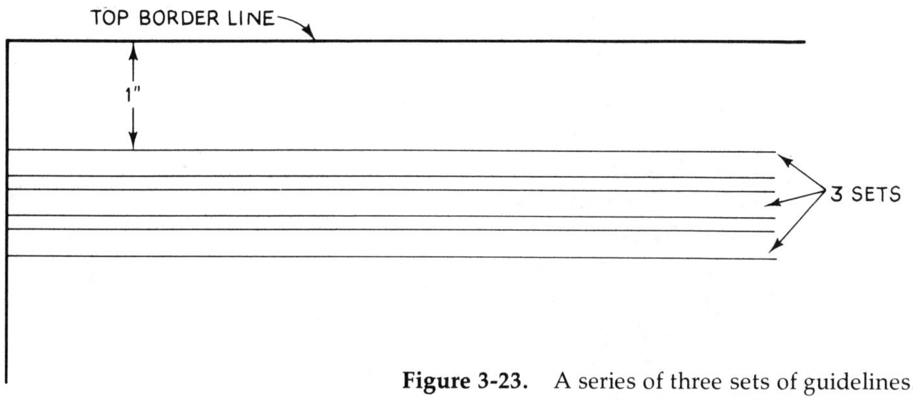

Figure 3-23. A series of three sets of guidelines.

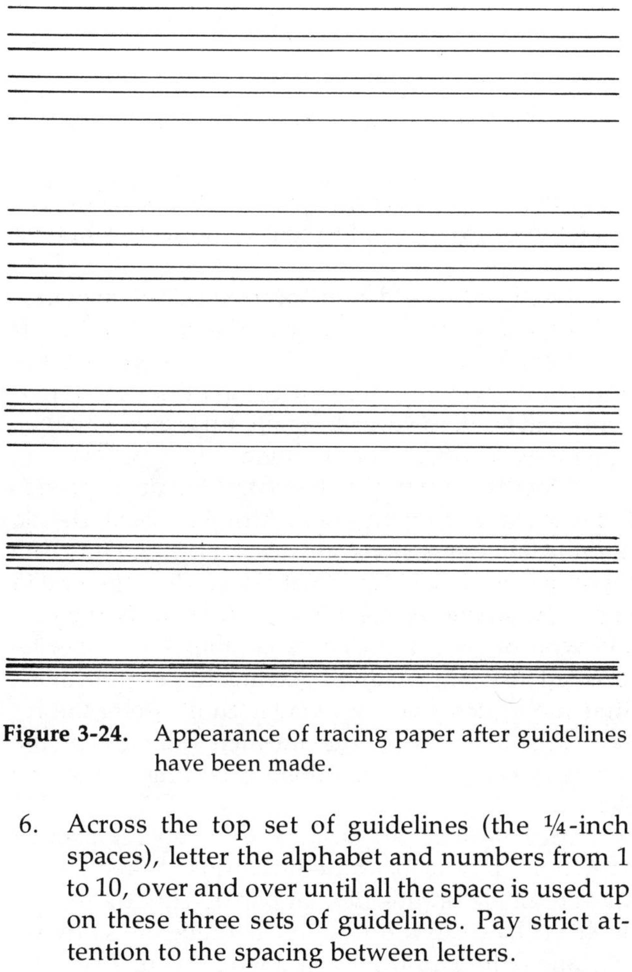

Figure 3-24. Appearance of tracing paper after guidelines have been made.

6. Across the top set of guidelines (the ¼-inch spaces), letter the alphabet and numbers from 1 to 10, over and over until all the space is used up on these three sets of guidelines. Pay strict attention to the spacing between letters.

   Go to the next set of guidelines and letter the alphabet and numbers from 1 to 10 until all the space is used up in this set. Continue on to each set until all guideline space is used up.

QUESTIONS

Answer the following questions according to data presented in this chapter. Write your answer on a separate sheet of paper.

1. In addition to achieving uniform height of letters, in what other respects is uniformity in lettering achieved?
2. What type of lettering aid is used in making uniform vertical strokes in lettering?
3. Why is neat lettering on electrical and electronic drawings very important?
4. Name an effective way of making guidelines.
5. How are the widths of the guidelines changed when one is using the Ames lettering guide?
6. What hardness of lead is recommended for guidelines on electrical and electronic drawings?
7. Although there are several styles of lettering in use, what style is used mostly on electrical and electronic drawings?
8. There are two types of lettering used with the style in Question 7; what are they?
9. The Ames lettering guide may be adjusted to draw guidelines for lettering from _____ inch to _____ inches in height.
10. The distance between words—while lettering—should be approximately equal to the _____ of the letters, while the spacing between sentences should be equal to twice the _____ of the letters.

3.3 Sketching

The subject of sketching is very important to the electrical/electronic draftsman, because many ideas and designs are usually expressed first in the form of a freehand sketch before they are later translated into finished drawings by the draftsman. The draftsman may also be called upon to go out and take measurements and sketch existing electrical systems so that they can be modified or updated. Therefore, it is essential that the electrical/electronic draftsman obtain the proper knowledge to quickly produce and interpret freehand sketches.

Before starting a sketch, study the system or apparatus to be sketched and decide upon what views will be necessary and how they will be arranged. They may be either orthographic or pictorial, depending on the function for which they are intended. For example, suppose that an electrical draftsman is required to furnish a sketch of an existing main distribution panel in a commercial store. The sketch could be isometric (Fig. 3-25) or orthographic (Fig. 3-26). Since electrical conduits enter the panel on several sides, it would probably be best—in this case—to

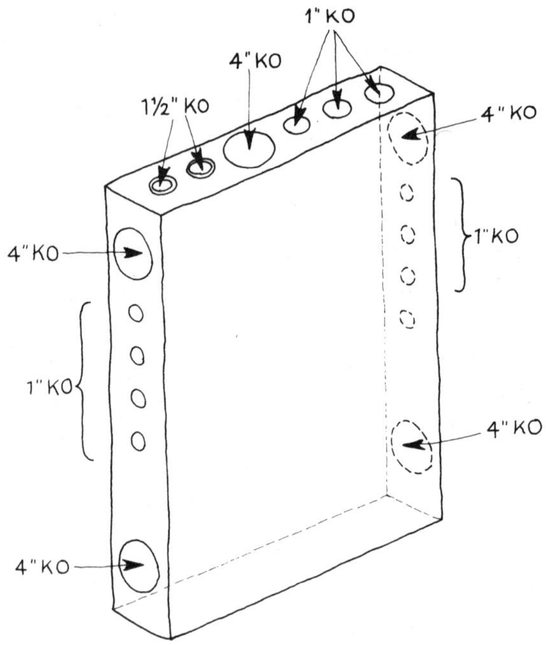

Figure 3-25. Example of an isometric sketch.

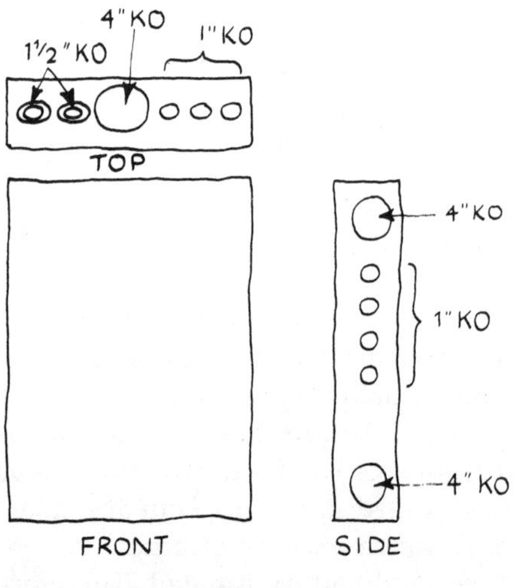

Figure 3-26. An orthographic sketch.

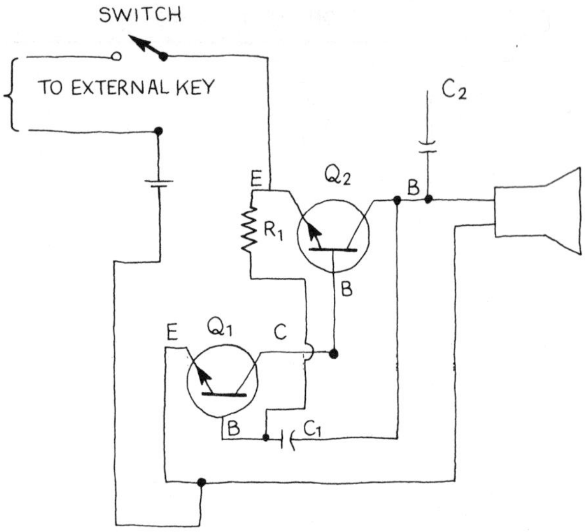

Figure 3-27. Sketch of a circuit board with components.

use orthographic views, as two of the sides in the isometric drawing would require the use of hidden lines.

On the same order, an electronic draftsman may be required to sketch the components on a circuit board as shown in Fig. 3-27 before preparing the finished drawing in ink, to ascertain that all components will fit on the board. Again, this could either be a pictorial drawing or an orthographic view.

Once the views have been decided upon, the object is sketched in with a clear firm line. Next, decide upon what dimensions are necessary and indicate these on the sketch. The final sketch (like the one in Fig. 3-28) should be complete, so that either the person who made the sketch or another can make the finished drawing without having to see the items after the sketch has been made. To illustrate the result, Fig. 3-29 shows the finished drawing of the sketch in Fig. 3-28—everything was complete on the sketch.

Most sketches are best made on paper imprinted with nonreproducible blue-line grids. When drawings are made on this type of paper, only the sketch itself will be reproduced on a blue-line copying machine—the grid itself will not show up. This type of paper also makes it easier to draw straight lines and secure good proportions, as each square of the grid can represent one inch, one foot, or any other desired size.

Besides the paper, the only other materials required for freehand sketching are some pencils, sharpening devices, and an eraser.

In order to develop a satisfactory sketching technique, attention should be paid to the direction,

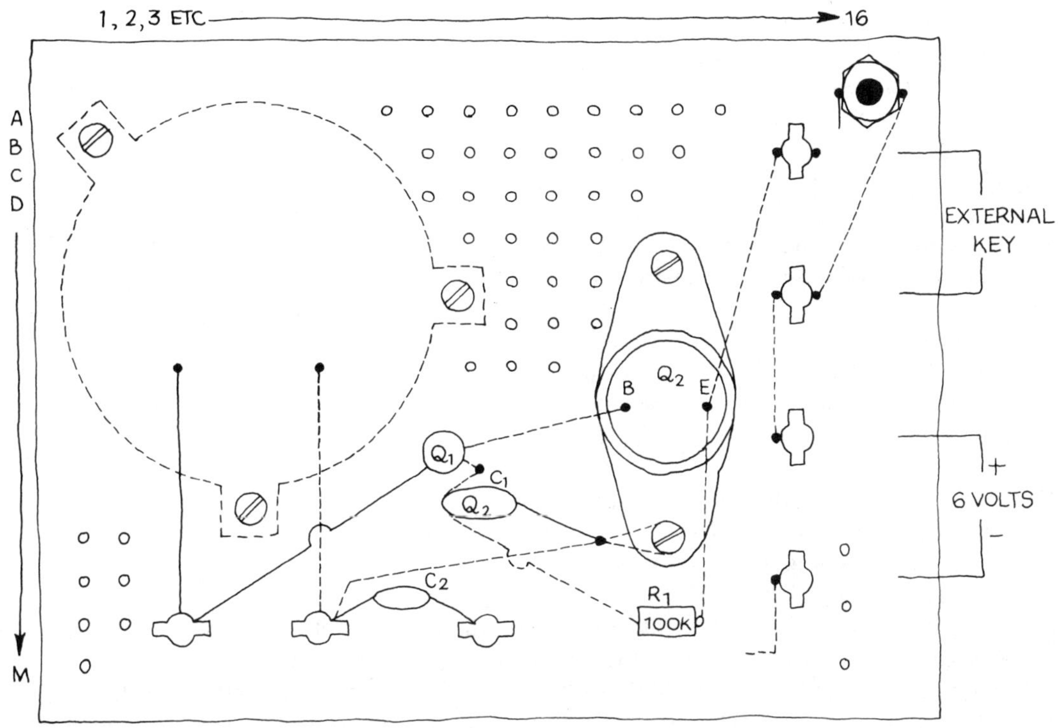

Figure 3-28. A complete sketch used as a guide for a finished drawing.

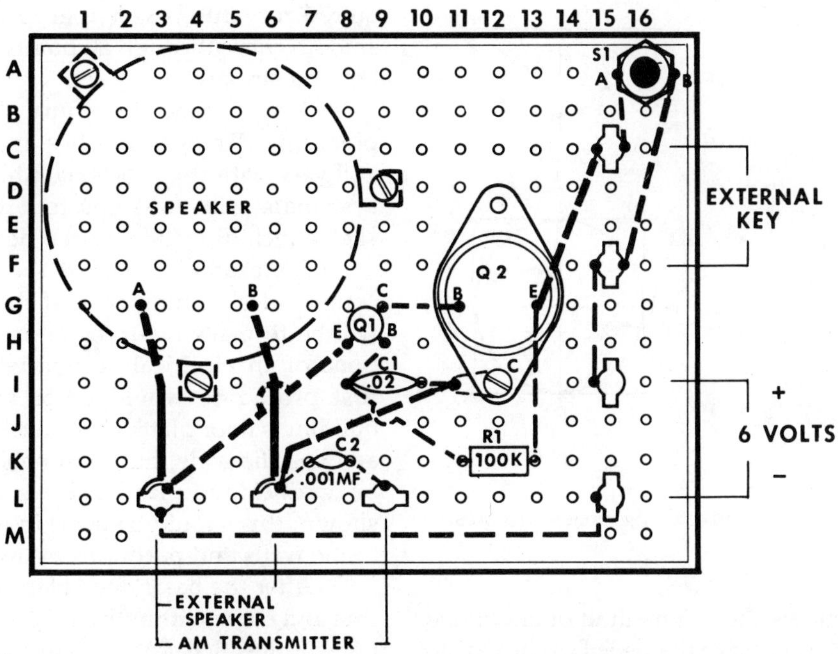

Figure 3-29. Finished drawing of the sketch in Fig. 3-28.

weight, and characteristics of the lines. Straight lines are usually drawn from left to right and from the top down, so that the person sketching is able to observe the point of the pencil at all times. It is sometimes easier to mark the extremities of a straight line with dots, then begin at one dot and draw the line towards the other dot while looking at it. A practice motion without making a mark is sometimes advisable to be sure that the hand can complete the desired stroke without error.

Circles and arcs require particular attention in sketching, because it is more difficult to produce a curved line of uniform weight than a straight line. Figure 3-30 illustrates an acceptable method of sketching circles and arcs.

In Fig. 3-30, first sketch the main axis (a) and mark off the radius (b) for small circles. Next construct a square (c) by drawing lines parallel to the main axis at the radius points. Finally, draw the circle (d) within the square, touching it at the midpoint of the sides. For larger circles, add two intersecting lines drawn at 45-deg angles to the horizontal. Mark off the radius on the lines and sketch an octagon. Draw the circle within the octagon, touching all sides of the octagon at their respective midpoint. See Fig. 3-31.

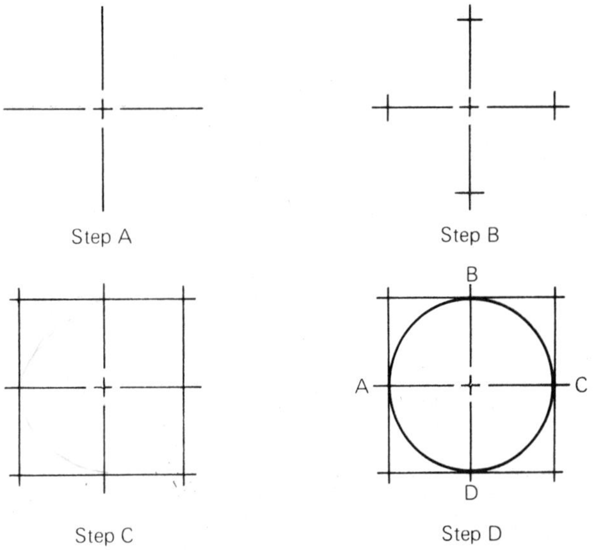

Figure 3-30. One method for sketching circles and arcs.

Figure 3-32 shows another method of drawing a circle. With this method grasp the pencil as shown, so that the radius of the circle will be the distance from the fingernail to the pivot finger. Then place the fingernail of the pivot finger on the point that is to be

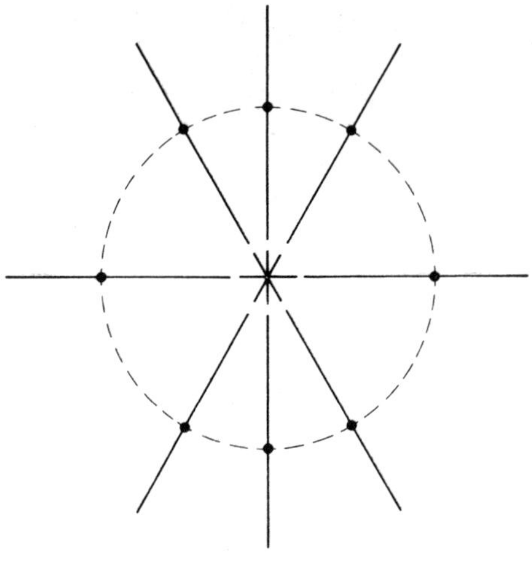

Figure 3-31. Completed sketch of circle.

the center of the circle and turn the paper, using the hand as a compass to draw the circle.

Still another method of drawing a circle is with straight lines. If you do not believe this, look at Fig. 3-33.

Slanted lines are made as shown in Fig. 3-34; that is, move the pencil from left to right. Sometimes it is better to turn the paper so that the slanted lines assume a horizontal position.

For any irregular curve (Fig. 3-35), locate a number of points to represent the shape of the required curvature, such as at ½-in. intervals. Sketch a smooth curve through all points located without overlapping strokes.

A systematic procedure should be followed in preparing electrical sketches in the field. The method will vary with the project conditions; however, being systematic will allow you to develop a clear accurate sketch including all required details.

For example, one of the most frequently encountered sketching jobs of the electrical draftsman will be to make a survey and record existing conditions of an electrical system within a building. The first procedure would be to make a sketch of the building's floor plan. This sketch need not be made to any specific scale, but reasonable proportions should be maintained. If windows and doors must be shown, this should be done in reasonable proportion to the walls and partitions of the building.

After the basic floor plan has been sketched, locate and draw within the floor plan all necessary electrical outlets, location of lighting fixtures, fuse panels, etc. If necessary, the drawing should then be dimensioned. A typical sketch of an existing electrical system is shown in Fig. 3-36.

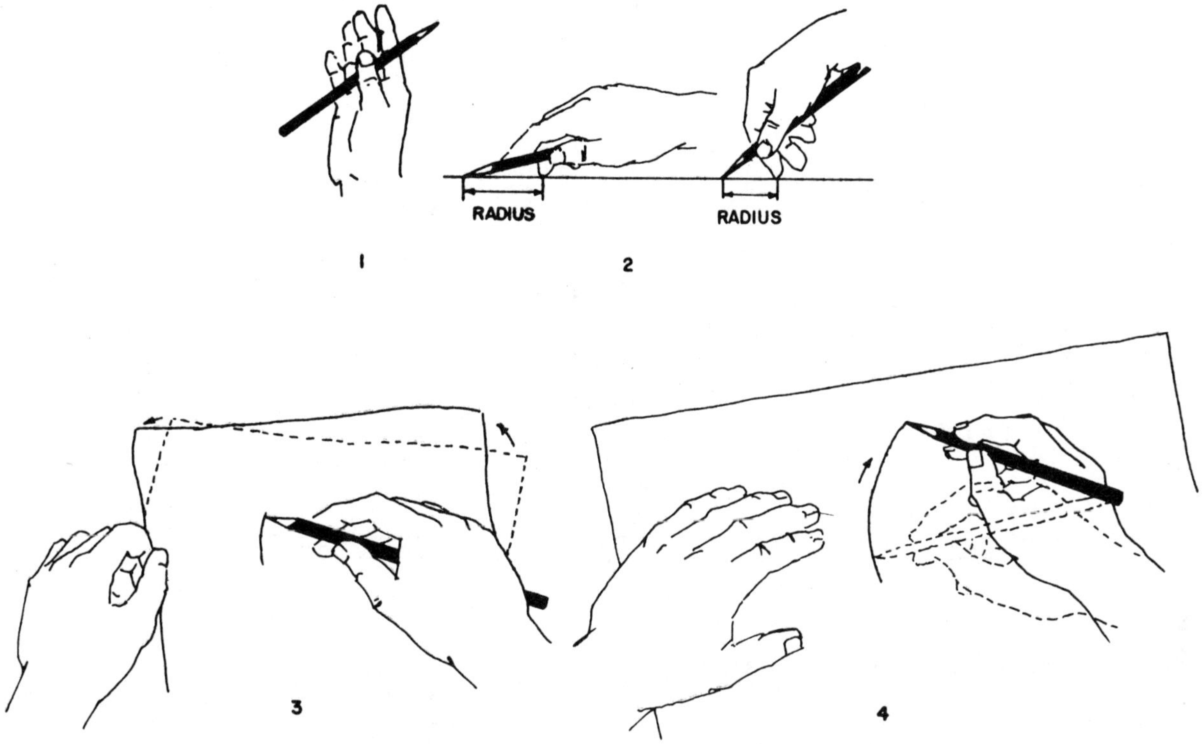

Figure 3-32. Another method of drawing a circle.

Figure 3-33. Method of drawing a circle by using straight lines.

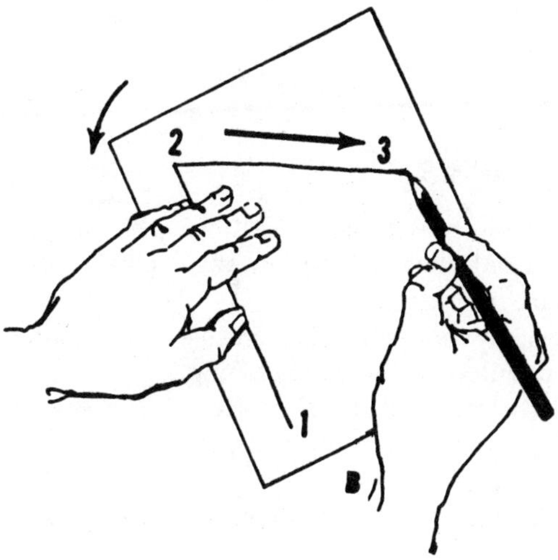

Figure 3-34. Method of sketching slanted lines.

Figure 3-35. Method of sketching irregular curves.

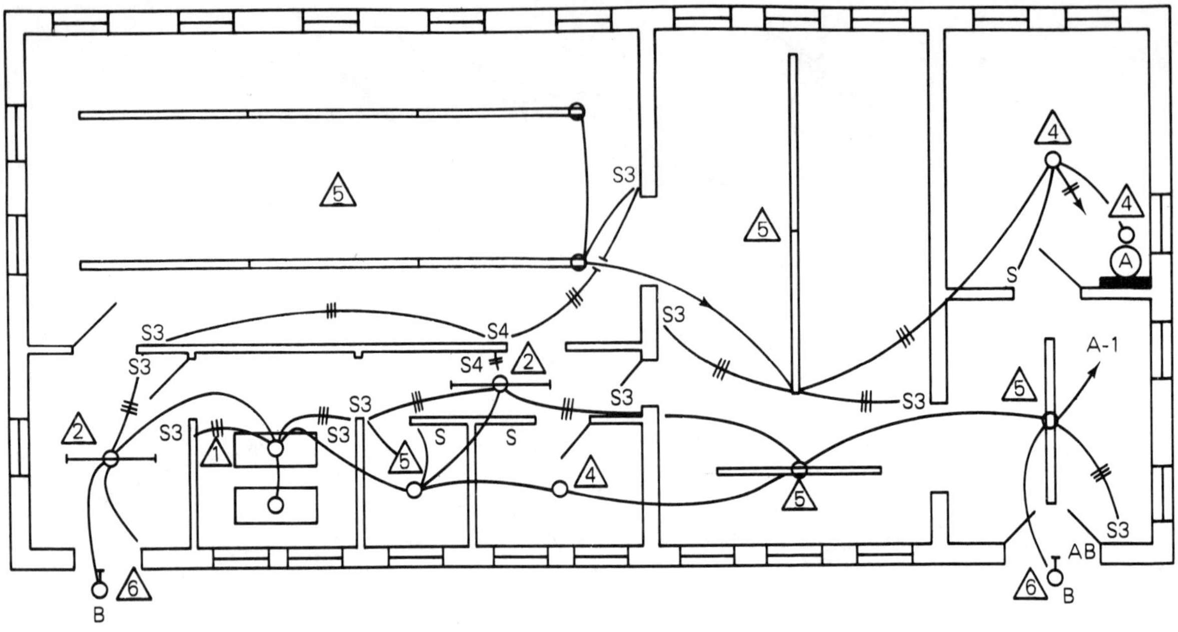

Figure 3-36. A typical sketch of an existing electrical system.

You can develop great skill in sketching with practice. Doing just a few minutes of sketching each day will help. Later in the book you will continue to develop your ability to sketch with practical applications.

SUMMARY OF 3.3

Freehand sketches may be either orthographic projections or pictorial, depending on the function for which they are intended.

It is usually better to use cross-section paper rather than unruled paper for sketching.

Although freehand sketches are seldom drawn to any certain scale, maintaining correct proportion is very important.

A soft lead (H or F) is recommended for sketching.

Only paper, pencil, and an eraser are needed for most sketching jobs.

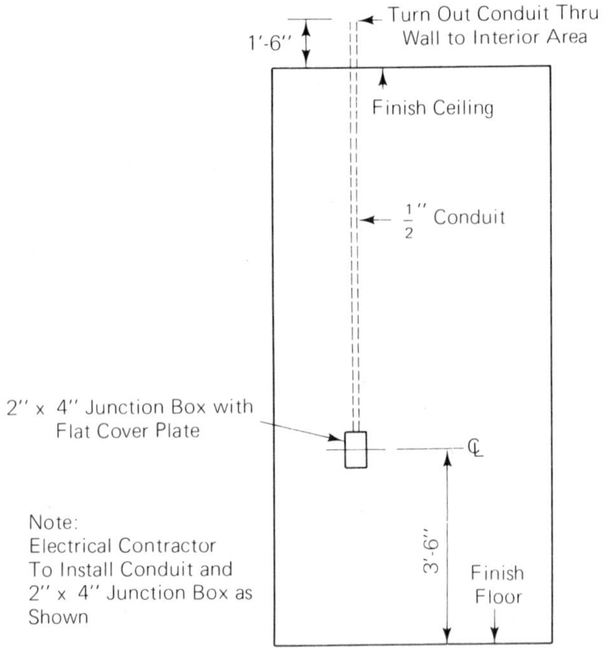

Figure 3-37. Drawing used in an assignment.

ASSIGNMENT FOR 3.3

1. Prepare a sheet of 11″ × 17″ tracing paper with border lines, as described previously in this chapter.
2. Sketch the drawing in Fig. 3-37 on the left-hand side of the sheet and the drawing in Fig. 3-38 on the right-hand side of the sheet. You may want to make a light vertical line to divide the sheet in half; then center your sketches in the center of each of these halves. Also be sure to leave room in the lower right-hand corner of the drawing for a 2″ × 3″ title block. Use the techniques learned in this section to make the sketches, particularly the circles. Your drawings should be approximately twice the size of the ones in this book.

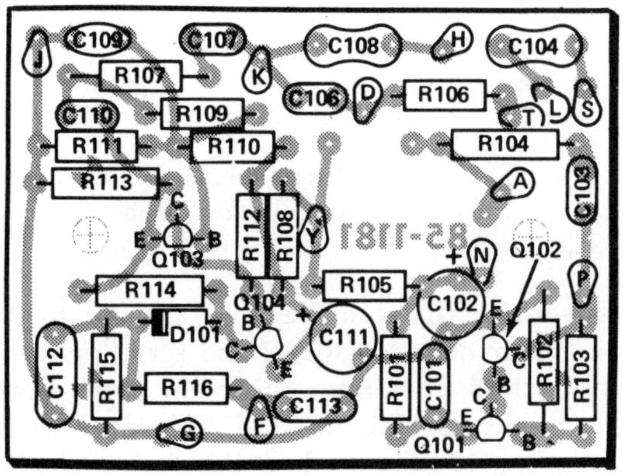

Figure 3-38. Drawing to be sketched in assignment.

3.4 Sheet Layouts

Electrical and electronic drawings normally prepared by electrical/electronic draftsmen for engineers, contractors, manufacturers, and others involved in the industry are very unique drawings. Most projects encompass all of the types of drawings that were described in Chapter 1. For example, a complete set of electrical construction drawings for an electrical system in a new building usually will consist of the following:

1. A plot plan (see Fig. 3-39) showing the location of the building on the property and all outside

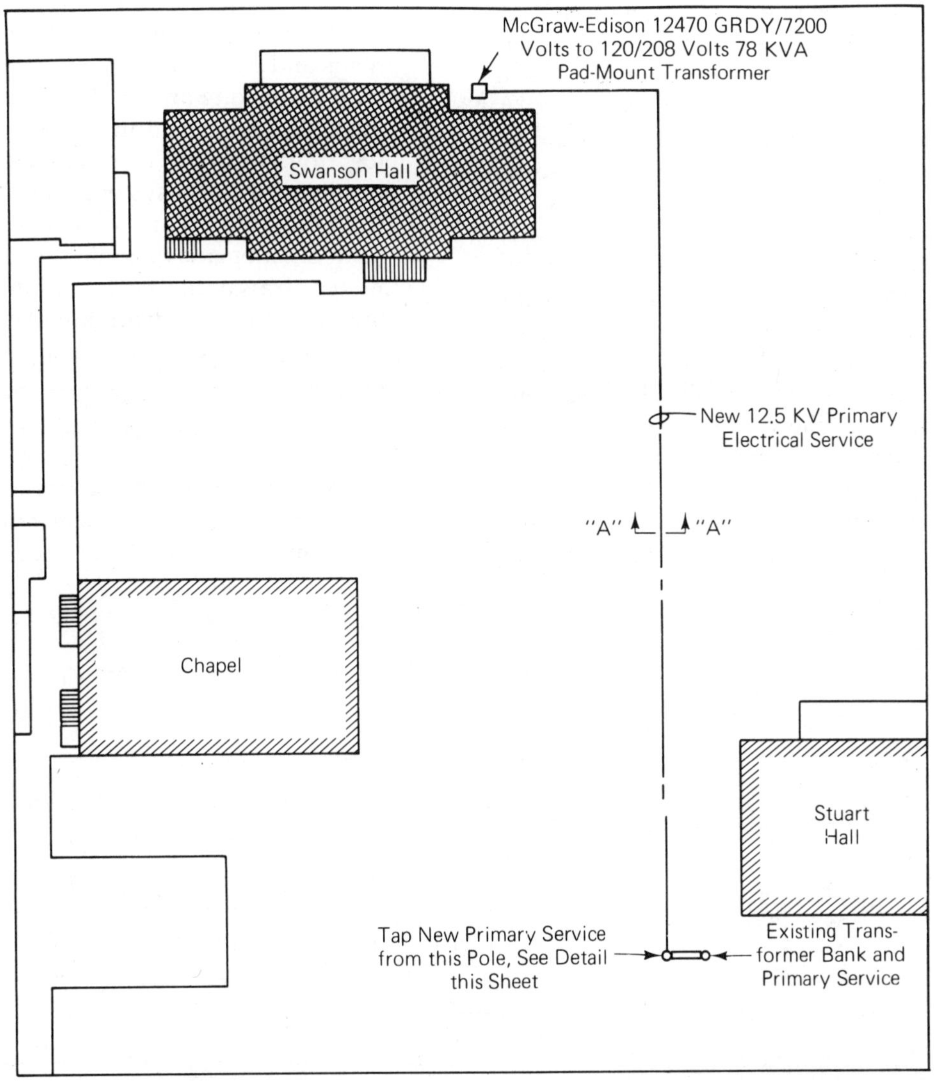

Figure 3-39. A typical building plot plan.

electrical wiring, including the service entrance. This type of plan is drawn to scale with the exception of the various electrical symbols, which must be enlarged to be readable. Figure 3-39 shows a typical building plot plan with related electrical wiring.

2. Floor plans showing the walls and partitions for each floor level. The physical location of all wiring with related outlets is shown for lighting, power, signal and communication, special electrical systems, and related electrical equipment. Again, the building partitions are drawn to scale, as are such electrical items as fluorescent lighting fixtures, panel boards, switch gear, and the like. The locations other electrical outlets and similar components are only approximated on the drawings, because they have to be exaggerated. To illustrate, a common duplex receptacle is only about three inches wide. If such a device were located on the floor plan of a building that was drawn to, say, a scale of ⅛″ = 1′0″, even a small dot on the drawing would be too large to draw the receptacle exactly to scale. Therefore, the symbol is used to indicate a duplex receptacle and is large enough to be clearly shown. When the electrician on the job is locating these outlets, he usually measures to the center of the circle (denoting the outlet) to determine the distance between the outlets. Where the dimensions are critical, dimension lines are used to show the workmen exactly where the outlets are to be located. See Fig. 3-40.
3. Diagrams are used in block, one-line, and schematic types to indicate such items as power-riser diagrams (Fig. 3-41); notice that this particular diagram shows the service-entrance conductors, metering, the main panel board, grounding, subpanels, and all related wiring, including circuits feeding the subpanels from the main panel board. The diagram in Fig. 3-41 can be termed a block wiring diagram. One-line diagrams like the one in Fig. 3-42 are commonly used for service equipment, motor controls, and the like, while schematic wiring diagrams (Fig. 3-43) are used to more clearly show the components and their proper connections for such systems as heating and cooling controls (shown in Fig. 3-43), control wiring for low-voltage switching, the electronic components and their connections for alarm systems like fire, burglar, etc. Communications are sometimes shown in the form of schematic diagrams. These diagrams are never drawn to scale; rather—as the term implies—they are diagrammatic only.
4. Control wiring schematic diagrams (Fig. 3-44).
5. Schedules, notes, and large-scale details on construction drawings (Fig. 3-45).

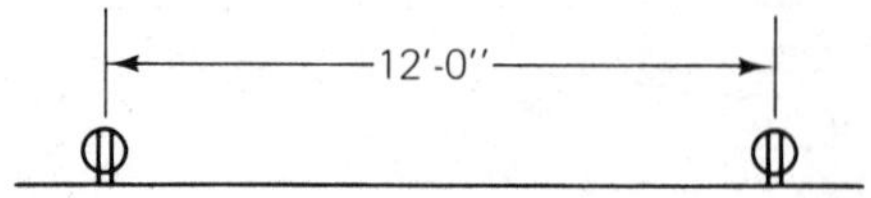

Figure 3-40. Drawing showing dimension lines indicating the placement of duplex receptacles.

In order to prepare an electrical drawing, one must first lay out the drawing sheet of an appropriate size to accommodate all of the necessary data. Care

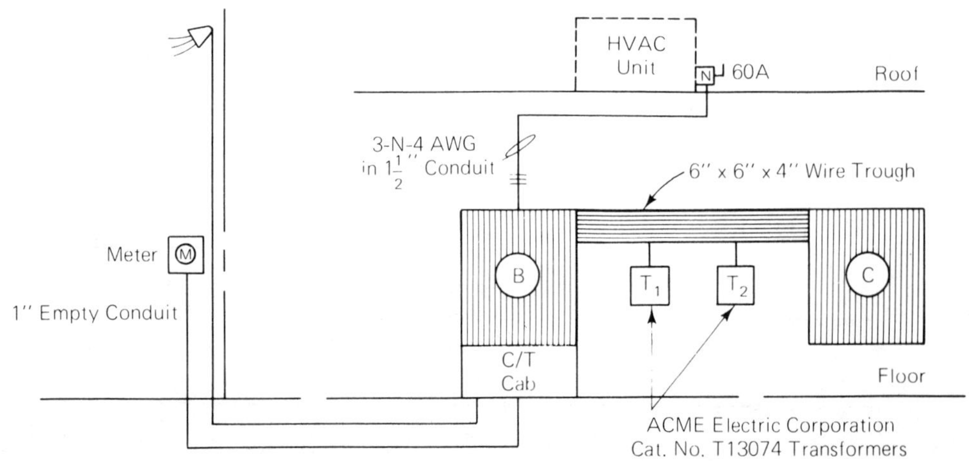

Figure 3-41. A power-riser diagram.

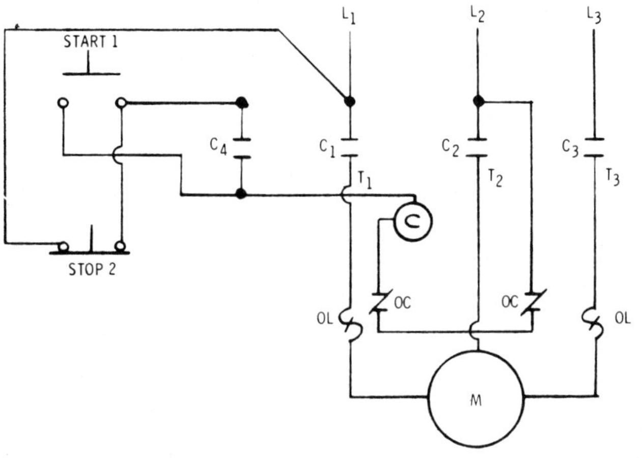

Figure 3-42. A one-line diagram.

should also be given to the arrangement of the data on the drawing sheet.

The size of the drawing sheet to be used for a given project depends on a number of variables—the most important, obviously, is the amount of data that must be drawn on the sheet. Other factors include the scale of the objects to be drawn and the standard size sheet that some firms use regardless of the amount of scale of the objects. To illustrate, the 24″ × 36″ drawing sheet is the most widely used size for architectural drawing throughout the United States. A few firms may use a slightly larger or smaller size, but the majority of architectural firms will use the 24″ × 36″ size. Therefore, when the electrical drawings are drafted for a building, the electrical draftsman must use the same size drawing paper so that they will fit neatly in the set of drawings comprising the construction documents for the entire project.

From this we can see that if only a small space is needed for the electrical data, much of the space on a 24″ × 36″ drawing sheet will not be used; if much information is required, several (24″ × 36″) sheets will have to be used.

Other sheet sizes are based on standards recommended by government agencies and the American Standards Association (ASA). These agencies have, over the years, undertaken the standardization of drafting practices that have been generally accepted by industry as a whole.

Although most industries have accepted 8½″ × 11″ as the basic multiple for flat sheet sizes, some industries (especially architectural and engineering firms) have standardized their drawing sheet sizes on the basis of the 9″ × 12″ multiple. See Tables 3-1 and 3-2 for sheet sizes currently used.

Notice from Tables 3-1 and 3-2 that the margins vary—depending upon the sheet size—from ¼ to ½

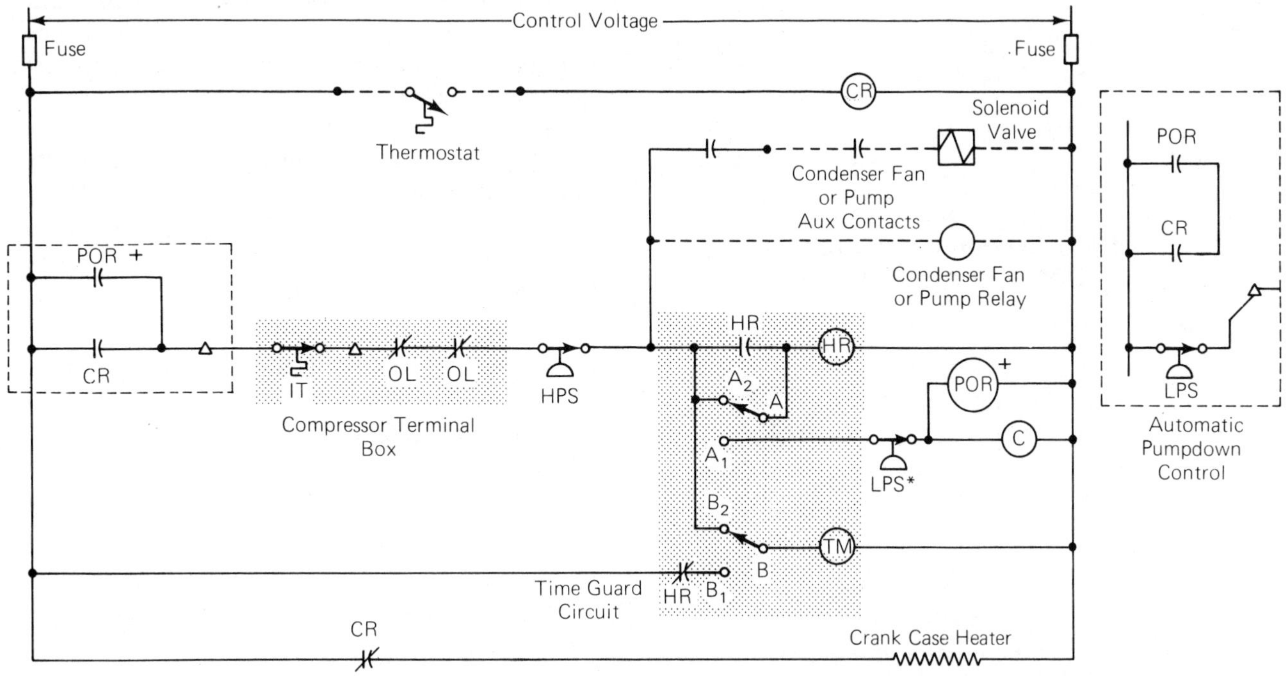

Figure 3-43. A schematic diagram of heating and cooling controls.

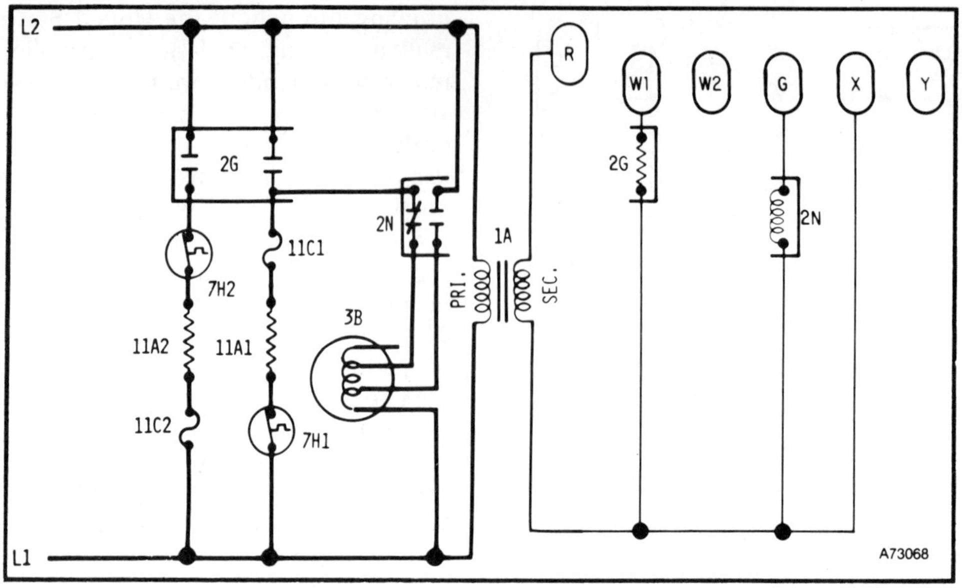

Figure 3-44. Control wiring schematic diagram.

Table 3–1
FINISHED FORMAT SIZES
(Based on 8½″ × 11″)

WIDTH	*LENGTH*	*AVERAGE MARGIN*
8½″	11″	¼″
11″	17″	⅜″
17″	22″	½″
22″	34″	½″
34″	44″	½″
28″	40″	½″

Table 3–2
FINISHED FORMAT SIZES
(Based on 9″ × 12″)

WIDTH	*LENGTH*	*AVERAGE MARGIN*
9″	12″	¼″
12″	18″	¼″
18″	24″	⅜″
24″	36″	½″
36″	48″	½″

in. When two or more sheets are to be bound together, it is recommended that at least a 1-in. margin be drawn on the left-hand side of the drawing, as shown in Fig. 3-46. This allows for stapling or binding the sheets together without hiding any of the data on the drawing sheets.

TITLE BLOCK

A title block is provided on all drawings for identification, dates, and the like. Location of this title block on the drawing is optional; however, the predominant practice in electrical drafting is to place the title block in the lower right-hand corner of the sheet. The dimensions of these title blocks also vary from firm to firm. However, for drawing assignments in this book using 11× 17-inch drawing paper, the space to be reserved for the title block should be a minimum of 3 inches long by 2 inches high.

In general, the contents of the title block should include the following:

1. Name of the project.
2. Address of the project.
3. Name of the owner or client.
4. Name of the architectural and/or engineering firm.
5. Date of completion.
6. Scale(s).
7. Initials of the draftsman, checker, and designer, with dates under each.
8. Job number.
9. Sheet number.
10. General description of the drawing.

Every architectural/engineering firm has its own standard form for drawing titles, and they are often preprinted directly on the tracing paper or else

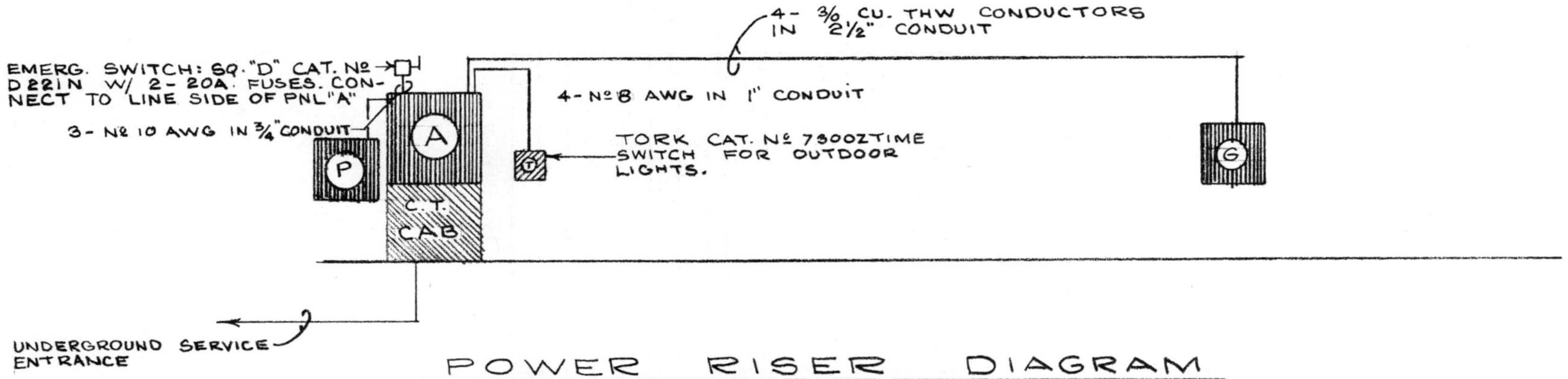

POWER RISER DIAGRAM

NO SCALE

PANELBOARD SCHEDULE

PANEL	TYPE	PANEL MAINS			BRANCHES					
Nº	CABINET	AMPS	VOLTS	PHASE	1P	2P	3P	PROT	FRAME	ITEM FED OR REMARKS
"A"	SURFACE	600	120/208V	3ϕ,4-W	21	—	—	20	70	LIGHTS
SQUARE "D" TYPE NQOB W/ MAIN CIRCUIT BREAKER					12	—	—	20	70	RECEPTS
					—	—	1	30	70	TIME CLOCK
					—	—	1	200	200	PANEL G
					—	—	1	200	200	PANEL P
P	SURFACE	200	120/208V	3ϕ,4-W						
SQUARE "D" TYPE NQOB FOR HTG & COOLING EQUIP.										

Figure 3-45. Schedules, notes, and large-scale details on construction drawings.

LIGHTING FIXTURE SCHEDULE						
FIXT. TYPE	MANUFACTURER'S DESCRIPTION	LAMPS Nº	LAMPS TYPE	VOLTS	MOUNTING	REMARKS
1	LITHONIA CAT.# 2RL 440 A12	4	40W F	120V	RECESSED	COOL WHITE LAMPS
2	LITHONIA CAT.# AF 296 PG	2	215W F		PENDANT	COOL WHITE LAMPS
3	LITHONIA CAT.# C-196	1	75W F		SURFACE	
4	LITHONIA CAT.# STC-196	2	75W F		SURFACE	
5	CRESCENT CAT.# ANG-220	2	20W F		WALL	
6	BENJAMIN CAT.# A-23	1	150W I		SURFACE	
7	BENJAMIN CAT.# 7614	1	200W I		SURFACE	
8	GOTHAM CAT.# 530	1	150W I		RECESSED	PAR-38 LAMPS
9	GOTHAM CAT.# R5006	1	100W I		RECESSED	
10	STONCO (2)CAT.# 30KL & (1) 3203B	2	150W I		SURFACE	PAR-38 LAMPS
11	STONCO CAT.# PRA11-250MM	1	250W M.V.		POLE	COLOR-CORRECTED LAMPS
12	PERFECLITE CAT.# P-873	2	25W	120V	SURFACE	CONNECT TO EMERG. PNL

PANELBOARD SCHEDULE										
PANEL Nº	TYPE CABINET	PANEL MAINS AMPS	PANEL MAINS VOLTS	PANEL MAINS PHASE	BRANCHES 1P	BRANCHES 2P	BRANCHES 3P	BRANCHES PROT	BRANCHES FRAME	ITEM FED OR REMARKS
G	SURFACE		120/208	3φ, 4-W	18	—	—	20	70	DUPLEX RECEPTS
SQUARE "D" TYPE NQOB M.L.O.					—	4	—	50	70	WELDER OUTLETS
					13	—	—	20	70	LIGHTS
					3	—	—	20	70	SPARES

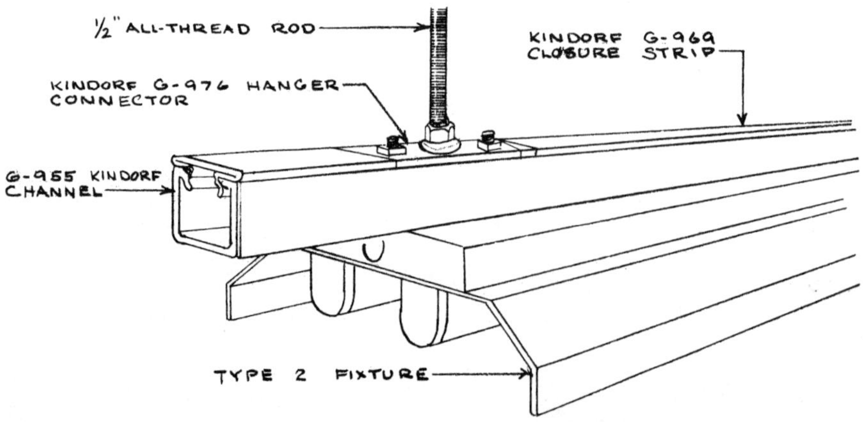

KINDORF MOUNTING DETAIL

NO SCALE

Figure 3-45. (cont.)

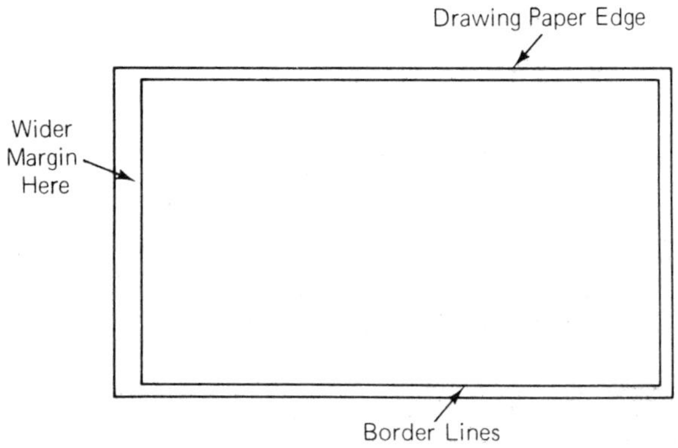

Figure 3-46. A 1-in. margin is recommended on the left-hand side of the drawing to allow for stapling and binding.

printed on stick-on paper which is placed on the drawing. A typical architectural title block is illustrated in Fig. 3-47.

Often the consulting engineering firm will be listed too, which means that an additional title will be applied to the drawing, usually next to the architectural title block.

The following steps describe the exact procedure to use when one is preparing a sheet of drawing paper for any given drawing.

Step 1. Clean drawing board and all drawing equipment. This is necessary in order to have a clean and neat finished drawing.

Step 2. Place a sheet of tracing paper—of the appropriate size—on the board, line it up with the T-square, and secure with drafting tape.

Step 3. Measure the correct distance (according to the size of paper) in from all sides of the paper, and make a very light dot to indicate these measurements.

Step 4. Draw a border line completely around the paper. Use the extra heavy full line for this border.

DRAWING THE TITLE BLOCK

Once the title block lines have been measured and drawn on the tracing paper in the proper form, the information should be lettered freehand in single-stroke capitals. However, the contents of the title block should first be written out, on a separate piece of paper, for two important reasons. The first reason is to make certain that all information required has been obtained, and the second reason is to give the draftsman an idea of the total space required for the information. The latter enables the draftsman to obtain good spacing for a neat appearance of the title block.

REVISIONS

Often electrical drawings must be partially redrawn or revised. Sufficient space should be left near the title block for dates and description of revisions. A space about two inches high and the width of the title block is usually sufficient (Fig. 3-48). A suggested title block for use on drawing exercises in this book is shown in Fig. 3-49.

LAYOUT FOR TRACING DRAWINGS

It should be clear from Chapter 1 that the architect plans and designs buildings; that the engineer or electrical designer ~~usually~~ lays out the complete electrical systems for the architect's buildings; and that the

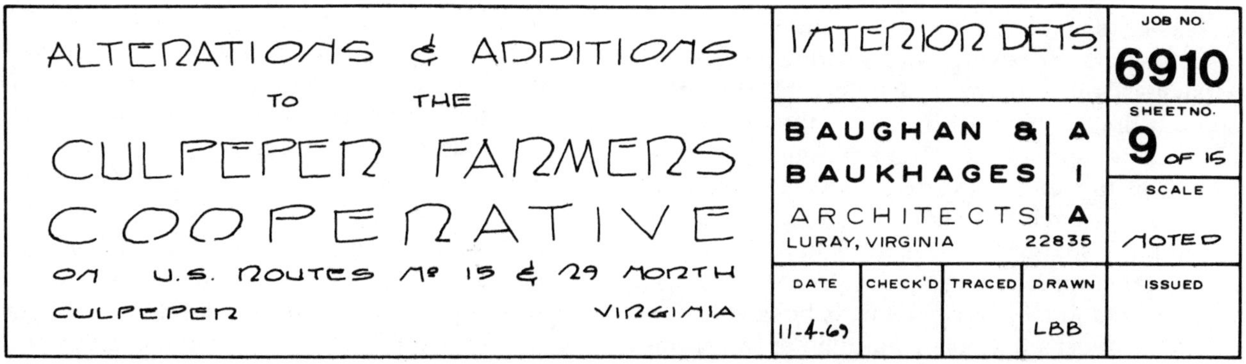

Figure 3-47. A typical architectural title block.

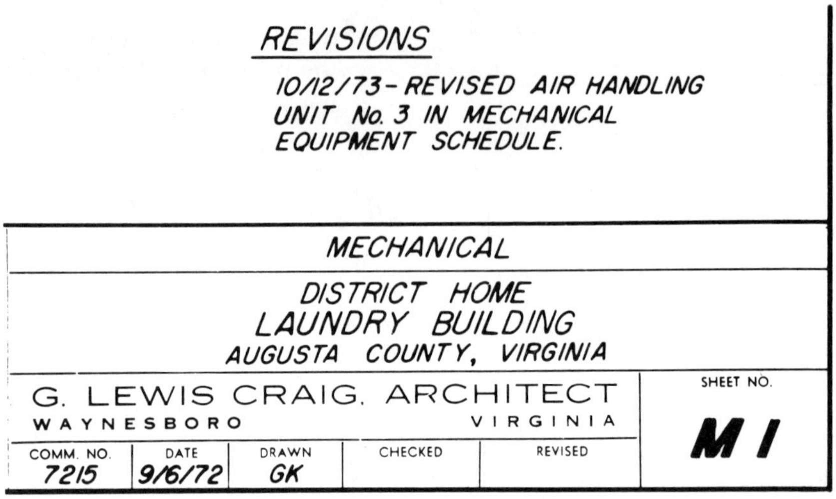

Figure 3-48. Sufficient space should be left near the title block for revisions.

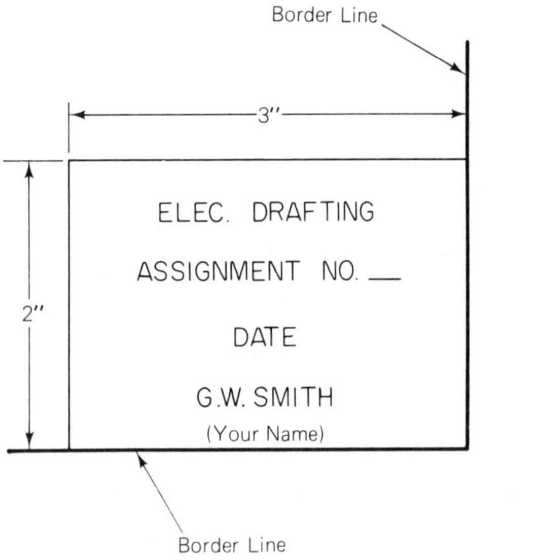

Figure 3-49. A title block for use on drawing exercises in this book.

electrical draftsman transforms the engineer's designs into working drawings. In preparing the electrical drawings, an outline of the architect's floor plan(s) is laid out first. To make this outline, the draftsman could scale the original architectural drawing, transfer the measurements to the new drawing sheet, and then redraw the architect's floor plan (or other part of the building) so that the electrical data could be placed in their appropriate places.

A faster and more accepted way, however, is to trace the architect's floor plan and then to insert the various outlets and circuits on the traced building outline. The following steps describe this procedure:

Step 1. Clean drawing board and all drawing equipment. This is necessary in order to have a clean and neat finished drawing.

Step 2. Take the print of the architect's floor plan and place it on the drawing board. If the print has been folded, rub out the creases as much as possible. Now line up the horizontal lines on the print with the T-square. When the print and the T-square are lined up perfectly, secure the print to the board with small pieces of drafting tape.

Step 3. Place a sheet of tracing paper over the print, line it up, and secure with drafting tape.

Step 4. Insert a 4H piece of lead into the lead holder so that a light solid line can be used throughout in tracing the architect's floor plan.

Step 5. With the use of a T-square or other type of straightedge, trace all necessary horizontal lines on the floor plan of the building. Now use a triangle in conjunction with a straightedge and trace all vertical lines. This drawing should show just enough of the building outline—including doors, windows, partitions, etc.—to identify the general areas so that the electrical equipment and outlets can be added later. Do not show dimension lines, wall finishes, or other details not necessary for the electrical work.

Step 6. Check over the tracing very carefully; make certain that no doors, windows, partitions, etc., are omitted. Now remove the tape securing the tracing paper and remove the tracing from the board. Do not remove the print, however, until you further check the tracing against the print for any omissions.

ASSIGNMENT

1. Clean the dust from your drawing board with your dust brush and make certain that all of your other drafting tools are clean and in good working order.

2. Gather all of your drawing assignments from previous assignments; place one sheet at a time on your drawing board and perform the following on each:
   a. Measure 1 inch in from the left side of the drawing sheet and make a very light dot to indicate this measure; then measure in from all of the other sides ¼ inch, making light dots to indicate these measures. Now draw a border line completely around the drawing paper as shown in Fig. 3-50. Use the extraheavy full line for this border.
   b. Following instructions given in the text and using Fig. 3-49 as a guide, draw a two-inch by three-inch title block in the lower right-hand corner of each drawing sheet and handletter in the required information—including the actual dates that each sheet was drawn.

On future assignments, each drawing sheet will be laid out in this manner (including title block and lettering) before the assignment is completed.

Figure 3-50. Draw a border line completely around the drawing paper.

Chapter 4

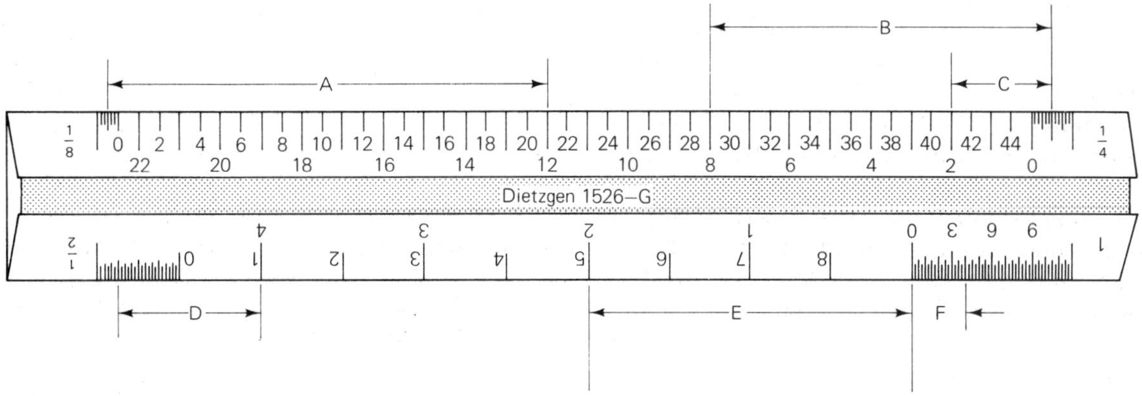

DIMENSIONING

Since it is the responsibility of the electrical draftsman to use the language of drafting to provide neat, detailed, and accurate drawings for workmen or technicians, a means of expressing accurate dimensions must be used. Without definite dimensions, it would be impossible, on most drawings, to indicate clearly any engineering intent or to achieve successful fabrication of the system or product. The ideal electrical or electronic drawings should indicate, beyond any question of a doubt, exactly what is required to install or fabricate the system in question. It should never be necessary for the workmen to have to calculate or assume any dimension in order to perform their work from drawings made by electrical draftsmen.

The dimensioning practices described in this chapter are those recommended by United States of America Standards Institute (USASI) and the U.S. Department of Defense—they are standard in almost all industries.

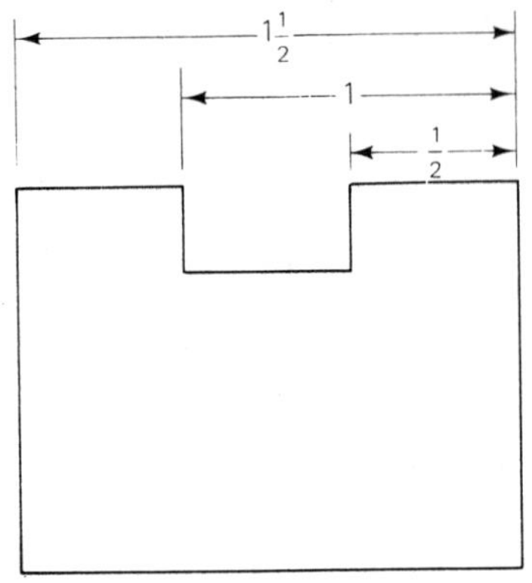

Figure 4-2. Where several parallel dimension lines are drawn, they should be the same distance apart, with the dimensions staggered so that they will not be crowded.

4.1 Fundamental Dimensioning Procedures

Dimensions on all drawings should appear in inches and fractional parts of an inch (or in feet and inches) unless the metric system is used. If the latter is used, a general note must be included on the drawings to indicate that dimensions are shown in millimeters.

Dimension lines are used with numerals and arrowheads to show the sizes or placement of an object or component on a drawing. These dimensions should be placed off the views as much as possible and should not be repeated in other views of the same object. The light full line is used for dimension lines, with a space left—usually in the middle—for the numerals that represent the size or dimension. These lines should be placed from 1/4 to 3/8 in. from the views, as shown in Fig. 4-1; when several parallel dimension lines are drawn, they should be the same distance apart, with the dimensions staggered so that they will not be crowded (Fig. 4-2).

Extension lines are used to connect a part of a view with its dimension line. Again, use the light full line and begin about 1/16 in. from the corners of the views and extend the extension line about 1/8 in. beyond the dimension line, as shown in Fig. 4-3.

Arrowheads are placed on the ends of the dimension lines, and all arrowheads on a particular drawing should be of uniform length—from 1/8 to 3/16 in. long. Arrowheads are made with two

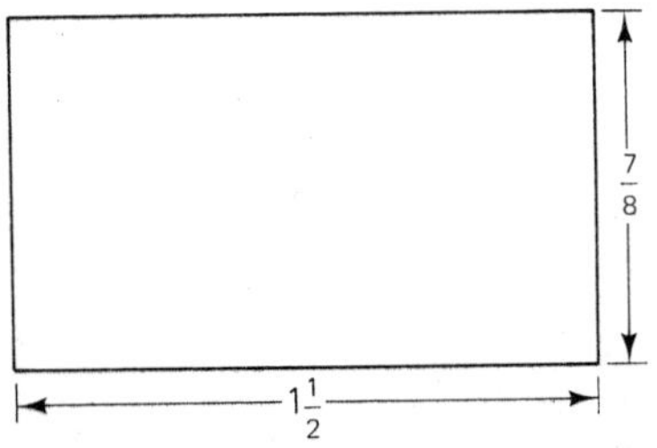

Figure 4-1. Dimension lines should be placed between 1/4 in. to 3/8 in from the views.

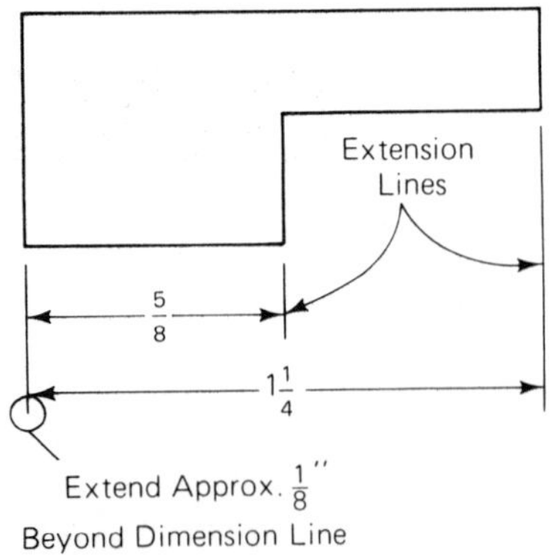

Figure 4-3. Extension lines should be extended about 1/8 in. beyond the dimension line.

freehand strokes from the barb to the point (Fig. 4-4). Sometimes a third stroke is used to fill in the other two, as shown in Fig. 4-5.

In some cases, there may not be enough room between the extension lines to insert the dimension lines and numerals. If so, the dimension lines may then be drawn outside of the extension lines and the arrowheads reversed to point inward (Fig. 4-6). This arrangement permits the fractions to be placed either inside or outside the extension lines so as to prevent crowding and also so that the length of the arrowheads can be the same as others on the drawing.

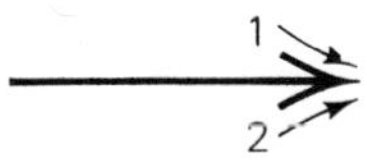

Figure 4-4. Arrowheads are made with two freehand strokes from the barb to the point.

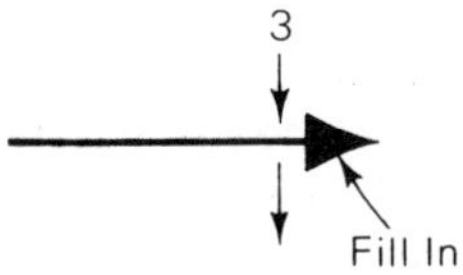

Figure 4-5. Sometimes a third stroke is used to fill in the other two when an arrowhead is made.

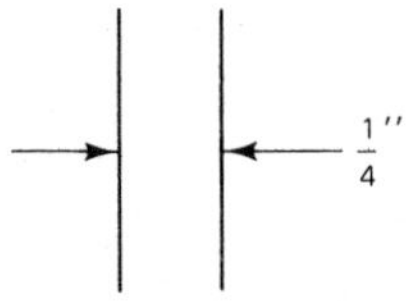

Figure 4-6. In tight places, the dimension lines may be drawn outside the extension lines and the arrowheads reversed to point inward.

If at all possible, dimension lines should not cross extension lines, as shown in the drawing in Fig. 4-7. Rather, the shorter dimension should be placed nearest the view, with the larger dimensions placed outside the shorter dimensions, as illustrated in Fig. 4-8. Also note that numerals in horizontal dimension lines should be placed so that they may be read from the bottom edge of the drawing (Fig. 4-9), while vertical dimension lines should be placed so that they may be read from the *right*-hand edge of the drawing. They should not be placed as shown in Fig. 4-10. Sloping dimension lines should also be drawn—

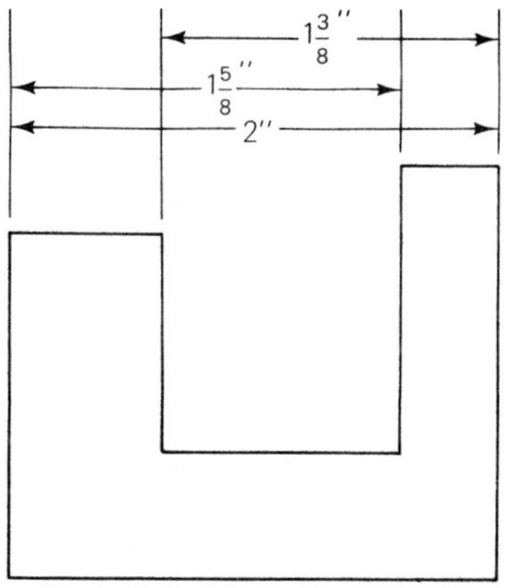

Figure 4-7. Dimension lines should not cross extension lines as shown here.

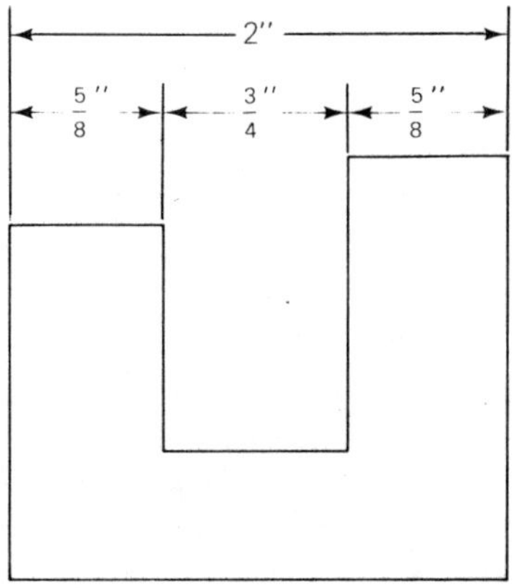

Figure 4-8. Proper method of placing dimensions on a drawing.

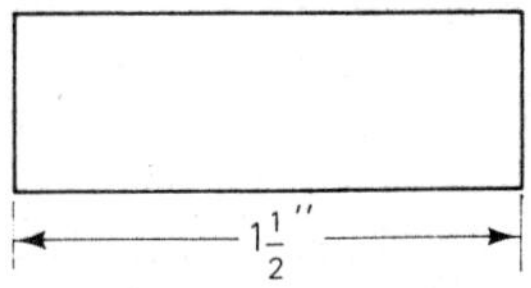

Figure 4-9. Numerals in horizontal dimension lines should be placed so that they may be read from the bottom edge of the drawing.

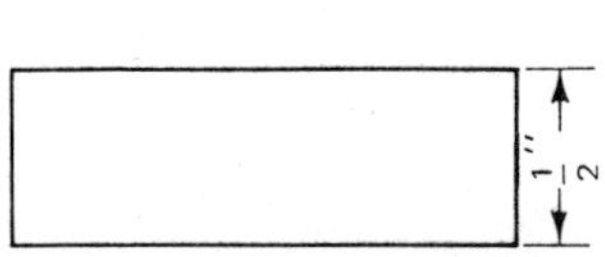

Figure 4-10. Vertical dimension lines should be placed so that they may be read from the right-hand edge of the drawing.

whenever possible—so that the numerals may be read from either the bottom or right-hand side of the drawing, as shown in the examples in Fig. 4-11.

Dimensions for the lengths of an object should be placed either above or below the view, while dimensions showing the height of an object should be placed next to the view, as shown in Fig. 4-12.

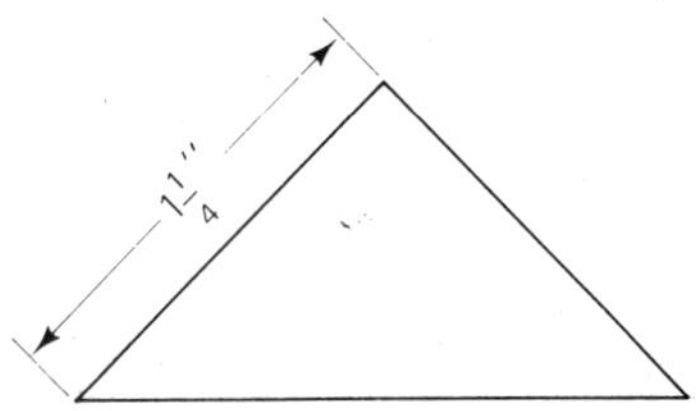

Figure 4-11. Sloping dimension lines should be drawn so that the numerals may be read from either the bottom or right-hand side of the drawing.

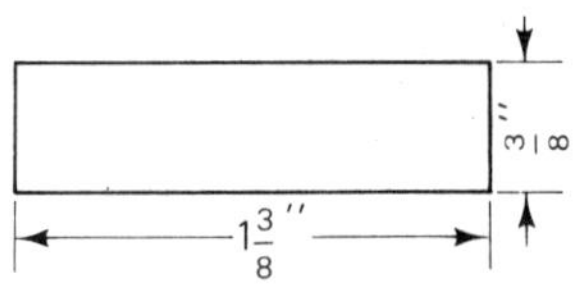

Figure 4-12. Dimensions for the lengths of an object should be placed either above or below the view, while dimensions showing the height of an object should be placed next to the view.

When arcs must be dimensioned on drawings, the center of the arc should be indicated by a small cross and the radius of the arc by a line drawn from the cross to the arc. The figures indicating the dimension should be arranged as shown in Fig. 4-13 for the radius of a conduit (pipe) bend or in Fig. 4-14, which shows the radius of the corners of an instrument chassis. The radius dimension is always followed by the letter R placed above the dimension line or next to the numerator in the case of a fraction. Only one arrowhead, however, is necessary; the point should touch the arc and is placed on the dimension line. Other examples of dimensioning arcs are shown in Figs. 4-15(a) and 4-15(b).

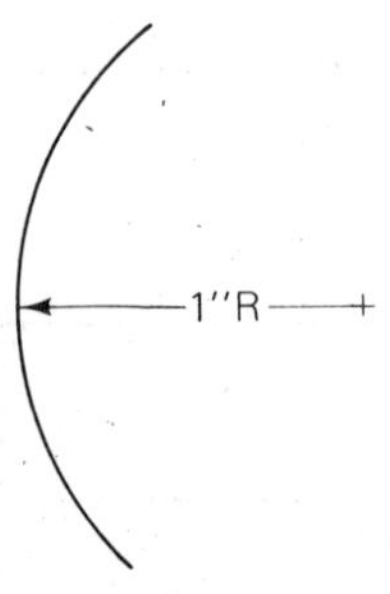

Figure 4-13. Method of arranging the dimensions for the radius of a conduit bend.

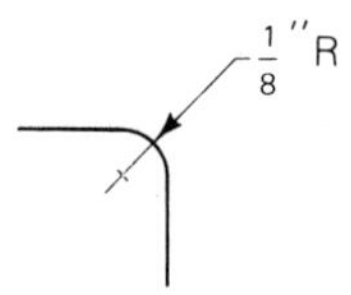

Figure 4-14. Method of arranging the dimension for the corner of an instrument chassis.

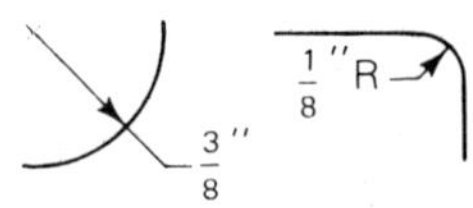

Figure 4-15. Other methods of dimensioning arcs.

The dimension line for an angle is an arc, examples of which may be seen in Fig. 4-16. These dimension lines are made either with a compass or, preferably, with a circle template. The point of the compass or the centerlines of the circle template are placed on the vertex of the angle, and the arc is then drawn from one side of the angle to the other. Except for large angles, the dimensions are placed to read from the bottom of the drawing. For small angles (Fig. 4-17), the arrowheads may be reversed to point inward.

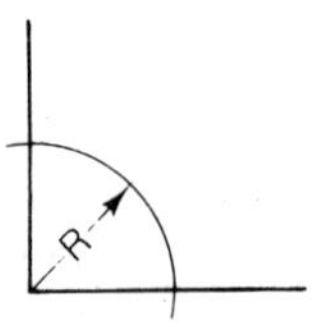

Figure 4-16. The dimension line for an angle is an arc.

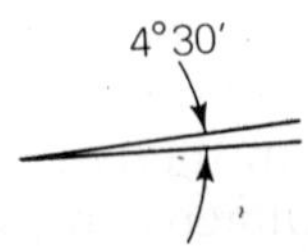

Figure 4-17. For small angles, the arrowhead may be reversed to point inward.

Drawings indicating the centerline of knockouts (holes) in such items as outlet boxes, panelboard housings, instrument chassis, etc., should appear in

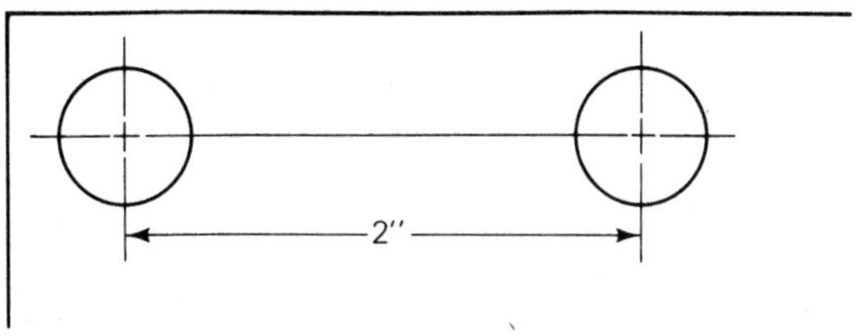

Figure 4-18. Drawings indicating the centerline of holes should appear in the view in which the hole appears as a circle.

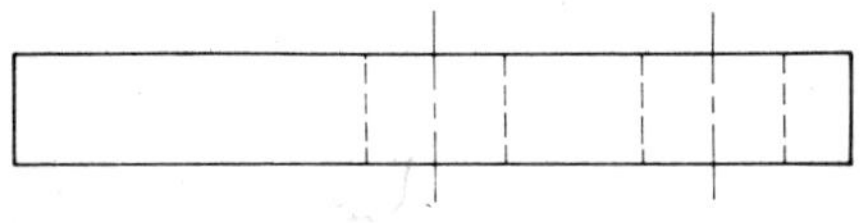

Figure 4-19. Example of views where holes appear other than as circles.

the view in which the holes appear as circles (Fig. 4-18) and not in veiws where these holes appear as other than circles, as shown in Fig. 4-19. The centerlines are extended away from the object to serve as extension lines for dimensioning.

Drilled holes are dimensioned with a note stating the diameter of the drill and perhaps the number of holes of one particular size to be drilled in the object. In most cases of this type, the diameter of the hole is expressed in decimal fractions for control and electronic equipment (Fig. 4-20) and in common fractions for most electrical equipment (see Fig. 4-21).

Pictorial drawings are dimensioned with extension lines in much the same manner as orthographic drawings, but the dimensioning of pictorial drawings (like isometric drawings) is somewhat more difficult, because less room is available on the single view. In dimensioning isometric drawings, the following rules should be followed:

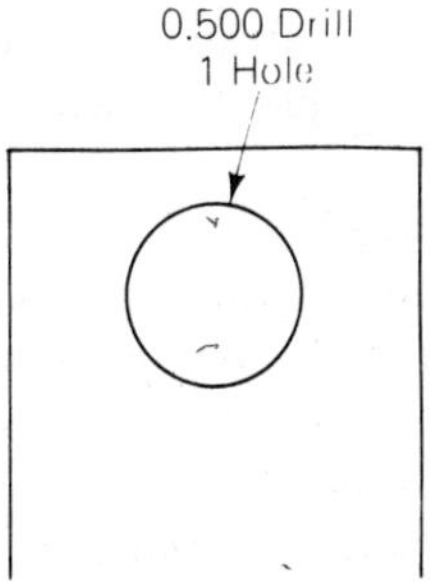

Figure 4-20. In most cases, the diameter of a hole is expressed in decimal fractions for control and electronic equipment.

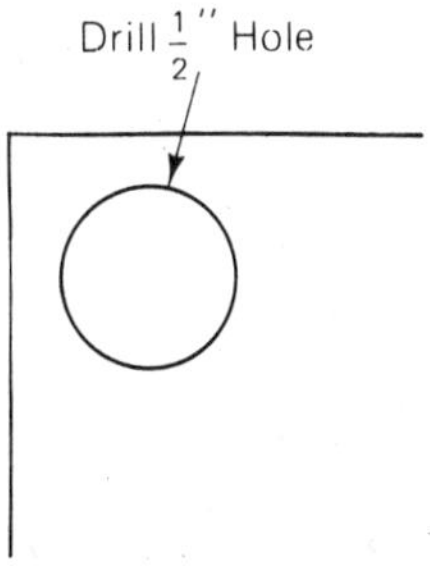

Figure 4-21. Common fractions are usually used for dimensioning electrical equipment.

1. Dimension lines and numerals should be kept off the view itself wherever possible.
2. The radius dimensions of an arc should be drawn from the center of the enclosing isometric square to the arc.
3. Visible surfaces should be dimensioned in preference to hidden surfaces.
4. Both extension lines for a dimension should be extended from isometric visible edges.
5. Vertical numerals should be used in dimensioning; numerals and arrowheads should be made in isometric.
6. Horizontal and vertical dimension lines on the orthographic views should be parallel to isometric axes in the isometric view.
7. Notes need not be lettered in isometric; rather, standard vertical single-stroke lettering may be used.

4.2 Using the Architect's Scale

While certain small electrical and electronic components can be drawn to full size on the drawing paper, drawings of buildings and certain other equipment obviously cannot. Therefore, the drawing is reduced in size so that all the distances on the drawing are drawn smaller than the actual dimensions of the building or piece of equipment, all dimensions being reduced in the same proportion. The ratio, or relation, between the size of the drawing and the size of the object being drawn is indicated on the drawing paper (1/8″ = 1′0″, for example), and the dimensions are the actual dimensions of the building or other object, not the distance that is measured on the drawing.

The most common method of reducing all the dimensions in the same proportion is to select a certain distance and let that distance represent 1 foot.

This distance is then divided into 12 equal parts, each of which represents one inch. If ½-inch divisions are required for the drawing (although they rarely are on most electrical drawings for building construction), these twelfths are further subdivided into halves. This method gives a usable scale that represents the common foot rule with its subdivisions into inches. When a measurement is laid off on the drawing, it is made with the reduced foot rule known as the architect's scale; when a measurement is taken on the building or object itself, it is made with the standard foot rule.

Architect's scales are available in many different degrees or scales; that is, 1 inch = 1 foot, ¼ inch = 1 foot, ⅛ in. = 1 foot, etc. However, the most common scales such as ½ in.= 1 foot and ¾ in.= 1 foot are in. = 1 foot, and ¼ in. = 1 foot. On enlarged details, scales such as ½ in. = 1 foot and ¾ in = 1 foot are also common. The following will illustrate how the various scales are used.

Figure 4-22 shows part of a building floor plan drawn to a scale of ¼ in. = 1 foot. The dimensions in question are found by placing the ¼-in. architect's scale on the drawing (as shown) and reading the figures. It can be seen that the dimensions read 3 feet 11 inches.

An architect's scale showing four degrees (⅛" = 1', ¼" = 1', ½" = 1', and 1" = 1') is shown in Fig. 4-23. The dimensions of the various lines are as follows: *A* equals 12 feet 6 inches on the ⅛-in. scale; *B* equals 8 feet 6 inches and *C* equals 2 feet 6 inches—both on the ¼-in. scale; *D* equals 1 foot 9 inches on the ½-in. scale; *E* equals 5 feet 0 inches on the 1-in. scale, and *F* equals 4 inches on the 1-in. scale.

Every drawing that is prepared in a given reduced scale should be plainly marked as to what the scale is. For example, the building floor plan in Fig. 4-22 should be marked

FIRST FLOOR PLAN
Scale: ¼" = 1'0"

4.3 Architectural Dimensioning

The common rules that apply to other types of drawings also apply to architectural drawings. However, only the experienced architectural draftsman can tell, with ease, which dimensions are necessary on an architectural drawing. The dimensions given must be clear and definite, and must tell the workmen the exact sizes of all parts of the building. Furthermore, they must check with one another from place to place and from plan to elevation to section, etc.

The electrical draftsman will not have too many drawings that will require architectural dimensioning, but in some cases—such as the placement of recessed lighting fixtures in the soffit of a building—some dimensioning may be required. Several points to be observed in architectural dimensioning are

1. Keep all outside dimension lines well away from the building lines. They should be located a minimum of ¾ in. from the building lines and should be approximately 5/16 to ⅜ in. apart.
2. Masonry openings should be marked "M.O." when an exact opening is required. When the opening is for a window or a door, a nominal dimension may be used.
3. Dimensions are usually to the centerlines of partitions or else are made to the outside walls. In any case, the wall thicknesses should also be shown.
4. Dimensions should be provided to the centerlines of columns in both directions.
5. Dimensions for openings are normally made to the centerline of the opening or else to the sides of the opening—as required.

Figures 4-24 and 4-25 show various methods of dimensioning architectural drawings, while Fig. 4-26 shows the application of architectural dimensioning on an electrical drawing. Notice that most of the dimensions on these drawings are from center to center of various objects. The reason is that lumber and other construction materials often vary slightly in size, making it impossible to indicate actual values of edges of structural members. This way, if structural members vary somewhat in size, the true location is always achieved.

SUMMARY

Architectural working drawings must give all the details necessary for the workmen to construct the building properly.

Nearly all working drawings are drawn to a reduced scale; that is, all dimensions are drawn smaller than the actual dimensions of the building.

When a measurement is laid off on the drawing, it is made with the reduced foot rule known as the architect's scale; when a measurement is taken on the building, it is made with the standard foot rule.

Most dimensions are given as center-to-center distances on architectural drawings because of the variation of the size of building materials.

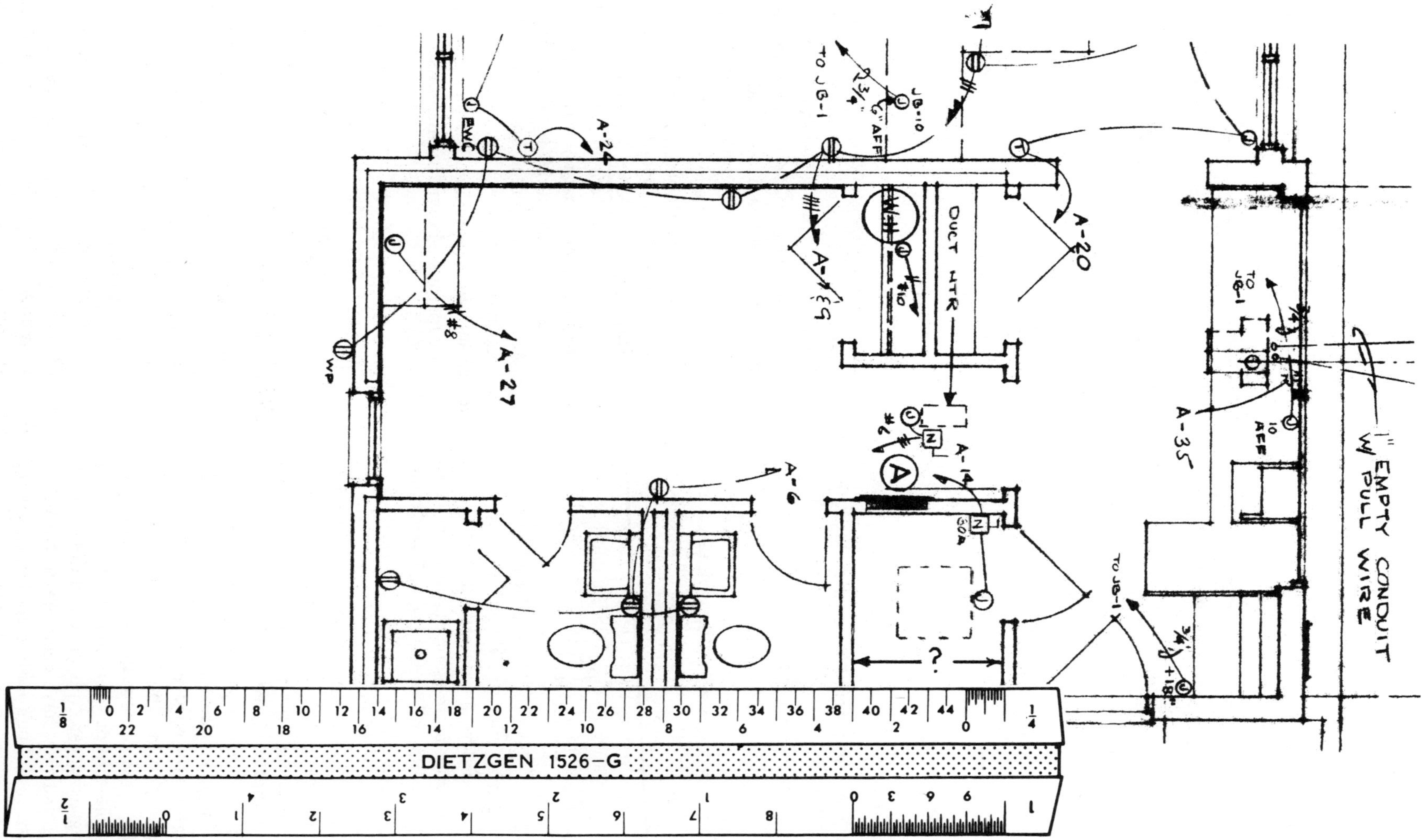

Figure 4-22. Method of using architect's scale to determine dimensions on building floor plans.

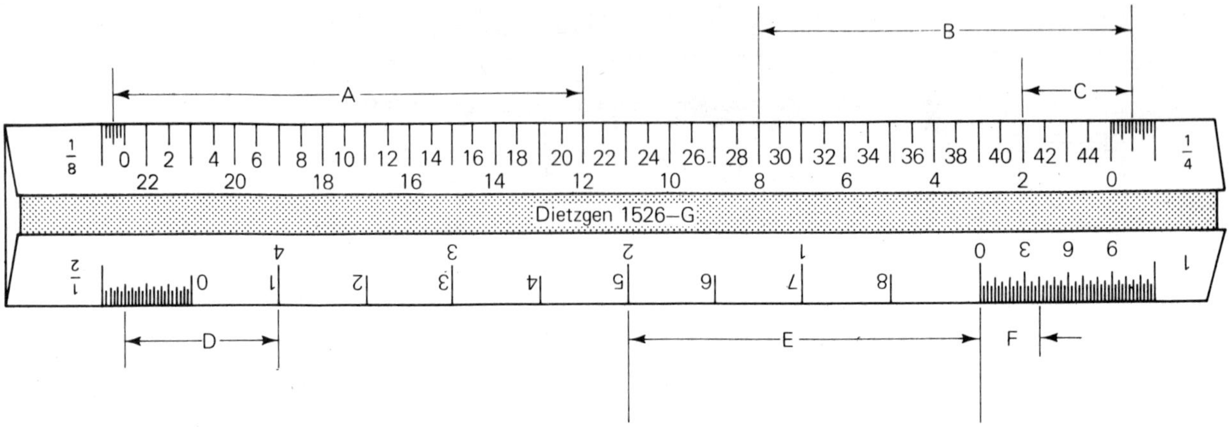

Figure 4-23. An architect's scale showing four degrees of dimensions.

ASSIGNMENT

1. Give the missing dimensions in the illustration in Fig. 4-27.

   A.______ B.______ C.______ D.______
   E.______ F.______ G.______ H.______
   I.______ J.______ K.______ L.______

2. Sometimes it becomes necessary to draw a part of a floor plan or other drawing to a larger scale than is originally shown on the drawing in order to show certain details more clearly. The electric service room shown in Fig. 4-28 is a good example. This floor plan is drawn to a scale of ⅛″=1′0″.

   Prepare a sheet of 11″ × 17″ tracing paper by securing it to your drawing board and preparing the borders as described previously; draw the electric-room floor plan to a scale of ¼″ = 1′0″. Include all of the panel boards and all notes indicated on the reduced-scale (⅛″) drawing. The enlarged drawing should be drawn in the approximate center of your tracing paper, room being left in the lower right-hand corner of the sheet for your title block, which will be inserted at a later date.

   Begin with a medium-hard lead, say, 5H; use the ⅛-in. architect's scale to measure all dimensions on the reduced-scale drawing in the figure, and then use the ¼-in. architect's scale to transfer these dimensions to your drawing. Once all lines have been transferred, the 5H lead being used, carefully check all dimensions to ensure that they have been drawn correctly. Then darken all lines with a 2H lead, making certain that you obtain clear sharp lines.

   With all line work completed, use your Ames lettering guide—set at number 3—and make the required guidelines, using 6H or harder lead. Then letter the notes as shown in Fig. 4-28. Now move the disc on the lettering guide to number 5 and make another set of guidelines directly under your drawing about one inch below the lowest line of the floor plan; this is for the drawing title. Letter this title: "Mechanical Room Floor Plan." Change the lettering guide back to 3, make another set of guidelines about ¼ inch below the title and letter in the scale; that is, "Scale: ¼″ = 1′0″. Also be sure to indicate all breaks in the adjoining partitions with a break line to "cut" them at the required location.

3. Using the ⅛″ architect's scale, find the missing dimensions shown in the illustration in Fig. 4-29.

   A.______ B.______ C.______ D.______ E.______

QUESTIONS

1. Dimensions of all electrical drawings should appear in ______and______parts of an inch unless the metric system is used.
2. What type of drawing line is normally used for dimension lines?
3. For what purpose are extension lines used?
4. Where should the dimensions be placed on a drawing to show the length of an object?
5. What is the dimension line for an angle?

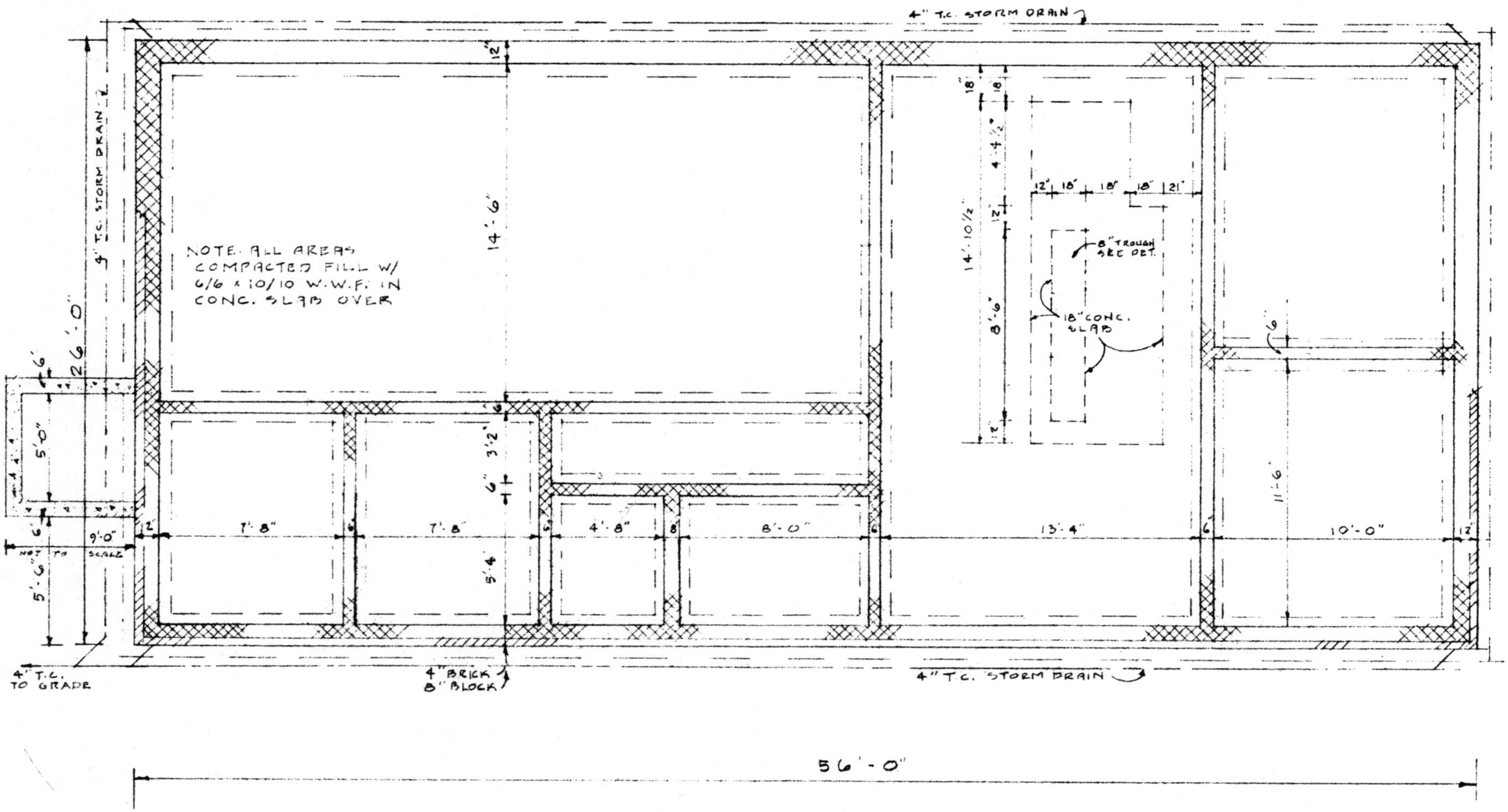

Figure 4-24. Method of dimensioning foundation plans for a building.

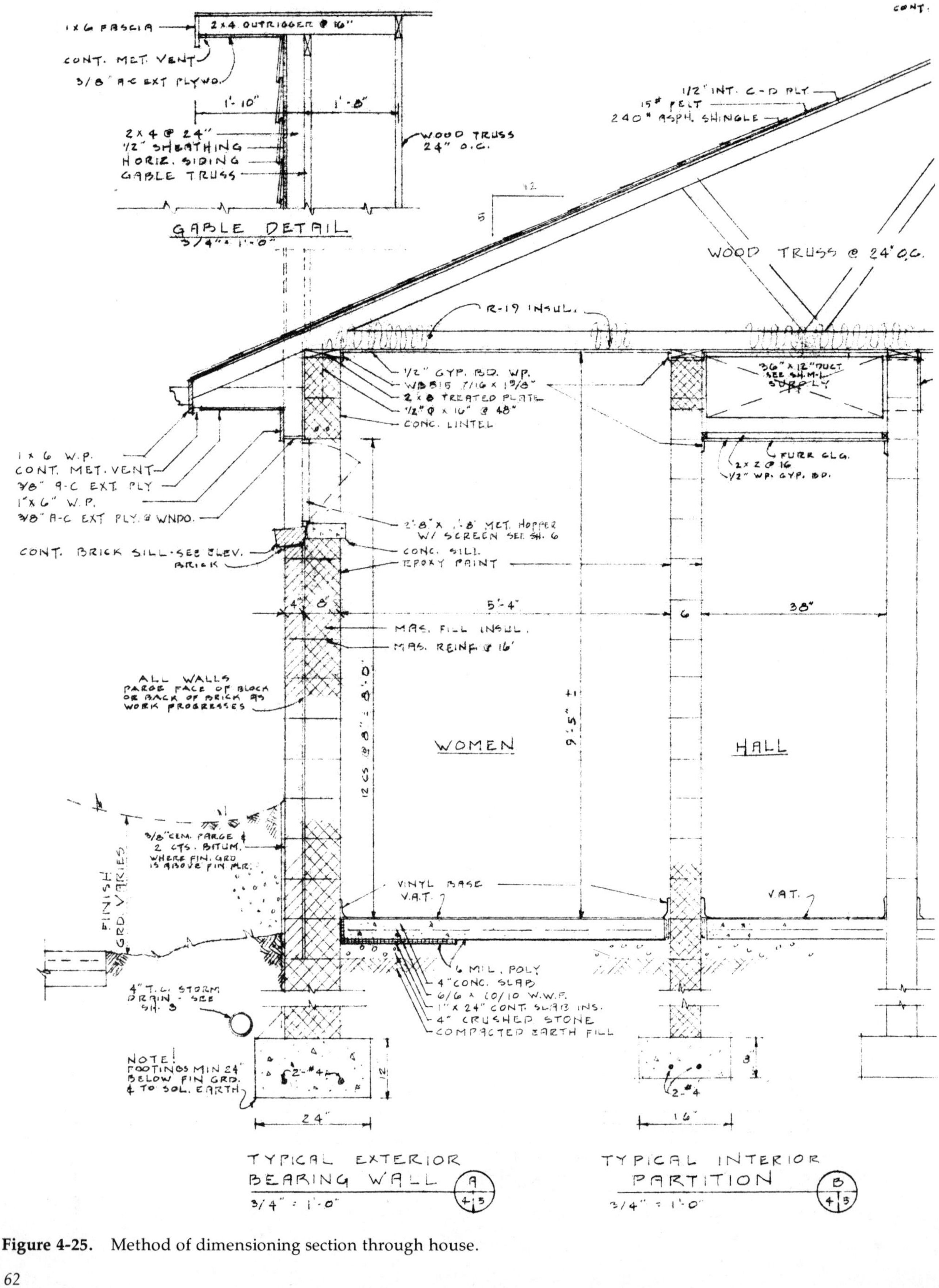

Figure 4-25. Method of dimensioning section through house.

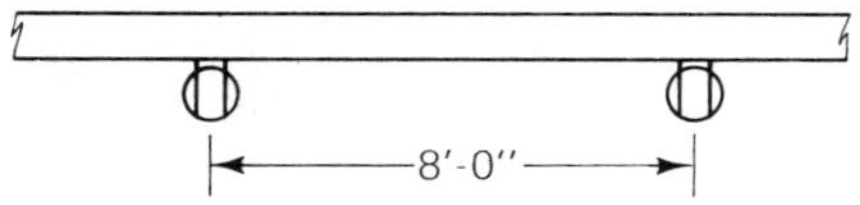

Figure 4-26. Method of dimensioning distance between electrical outlets.

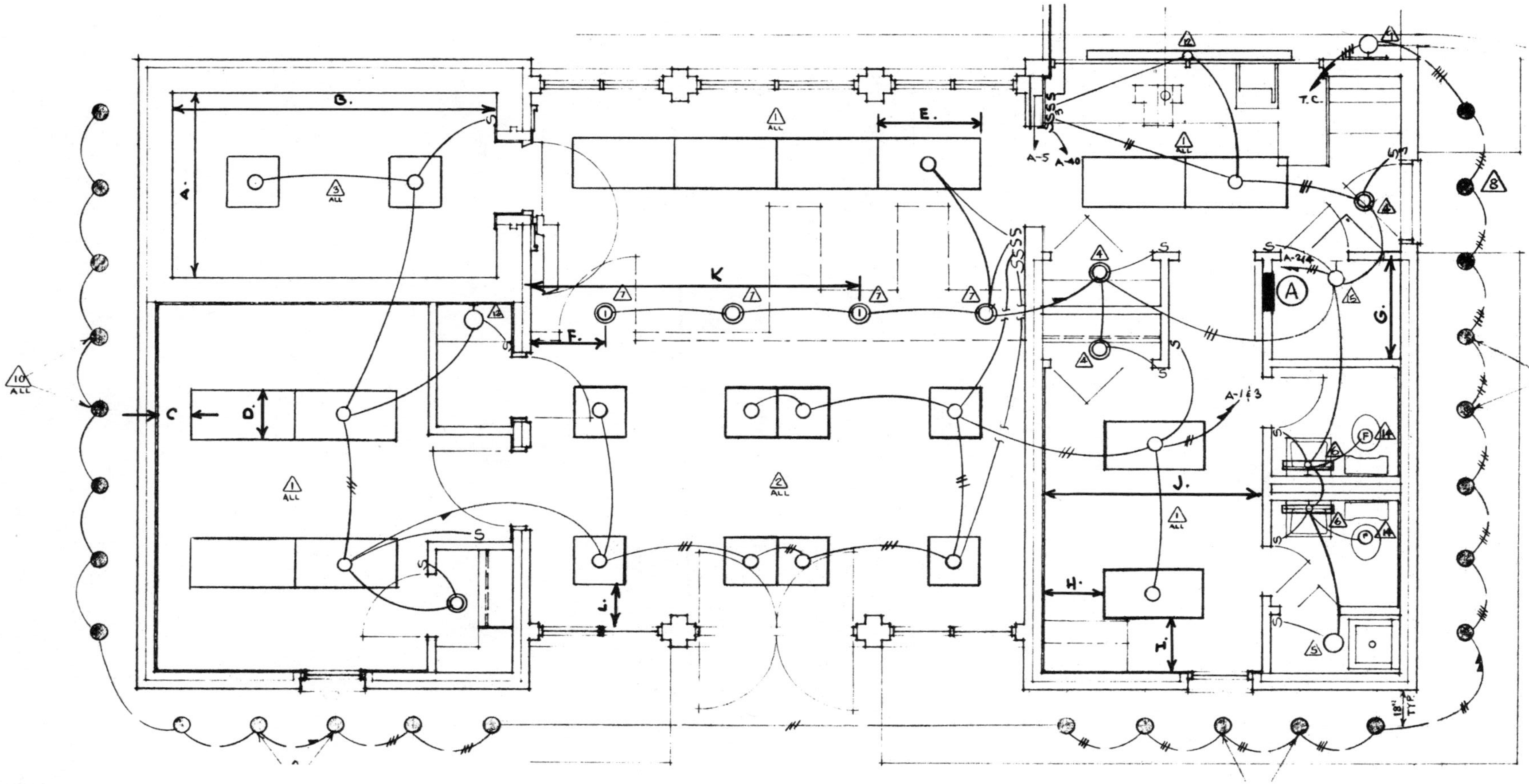

Figure 4-27. Drawing to be used for assignment.

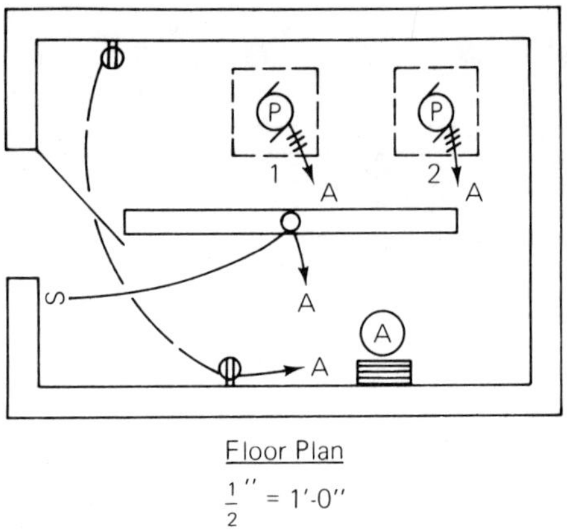

Figure 4-28. Electric service room for assignment 2.

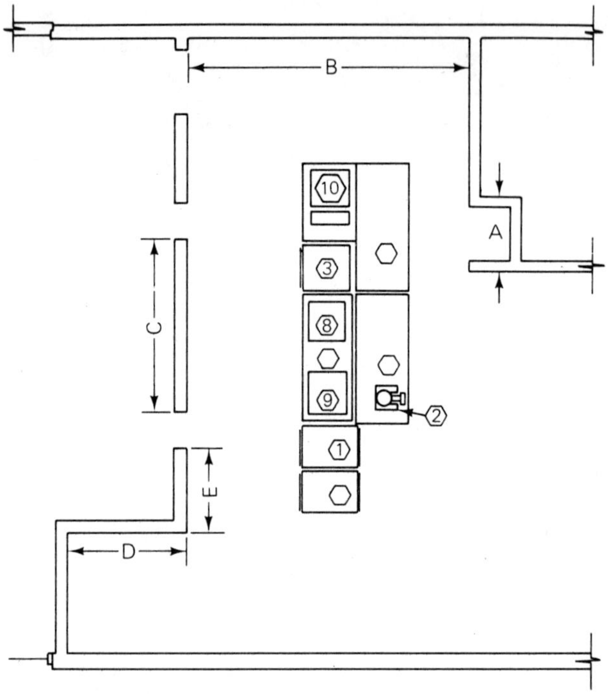

Figure 4-29. Drawing for assignment.

6. Name three scales commonly found on the architect's scale.
7. What minimum distance should dimension lines be located from building lines on architectural drawings?
8. When a measurement is laid off on an architectural drawing, it is made with the reduced foot rule known as the ______scale.
9. Why are most dimensions on architectural drawings given as center-to-center distances?
10. When a measurement is taken on an actual building, it is made with the ______ ______rule.

Chapter 5

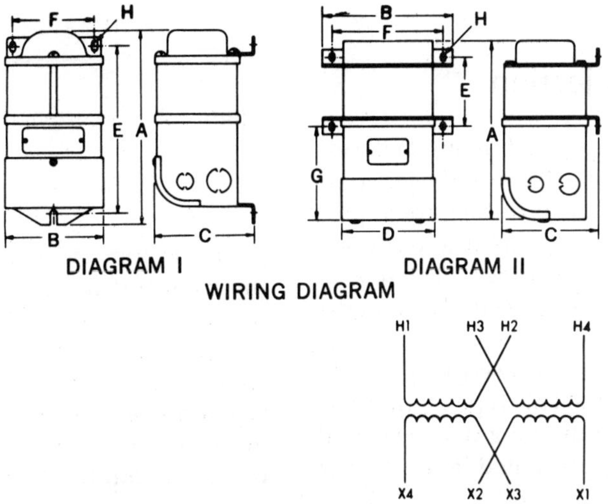

ORTHOGRAPHIC PROJECTION

The main objective of this chapter is to explain an orthographic-projection drawing, how it is laid out on the drawing sheet. The principal parts of an architectural drawing (as applied to electrical drafting) will be covered, as well as the actual procedures for making a working drawing in both the electrical and electronic fields.

5.1 Orthographic Projection

Although there are several types of drawing methods used in the preparation of electrical and electronic drawings, the views known as an orthographic projection will be the most commonly used to draw objects that are to be constructed by others. The reason is that although pictorial views of objects give a realistic appearance, only drawings with exact information concerning shape and size can be used to properly construct an object or a system. This type of information is usually best given in a drawing where several related views of an object are presented in the proper manner.

The basic elements of a working drawing, then, include projection methods for presenting views and sections, dimensioning methods, lines, symbols, and notes. The orthographic-projection method enables you to draw an object that has depth on a sheet of drawing paper, which has only length and width. A brief discussion of orthographic projection is necessary, however, before any drawing is attempted.

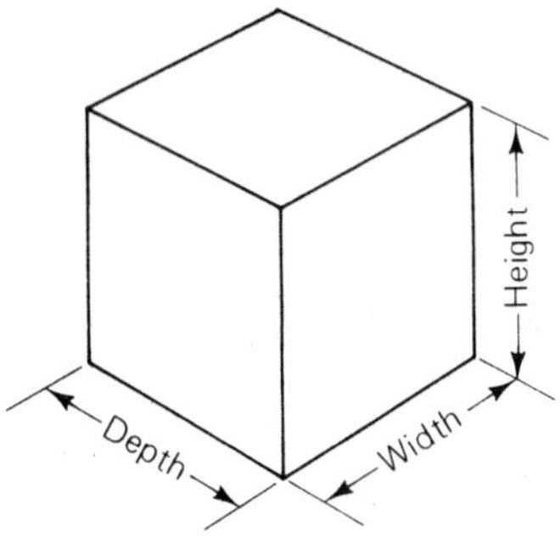

Figure 5-1. Pictorial drawing of a cube.

The pictorial drawing of a cube in Fig. 5-1 is a single-view drawing that shows how the object appears to the viewer. The words *length, width,* and *height* are used to show the basic dimensions on the drawings. If this same object were converted to an orthographic projection, it would appear as shown in Fig. 5-2. Here, only three separate views are required,

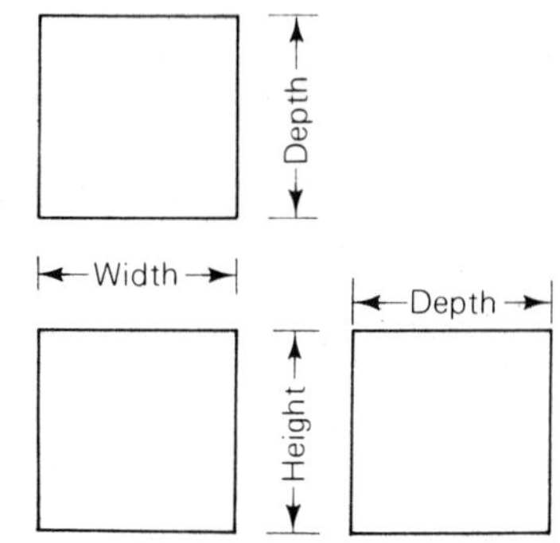

Figure 5-2. An orthographic projection of the cube in Fig. 5-1.

since the top and bottom are the same dimensions, the back and front are the same, and both sides are the same.

In both figures (Figs. 5-1 and 5-2) the width represents the horizontal distances on the drawing in the top and front views. The height refers to the vertical distances on the drawing in the front and side views, while the depth refers to the vertical distances in the top view and the horizontal distances in the side view.

An orthographic projection of a building would consist of the views illustrated in Fig. 5-3. One drawing shows the building as though the observer were looking straight at the front; one as though the observer were looking straight at the left side; one as though the observer were looking straight at the right side; and one as though the observer were looking straight at the rear of the building. These drawings may be referred to as *building elevations*.

Referring again to Fig. 5-3, assume that a horizontal cut is made through the building along a line already indicated as A-A on the front elevation. Then imagine that the top part is removed and a drawing is then made as if the artist were looking straight down at the remaining part. This drawing is known as a floor plan. It indicates the outside wall lines, interior partitions, windows, doors, and similar details. The final drawing would appear as shown in Fig. 5-4.

If the building should have more than one floor, horizontal cuts may be taken at varying distances from the ground in order to show the other floors on the drawing.

The instrument chassis in Fig. 5-5 is the kind of object that will most often be encountered in electronic drafting. To describe the correct shape and size of this chassis, at least three views are required: top, front, and side. If all sides had, say, knockouts (holes) of different sizes, then perhaps five or six different

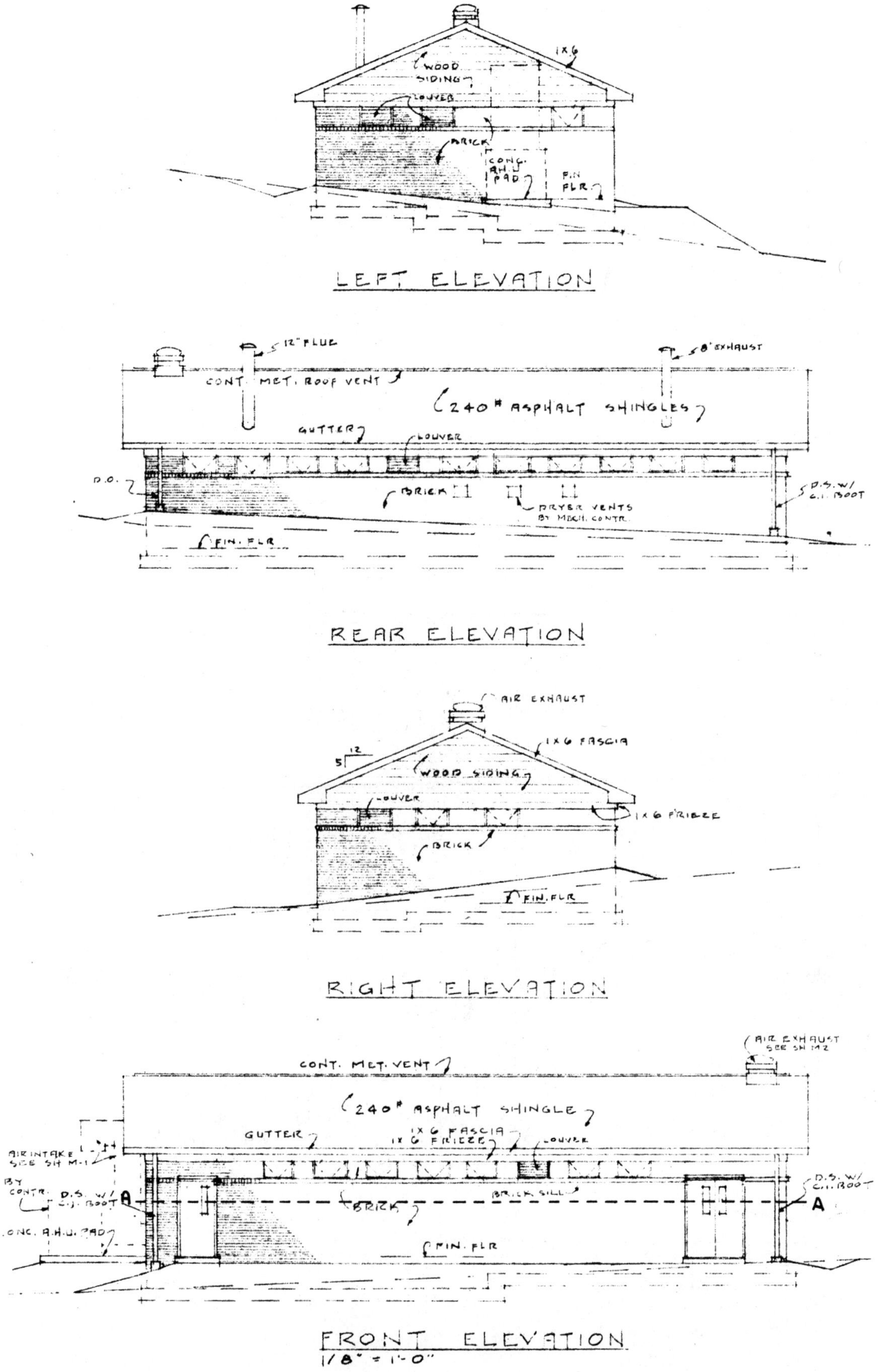

Figure 5-3. An orthographic projection of a building.

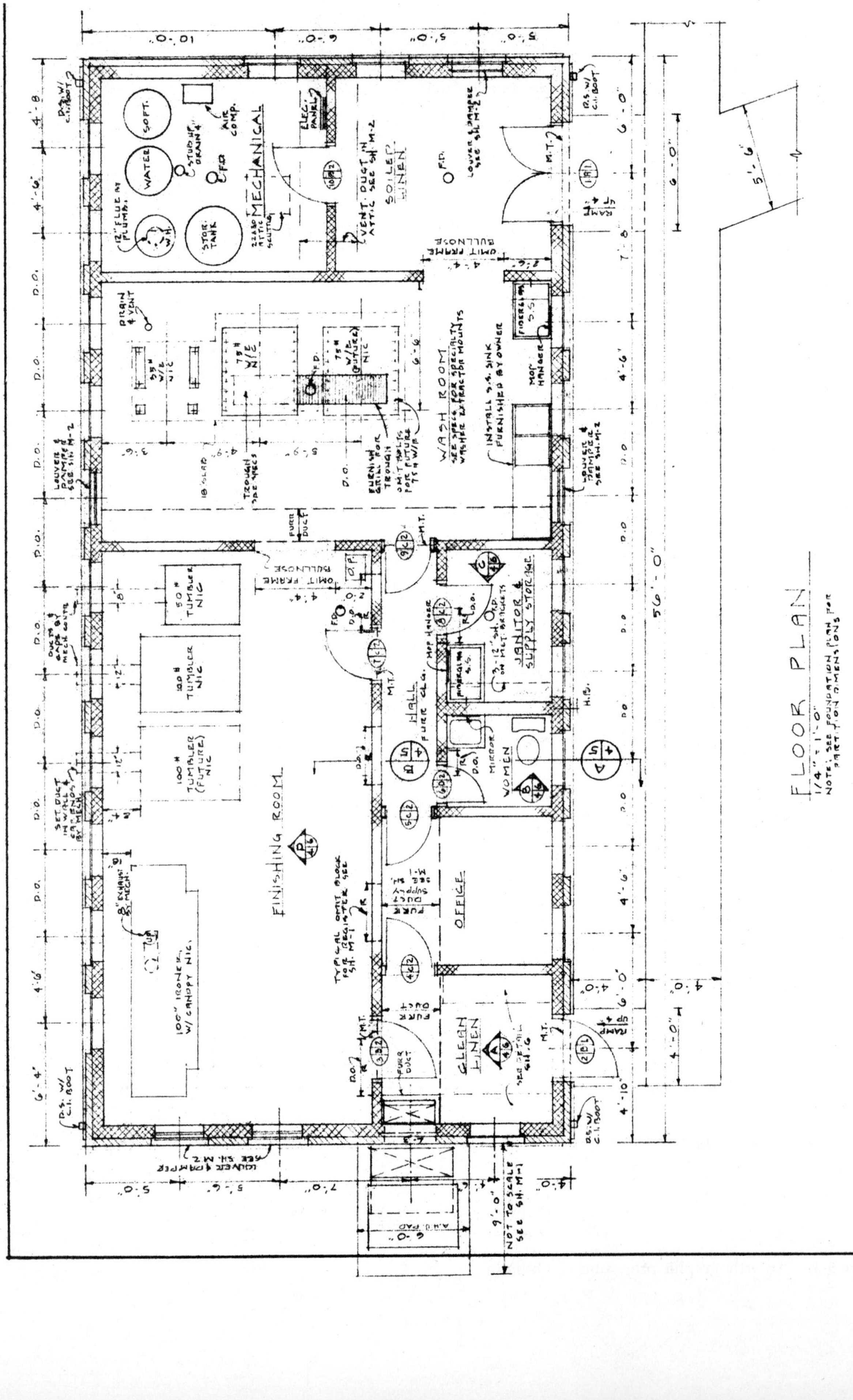

Figure 5-4. Floor plan of a building.

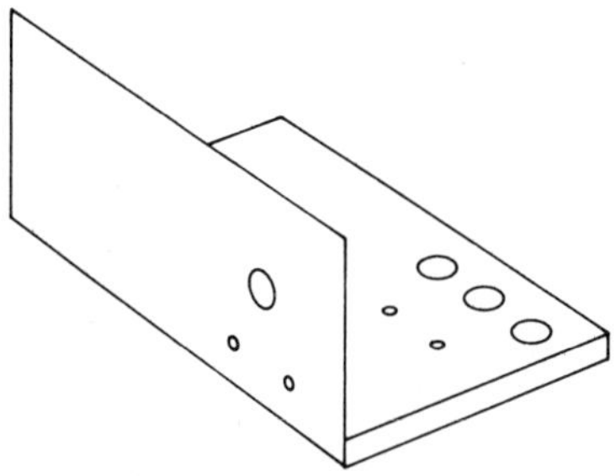

Figure 5-5. Pictorial drawing of an electronic instrument chassis.

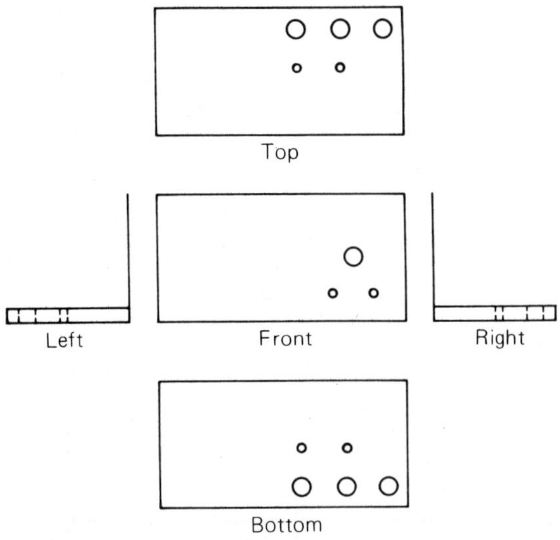

Figure 5-6. An orthographic projection showing all six sides of the chassis in Fig. 5-5.

views would be required. An orthographic-projection drawing showing all six sides of this chassis is shown in Fig. 5-6.

Once all of the necessary views are drawn, the various views should be dimensioned by methods described in Chapter 4.

5.2 Procedures for Making an Electrical Drawing

Figure 5-7 shows a sample floor plan from a set of electrical working drawings. First, notice that the entire drawing sheet has a border line around its perimeter. This is to square and confine the drawings. Within this border is a floor plan of a residence titled "Basement Floor Plan—Electrical"; note that it is drawn in a scale of ¼″ = 1′0″ (although the whole drawing has been greatly reduced for this book). The symbol list is in the upper right-hand corner of the drawing, and a lighting fixture schedule (to identify the various lighting fixtures) appears on another sheet of the drawings (see Fig. 5-8).

The basement floor plan of this building shows all outside walls, interior partitions, windows, doors, toilet fixtures, and stairwells (stairways). All of the electrical outlets are indicated by the appropriate symbol from the list, and the circuits are indicated by solid and broken lines. A note on the right-hand end of the floor plan tells the electrical workman what type of electric wall heater is required.

The printed hands with a figure pointing to a particular part of the drawing is one of several methods to indicate a revision of the original drawings. In this particular instance, details of the revisions were explained in an addendum issued in written form after the drawings were released. In other cases, drawings or notes may appear on the drawings.

A title block appears in the lower right-hand corner of the drawing sheet and contains sufficient information for identification of the project and other pertinent data. You will find that most electrical and electronic drawings will have a title block in the lower right-hand corner of the drawing sheet, the size of the block varying with the size of the drawing and also with the information required. For 11″ × 17″ drawing sheets—like the size used for assignments in this book—the space to be reserved for the title block should be a minimum of three inches long and one inch high.

In general, the title of an electrical drawing should contain the following:

1. Name of the project.
2. Address of the project.
3. Name of the owner or client.
4. Name of the architectural and/or engineering firm.
5. Date of completion of the drawings.
6. Scale(s).
7. Initials of the draftsman, checker, and designer, with dates under each.
8. Job number.
9. Sheet number.
10. General description of the drawing.

Figure 5-8 shows the main floor level of the residence in Fig. 5-7. Again, lighting fixture outlets, receptacles, special outlets for kitchen appliances, and the like are shown on the floor plan by means of elec-

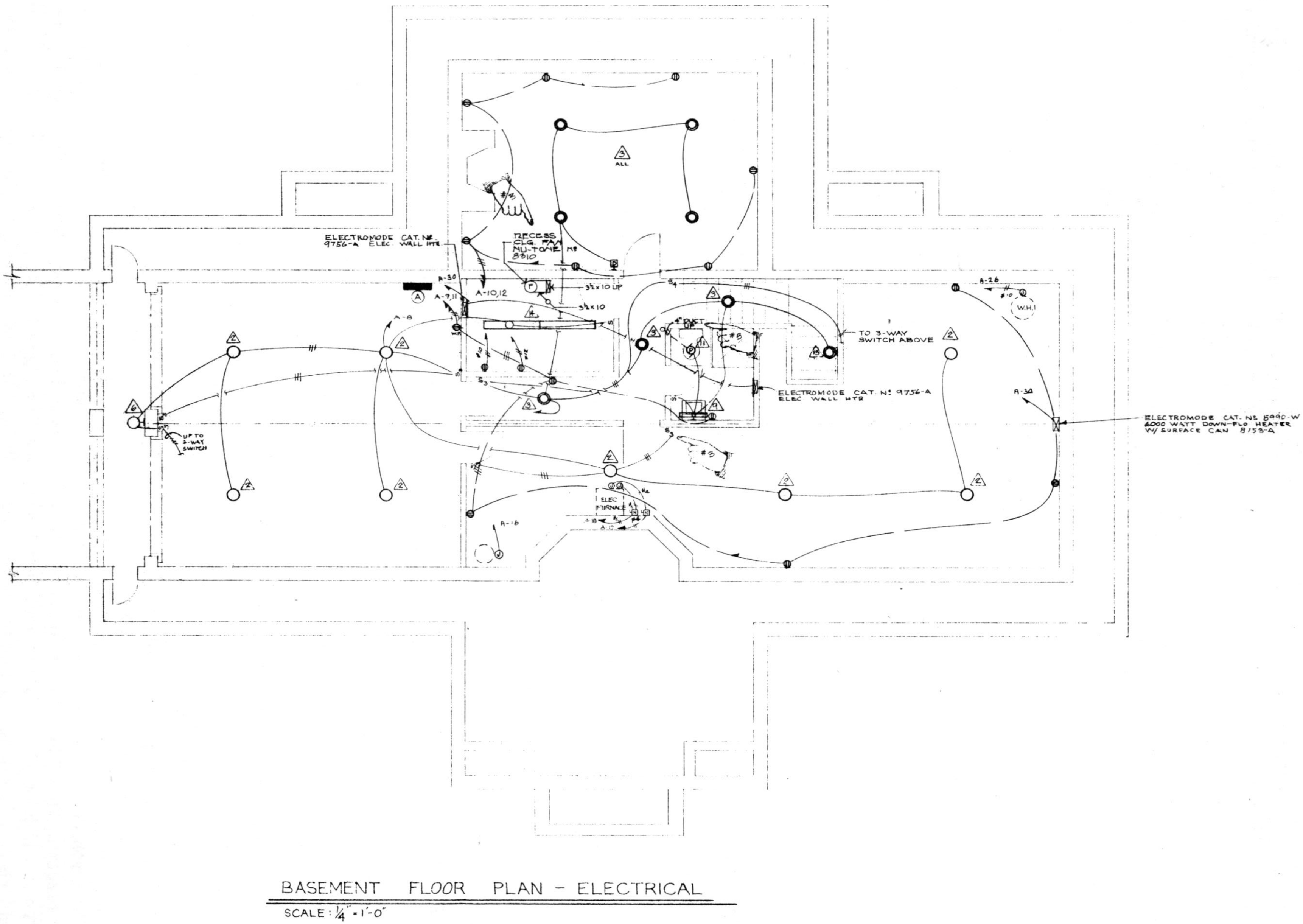

Figure 5-7. A sample floor plan from a set of electrical drawings.

LIGHTING FIXTURE SCHEDULE

FIXT. TYPE	MANUFACTURER'S DESCRIPTION	LAMPS Nº	LAMPS TYPE	VOLTS	MOUNTING	REMARKS
1	MARKSTONE CAT. Nº. W5509	1	100W I	120	RECESSED	GASKETED
2	P&S CAT. Nº. 110	1	100W I	↓	SURFACE	
3	MARKSTONE CAT. Nº. WAL-F10-6-FS	1	100W I	↓	RECESSED	
4	LITHONIA CAT. Nº. S120PH	1	20W F	↓	IN COVE	W/ WARM-WHITE LAMPS
5	MARKSTONE CAT. Nº. EBF3	1	50W I	↓	RECESSED	R20 LAMPS
6	TO BE SELECTED BY OWNER - ALLOW $125.00			↓		
7	TO BE SELECTED BY OWNER - ALLOW $125.00			↓		VERIFY LOCATION W/OWNER
8	TO BE SELECTED BY OWNER - ALLOW $125.00			↓		
9	CRESCENT ANG 220	2	20W F	↓	WALL	6" ABOVE MIRROR
10	TO BE SELECTED BY OWNER - ALLOW $75			↓		
11	NUTONE CAT Nº 8662	1	100W	↓	RECESSED	RUN 4" DUCT & TERMINATE W/ 6×6" SOFFIT GRILLE
12	TO BE SELECTED BY OWNER - ALLOW $50			↓		VERIFY LOCATION W/ OWNER
13	MOLDCAST CAT Nº V-372	2	60W I		RECESSED	18" ABOVE LANDING
14	MOE CAT CAT Nº M-315	1	40W F		SURFACE	
15	ALKCO CAT. Nº SS-213-S	2	13W F		SURFACE	UNDER CABINET

Figure 5-8. A typical lighting fixture schedule used to identify the various lighting fixtures on a drawing can be seen on this drawing sheet.

trical symbols. They are connected by broken and solid lines to indicate the circuits. The types of lighting fixtures are indicated by a numeral inside a triangle adjacent to a fixture or a group of lighting fixtures and are then identified in the lighting fixture schedule appearing to the right of the floor plan. A panelboard schedule also appears above the lighting fixture schedule to give necessary data for the selection and installation of the panel board—in this case, panel "A."

Although the descriptions of the drawings in Figs. 5-7 and 5-8 were very brief, they should be sufficient to give the reader a general knowledge of how an electrical drawing for building construction is laid out. To summarize, the contents of the drawings are enclosed in border lines around the entire drawing; a title block appears on the drawing (usually in the lower right-hand corner) to identify the drawing's contents and the name of the project; the floor plan of the building usually appears on the left-hand side of the drawing sheet, while notes, schedules, lists, etc., appear on the right-hand side of the drawing; the design and layout of the electrical system are indicated on the drawings by lines, symbols, and notation.

The draftsman who is employed by an electrical equipment manufacturer will be required to prepare drawings of the company's equipment; some will be pictorial for catalogs, and others will be orthographic views, such as the ones in Fig. 5-9. The orthographic views in this figure show the front and side views of two types of electrical transformers and show the location of the knockouts for wiring connectors or conduit, the connection plate, and other details. The dimensions (represented in this drawing) are indicated by letters—A, B, C, etc.—rather than numerals. The reason for this is that transformers of different sizes can be represented by the one drawing, and then the dimensions of each size can be shown in a separate schedule—the letters being used as references.

This drawing also shows a simplified wiring diagram of the connections; that is, the H (1, 2, 3, and 4) points represent the line side or high-voltage side of the transformer, and the X's represent the load or low-voltage side.

The orthographic view in Fig. 5-10 is that of an electronic circuit board, showing the various components of an electronic instrument (designated by the combination of letters and numerals); the printed circuits are shown in a lighter shade to signify that these are on the opposite side of the circuit board from the components. The components themselves are identified by a separate schedule like the one in Fig. 5-11.

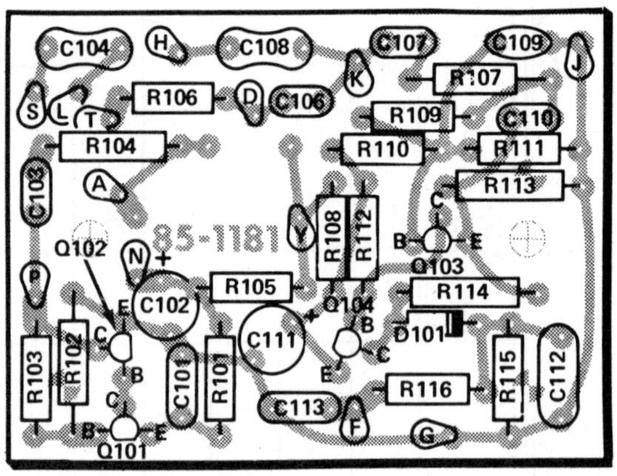

Figure 5-10. An orthographic view of an electronic circuit board.

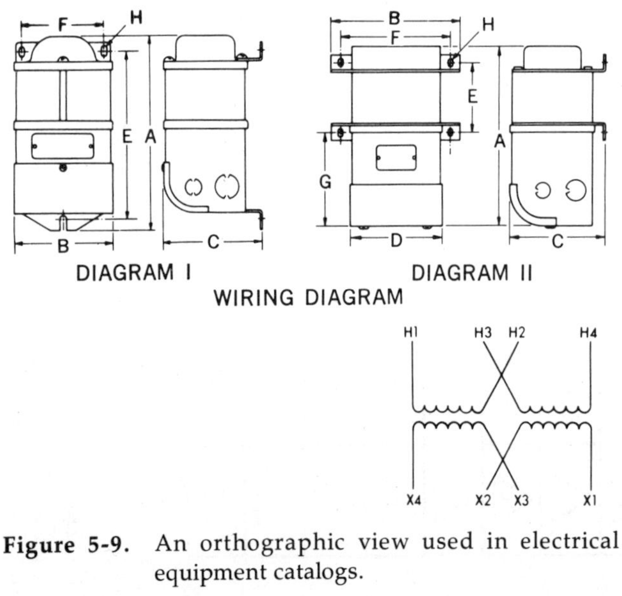

Figure 5-9. An orthographic view used in electrical equipment catalogs.

The electronic drawing just mentioned is in a semipictorial view—showing in detail how the components are arranged on the circuit board—although it is still classified as an orthographic drawing, since it is drawn in only one plane.

Another type of orthographic drawing that often accompanies a drawing such as the one in Fig. 5-10 is shown in Fig. 5-12. This is a schematic diagram, but since it appears in one plane, it is classified as an orthographic view.

All types of drawings encountered in electrical and electronic drafting will have many of the same characteristics, in that all drawing sheets should be enclosed in border lines, title blocks or title strips should appear on all sheets, and the objects will be drawn with lines and symbols.

KEY No.	PART No.	PARTS Per Kit	DESCRIPTION
Composition (Cont'd.)			
A1	1-104	3	100 kΩ, 5% (brown-black-yellow-gold)
A1	1-77	1	390 kΩ, 5% (orange-white-yellow-gold)
Film			
A2	4-58	2	150 Ω, 5% (brown-green-brown-gold)
A2	4-61	1	560 Ω, 5% (green-blue-brown-gold)
A2	4-63	2	2700 Ω, 5% (red-violet-red-gold)
A2	4-38	3	9100 Ω, 5% (white-brown-red-gold)
CAPACITORS			
B1	21-150	1	820 pF ceramic
B2	20-137	1	1800 pF mica
B2	20-150	1	3300 pF mica
B3	27-68	1	.0033 μF Mylar* (3300)

Figure 5-11. A schedule used for identifying the various components used in the drawing Fig. 5-10.

Further data would include alignment lines or arrowheads to align the drawing on the drawing board for revisions or for microfilming. These should be located at the horizontal and vertical center of the overall sheet at each of the four sides and in the margin between the outside edges of the drawing paper and the border lines, as shown in Fig. 5-13. A revision block as well as room for notes and schedules should also be provided.

Regardless of the type of drawing, proper planning is essential for a good layout. It is not good enough just to be able to get the various views and notes on the sheet; they should be arranged in a neat and logical manner. For example, when the necessary views have been decided upon, they should be laid out with adequate space for dimensions without crowding. If there is not enough room on the sheet for schedules and notes, use a separate sheet for the latter, but enough space must be provided for a title block, revisions, and similar necessary data on every sheet.

The drawing title (usually found in the title block) should clearly define the type of drawing; for example, a floor plan, a single-line diagram, a

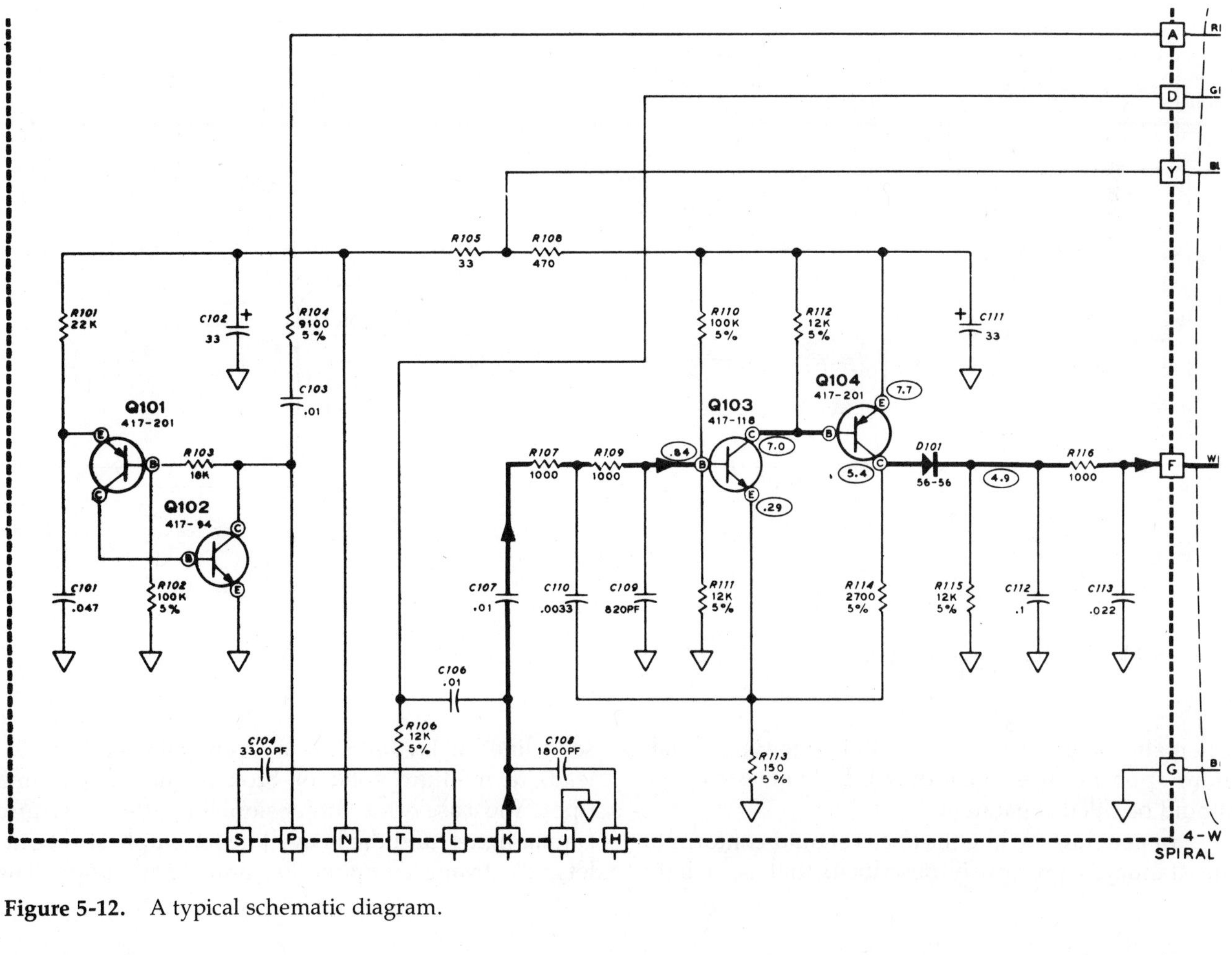

Figure 5-12. A typical schematic diagram.

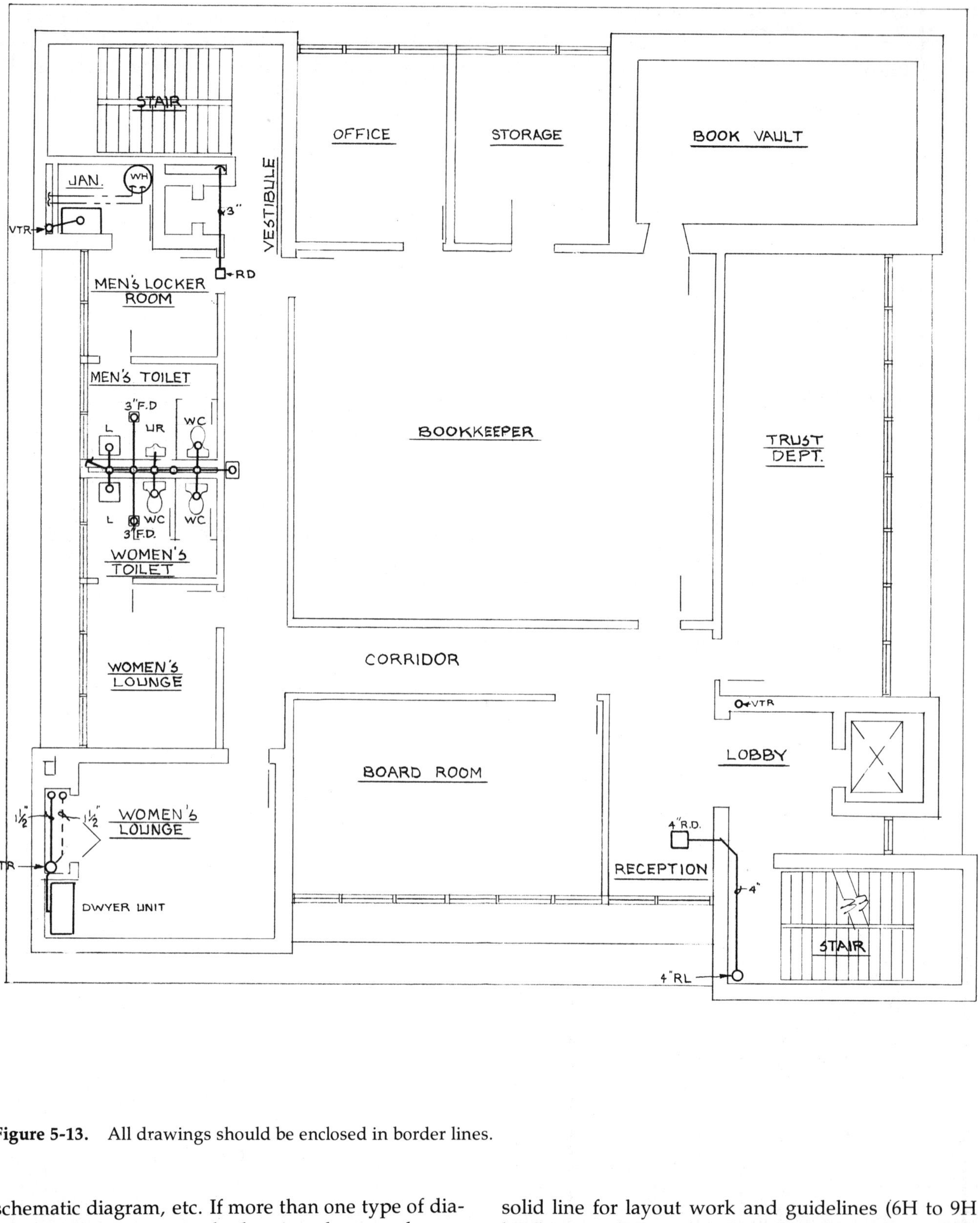

Figure 5-13. All drawings should be enclosed in border lines.

schematic diagram, etc. If more than one type of diagram appears on a single drawing sheet, each type should be titled separately.

In general, the line thickness should conform to the standards previously described; that is, a light solid line for layout work and guidelines (6H to 9H lead), a medium solid or broken line—depending upon the case—for dimension lines, leaders, and similar items, and the heavy line for emphasis of outlets, electronic components, and circuit lines. The

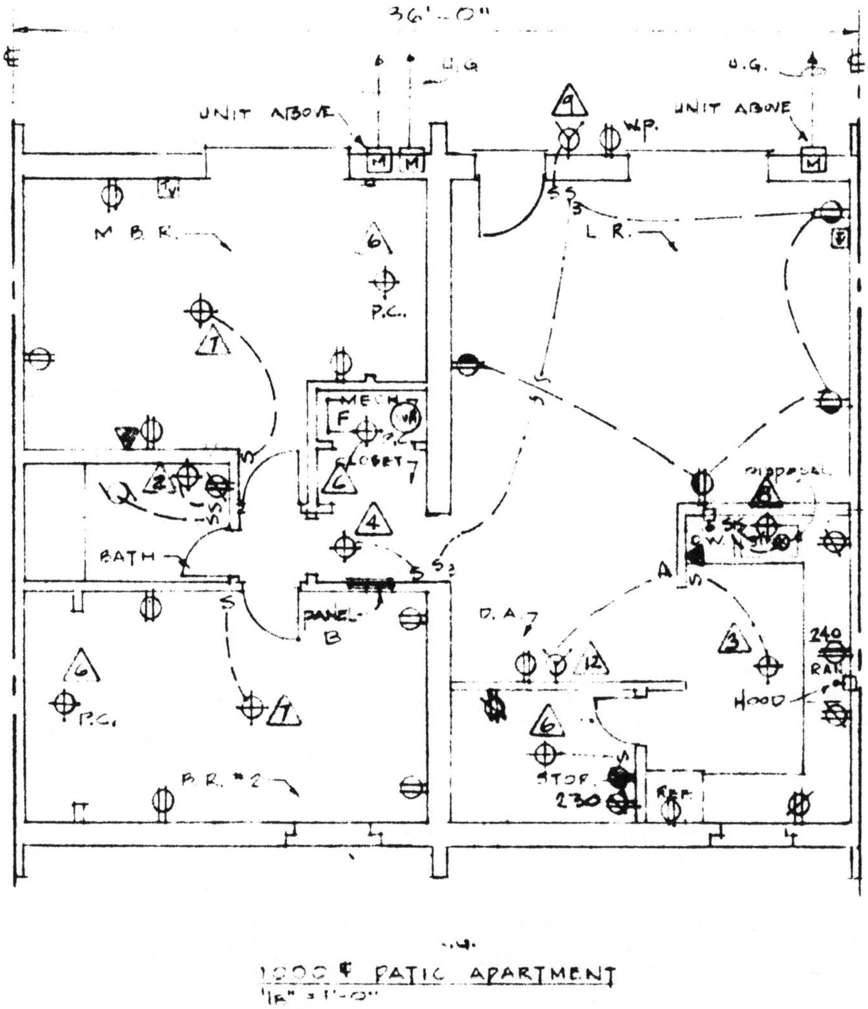

Figure 5-14. A building floor plan with both the lighting and power circuits shown in one view.

border lines and lines enclosing the title block, schedules, etc., are usually made with the extraheavy solid line.

Use standard electrical and electronic symbols as described in Chapter 6 where possible, and if these symbols must be modified to solve a particular need, always identify the symbol by note or symbol list. The size of the symbols will vary—especially on electrical drawings for building construction drawn to different scales—but they should always be in proportion to the other symbols used in the drawing.

Abbreviations should conform to the American Standard Abbreviations for Use on Drawings, Z32.13 and approved by the ASA. In the absence of a specific abbreviation in the list just mentioned, a special abbreviation may be devised, but it should be explained by note somewhere on the drawings. Obviously, if the word appears only a few times on a drawing, it would be simpler to write the word out rather than abbreviate.

In preparing drawings in which circuit lines appear, use as few crossovers in the lines as possible; if several lines happen to meet at one point, try to make another arrangement. One solution to the problem would be to use more drawing sheets to convey the data in a clearer method. To illustrate, the electrical construction drawing in Fig. 5-14 shows a building floor plan with both the lighting and power (receptacles, etc.) circuits on one plan. Note that the numerous lines on this drawing make the tracing of them difficult. A better arrangement appears in Figs. 5-15 and 5-16. Here, a separate drawing is used for the lighting and one for the power—making fewer lines on each sheet and, in turn, making the drawings less confusing.

Sometimes it is easier to interrupt a line to a

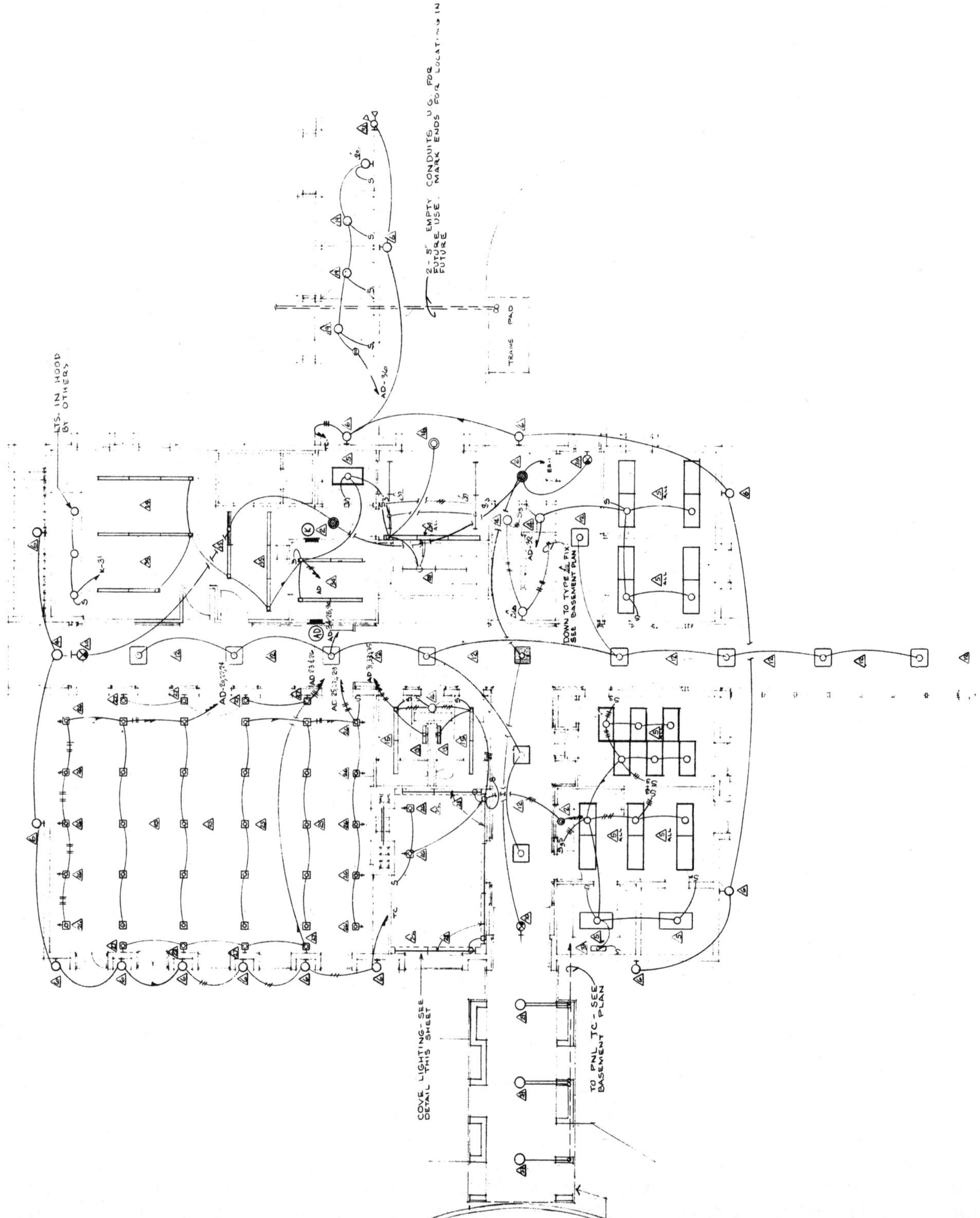

Figure 5-15. A better arrangement of a floor plan, showing only the lighting layout in one view.

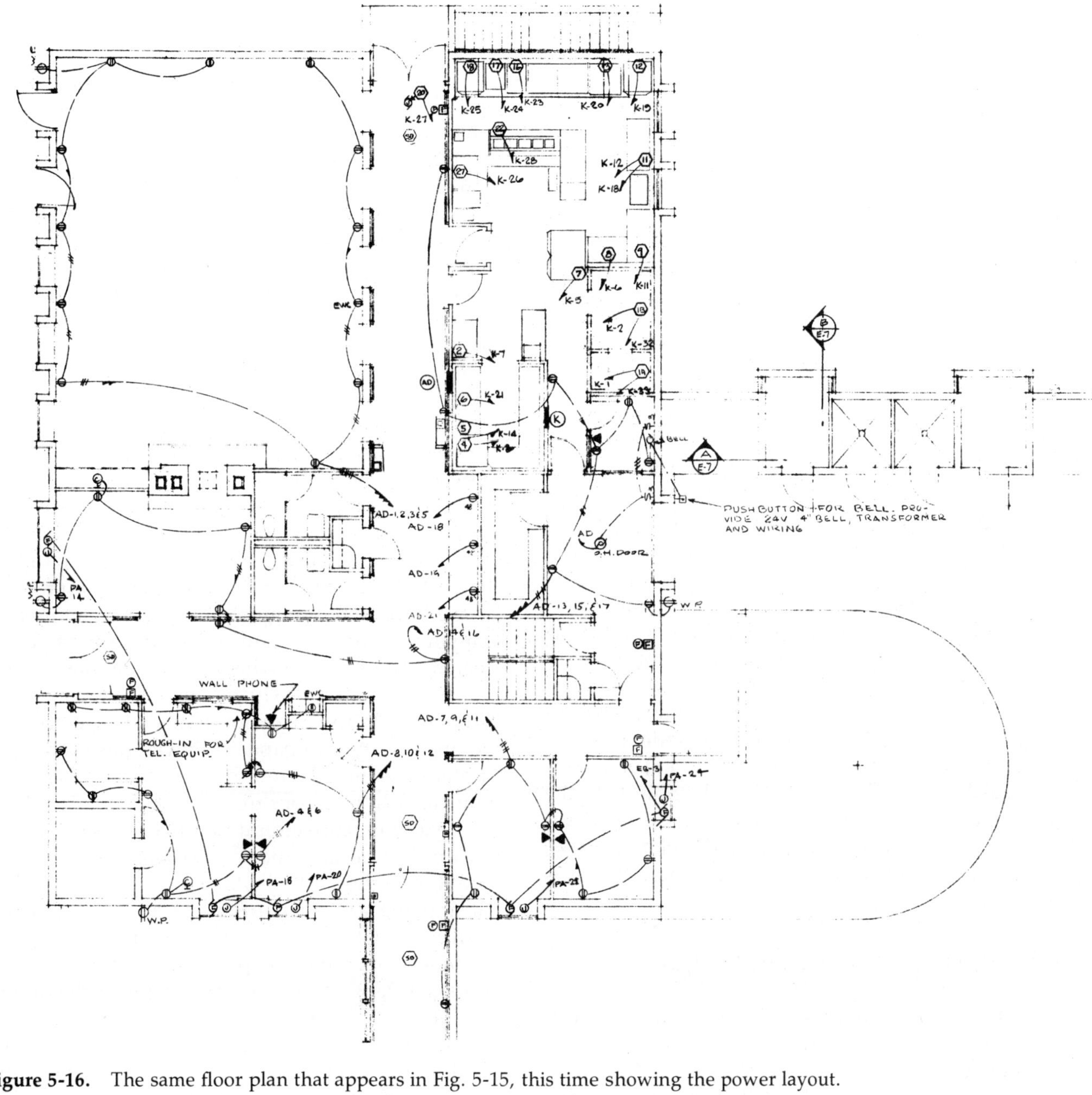

Figure 5-16. The same floor plan that appears in Fig. 5-15, this time showing the power layout.

piece of electrical equipment or component and note its destination. For example, the circuit homerun from the last outlet on the circuit to the panel board in Fig. 5-17 is not drawn out completely. Rather, the line is interrupted shortly after it leaves the last outlet, an arrowhead is drawn at the interruption point, and a note "to panel A" is lettered. This tells those reading the drawings that the circuit continues on and is connected to panel board A.

In preparing electrical and electronic drawings, the draftsman must learn the meaning of such words as *ground, capacitance, voltage, wattage, ohms,* etc., before he can intelligently prepare correct drawings. The beginner can learn some of the meanings of the basic terms in Appendix A of this book, and when they are understood, he can refer to the American Drafting Standards Manual published by A.S.M.E. for the standard practices relating to these terms.

Electrical and electronic symbols will be completely covered in Chapter 6 to show the draftsman how the various electronic components and electrical equipment are shown on working drawings. How-

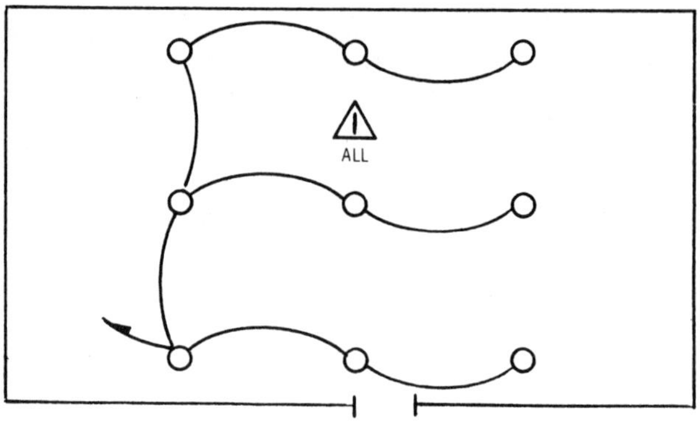

Figure 5-17. Drawing showing circuit homeruns from the last outlets on circuits to the panel board.

ever, there are other basic practices that the draftsman should learn to be proficient in his field. Two books published by Reston Publishing Company will help the draftsman obtain this information: *Basic Principles of Electricity* and *Basic Principles of Electronics* by Vester Robinson.

An example of what a draftsman should know includes his knowledge that resistance is expressed in *ohms* (Ω) for values beginning with 1 and going to 999, and in *megohms* (MΩ) for resistance values over 1,000,000. For values of 100–999,000 ohms, the abbreviation K for kilohms (kΩ) is sometimes used. Thus, 25,000 ohms would be written 25 kΩ or 25 K.

Further, the draftsman should know that the basic value in expressing capacitance is the micromicrofarad (μμF) for values up to and including 9999; and for values over 10,000, the microfarad (μF) is used.

5.3 Reproduction of Drawings

With the few assignments you have done thus far, you should be able to see that a great amount of work is required to prepare even the simplest of drawings. For this reason, the original drawing of an electrical or electronic project is often very valuable and must be guarded against wear or becoming lost. One simple solution to this problem is to make copies of the original drawings for distribution to those who need them for reference.

Blueprinting has been the accepted method of reproducing drawings from tracing paper for many years. Basically, the process of making a blueprint consists of placing a special paper that has been coated with chemicals sensitive to sunlight in a glass frame with its coated side up. The original drawing on tracing paper is laid over the coated paper, and both are held together by the glass frame.

The entire assembly is then subjected to sunlight for a few minutes, and then the coated paper is thoroughly washed in clean water and hung up to dry. If the exposure has been timed correctly, the coated surface of the paper is now a clear, dark blue color except where it was covered by the lines on the drawing; the latter are perfectly white.

An electric blueprinting machine is also used for making blueprints. In place of sunlight, the coated paper along with the original drawing passes around a glass cylinder containing electric lamps. The speed at which the paper travels may be adjusted to suit the quality of the drawing, as far as the intensity and size of the lines and the depth of the background are concerned.

Most drafting rooms now use the reproduction method known as *whiteprinting*. This method consists of exposing chemically treated paper and the original drawing to a high-intensity discharge or fluorescent lamp enclosed in a suitable housing. The ultraviolet light from either of these two light sources reduces the part of the sensitized surface that is unprotected by the lines of the original drawing into an invisible compound. After exposure, the print (or chemically treated paper) is then subjected to ammonia vapors, which develop the sensitized lines. The finished print is true to scale and ready for immediate use. Depending upon the type of paper used with this process, the printed lines may be red, brown, blue, or black on a white to light blue background.

Another convenient method of reproducing drawings (especially the smaller ones) is photocopying. This method actually takes a picture of the original drawing and then produces as many prints as desired—just like a Xerox or similar copying machine. This type of copier also has the advantage of making reductions or enlargements of the original drawing.

Over the past decade or so, microfilming has rapidly moved into the drafting industry for reproducing and preserving drawings. Because of the small size of the film, much space is saved, as well as shipping charges for the drawings.

The first step in microfilming a drawing is to convert the original drawing to a microfilm frame by means of a special camera mounted on a frame over a platform such as a table. The camera reduces a drawing so that it will fit on the microfilm; the reduction can be varied by changing the height of the camera above the drawing.

After the film is exposed, it is passed through a processor, which develops it, and it is then mounted on some type of card or frame.

Once the film has been mounted, the image can

be blown up for viewing by means of a special viewer. Some viewers will even make full-size prints from the microfilm.

Since a great amount of reduction is obviously used in the microfilm process, drawings of the highest quality should be used, with every line sharp and the lettering of sufficient size.

SUMMARY

Orthographic drawings enable an object to be drawn with depth on a sheet of drawing paper, which has only length and width. It is probably the type of drawing most often used in electrical and electronic drafting.

Proper planning is essential for a good layout on any type of drawing. The views should be laid out with adequate space for dimensions, notes, title block, revisions, schedules, and legends.

All drawing should have a title to define the type of drawing being made; if more than one type of diagram appears on a sheet, each should have a separate title.

In preparing drawings where circuit lines appear, avoid as many crossovers in the lines as possible. Use more drawing sheets if necessary, as long as it is practical.

The draftsman must know and understand certain trade terms before he can intelligently prepare correct electrical and electronic drawings.

To protect the original drawing and to have enough copies for those involved in a project, the original drawing usually is copied by one of the following methods:

1. Blueprinting
2. Whiteprinting
3. Photocopying
4. Microfilming

ASSIGNMENT

Prepare a sheet of 11″ × 17″ tracing paper with title block, border lines, etc. Then make an orthographic-projection drawing of the pictorial drawing in Fig. 5-18. Draw the object to a scale of 1/4″ = 1′0″ on your drawing sheet, using the dimensions in the figure. In doing so, the top, front, and side views should give a complete picture of the object. The views should be arranged on your drawing paper so that the front view is drawn approximately in the center of the sheet (from top to bottom), and the left side of the object should be about 1½ inches in from the lefthand border line. The top view should be directly above and in line with the front view, and the side view should be along the right side of the front view—in line with it and with about 1½ in. of space between them. This arrangement helps one to see the relationship between the different views.

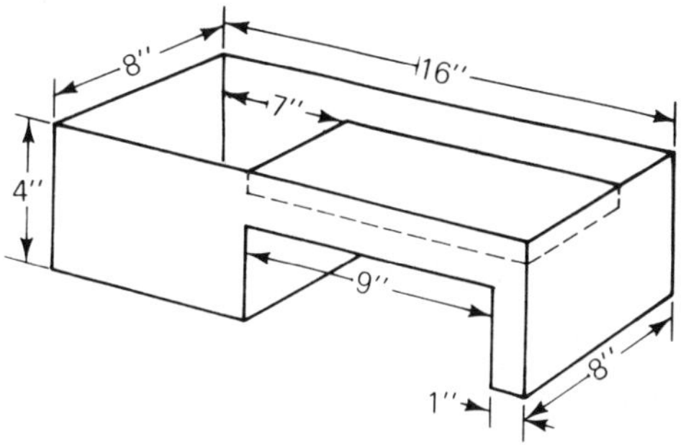

Figure 5-18. Pictorial drawing for use in written assignment.

The object itself should be drawn with the heavy solid line. Then dimension the object by methods described earlier in this book.

When the object itself has been correctly drawn, transfer the schedule shown in Fig. 5-19 to the right-hand side of your drawing sheet, making the size of the schedule lines large enough to accept lettering guidelines made with the Ames lettering guide set at position 3.

A1	1-1	2	47 Ω (yellow-violet-black)	R305, R308
A1	1-3	1	100 Ω (brown-black-brown)	R301
A1	1-4	2	330 Ω (orange-orange-brown)	R309, R316
A1	1-11	1	1500 Ω (brown-green-red)	R317
A1	1-44	3	2200 Ω (red-red-red)	R302, R306, R312
A1	1-16	2	4700 Ω (yellow-violet-red)	R311, R314
A1	1-24	1	33 kΩ (orange-orange-orange)	R315
A1	1-115	1	47 kΩ, 5 percent (yellow-violet-orange)	R304
A1	1-142	2	560 kΩ (green-blue-yellow)	R313, R318

Figure 5-19. Schedule for use in written assignment.

Have a reproduction of your drawing made by one of the methods described in this chapter and examine it for the following.

1. Are all lines clear and readable?
2. Is there enough contrast between the different types of lines used?
3. Are there smudge spots caused by erasing on your drawing?

QUESTIONS

1. What type of drawing is used more than any other in electrical and electronic drafting?
2. Electrical and electronic drawings are made with three different procedures; what are they?
3. Why should every drawing sheet contain a title block?
4. Name the 10 descriptive items that should appear in the title block of electrical drawings.
5. How should various views appear on drawing sheets?
6. Where is the drawing title usually located on a drawing?
7. To what standard should abbreviations conform?
8. Name three methods of reproducing drawings.
9. Why is it necessary for the draftsman to understand certain trade terms before he can correctly prepare electrical and electronic working drawings?
10. Why are original drawings normally copied?

Chapter 6

ELECTRICAL AND ELECTRONIC DRAFTING TECHNIQUES

The main objectives of this chapter are to teach the reader the various graphic symbols used in electrical and electronic drafting and to apply these symbols to working drawings. The instructions given on the use of templates will facilitate the drawing of these symbols, and the data on schedules and wiring diagrams will equip the electrical/electronic draftsman with other time-saving techniques for preparing high-quality drawings.

6.1 Electrical and Electronic Graphic Symbols

The purpose of a working drawing—as applied to the electrical/electronic industry—is to show how a certain object, piece of equipment, or system is to be constructed, installed, modified, or repaired. An electronic instrument, for example, usually has drawings and specifications showing the mechanical arrangement of the chassis and housing, and a schematic diagram showing the various components, the power supply, and the connection between each. An electrical drawing of a system used in a building normally shows the location of all lighting, power, and special outlets; the number and sizes of wire (conductors) feeding these outlets; service equipment, panel boards, communication equipment, and other information for the proper construction of the system.

In the preparation of working drawings for the electrical and electronic industries, symbols are used to simplify the work of those preparing the drawings. To illustrate this fact, look at the pictorial drawing of a lighting fixture in Fig. 6-1. Although this drawing clearly indicates the type of the lighting fixture and how it is connected to an outlet box, such drawings take hours to complete, costing more money than could possibly be allotted for a working drawing. However, by using a symbol for this lighting fixture, a drawing like the one in Fig. 6-2 can be prepared in only minutes; for those who are familiar with blueprint reading, the drawing in Fig. 6-2 gives exactly the same information as the one in Fig. 6-1.

Most engineers, designers, and draftsmen use symbols adopted by the United States of America Standards Institute (USASI) for use on electrical and electronic drawings. However, many designers and draftsmen frequently modify these symbols to suit their own particular requirements for the type of work they normally encounter. For this reason, most

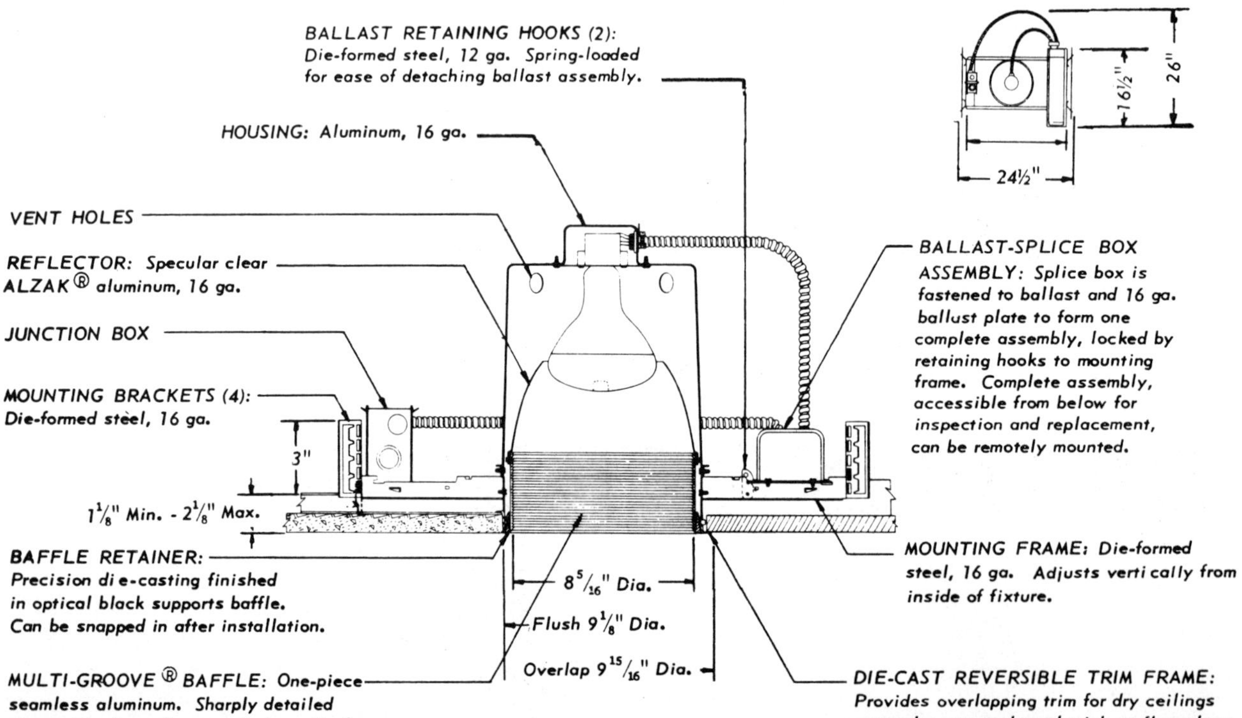

Figure 6-1. Pictorial drawing of a lighting fixture.

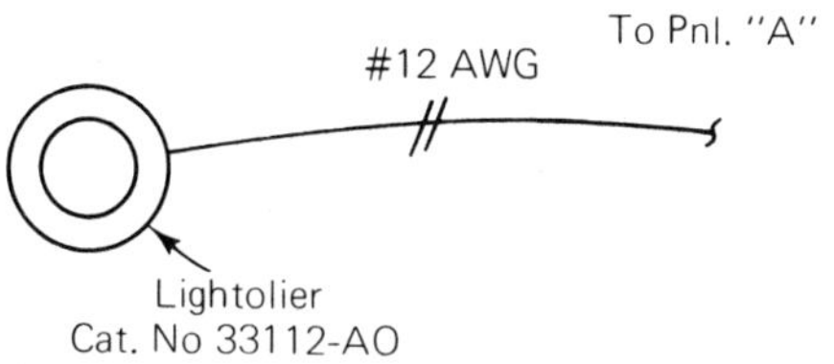

Figure 6-2. An electrical drawing using symbols to indicate lighting fixtures.

drawings have a symbol list or legend drawn and lettered on each set of working drawings. The symbols are made with straightedges, templates, and French curves, and then an explanatory note is lettered adjacent to the symbol.

Selecting the proper symbol for a given item should be given careful consideration. Ideally, the selected list of symbols should be

1. Easy to draw.
2. Easily interpreted by workmen.
3. Sufficient for most applications.

Obviously, item 1 above affects the draftsman's board time. Certainly, all designers and draftsmen in this day and time would not consider using symbols for which templates are not available—if it is at all possible to avoid them. But even with templates, certain modifications of the standard symbols can reduce board time considerably. For example, the symbol  was once very popular in representing a motor on an electrical drawing. This symbol depicted the round rotor of the motor, and the two slanted lines represented the brushes. Although this symbol would take only a few seconds to draw on a working drawing (with proper templates), the modified symbol would take less than half the time to draw. If many such symbols were required on a set of working drawings, much time could be saved by using the latter symbol.

The next item, "Easily interpreted by workmen," affects not only labor on the job, but all the time of the draftsman and other office personnel. To illustrate, a misinterpretation of a given symbol on a set of working drawings can mean a costly change to the engineer, the contractor, and the like; if the workmen cannot understand the meaning of a symbol, it means lost time by the workmen as well as time wasted in calling the draftsman to get an interpretation.

The final item, "Sufficient for most applications," means just what it says. It would be ridiculous to spend hours or days organizing what you consider to be a good symbol list, and then to find out that the ones you modified were only applicable to one out of ten projects. The list in Fig. 6-3 represents a good set of electrical symbols for guilding construction, and the list in Fig. 6-4 should suffice for the majority of the draftsman's needs on electronic drawings. Both meet the previously mentioned criteria. If additional data are occasionally needed on a given project, notes may be used on the drawings, or a new symbol can be devised to cover the situation at hand.

It is evident from these lists that many of the symbols have the same basic form, but their meanings are different because of some slight difference in the symbol.

For example, the outlets symbols in Fig. 6-4 have the same basic form—a circle; however, the addition of a line or a dot to the circle, or perhaps a letter or note, gives each an individual meaning. Therefore, a good procedure to follow in learning the different symbols is first to learn the basic form and then to apply the variations for obtaining different meanings.

Some of the symbols listed utilize abbreviations to obtain their meaning, such as "WP" for weatherproof and "E" for emitter, "C" for collector, etc. Others are simplified pictographs, such as 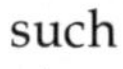 for a double floodlight fixture or  for a loudspeaker or  for a general antenna.

In some cases, the symbols are combinations of abbreviation and pictographs, such as  for fusible safety switch, and  for a PNP transistor.

SUMMARY OF 6.1

The purpose of an electrical or electronic working drawing is to show how a certain piece of equipment, object, or system is constructed, installed, modified, or repaired.

Symbols are used to simplify the draftsman's work and to save time.

A good procedure to follow in learning symbols is first to learn the basic form and then to apply the variations for obtaining different meanings.

6.2 Electrical Templates and Their Use

A variety of special templates has been developed to simplify the job of the electrical/electronic draftsman and to minimize many tedious and time-consuming operations. Most templates are made of thin pieces of transparent plastic out of which different shapes have been punched. The draftsman then uses these templates to trace objects of the required shape onto the drawing.

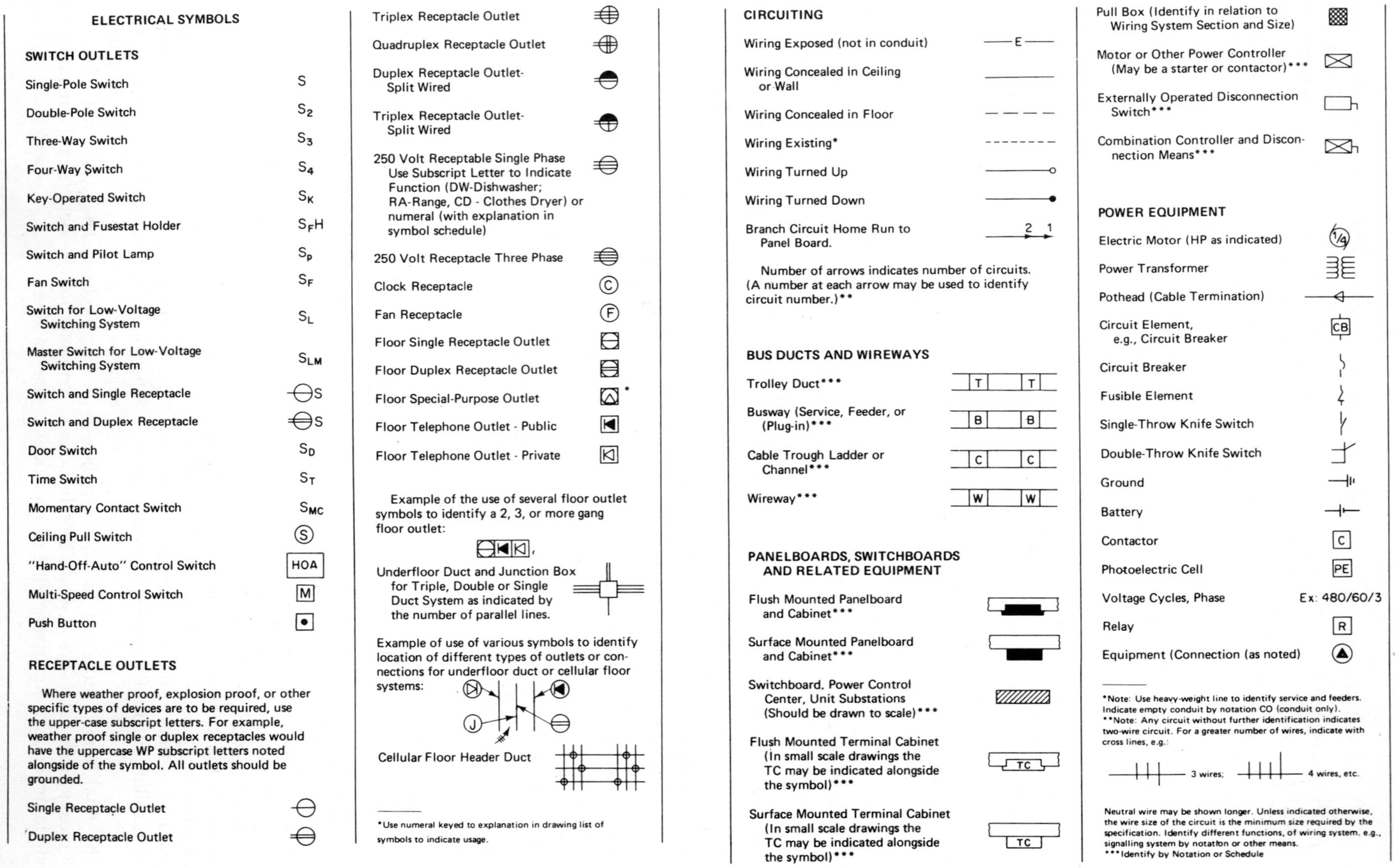

Figure 6-3. Electrical symbols used for building construction drawings.

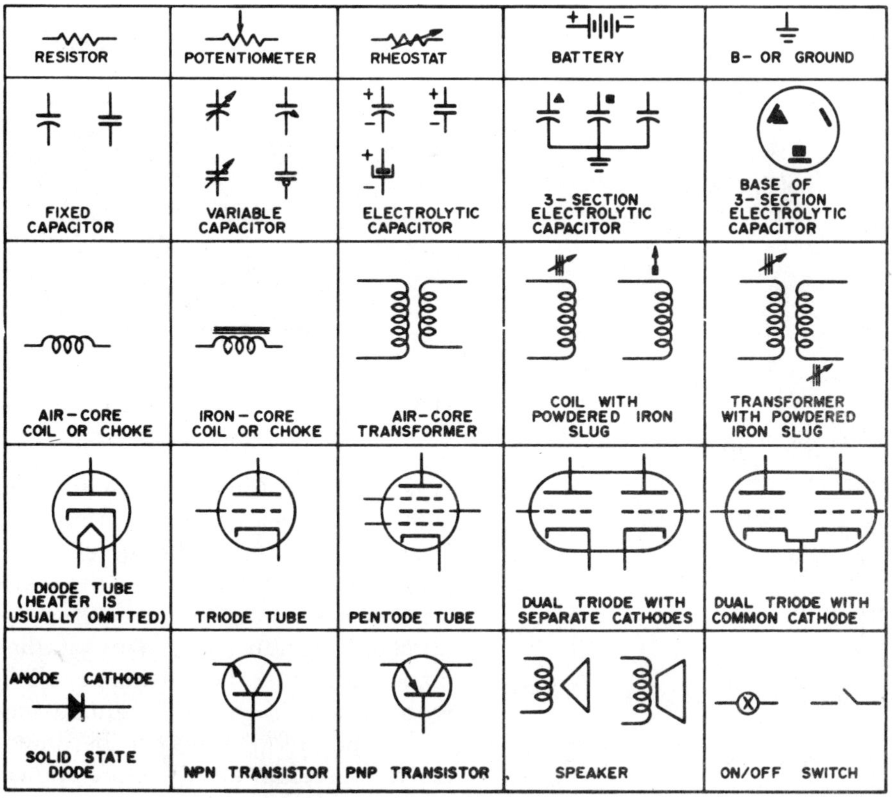

Figure 6-4. Symbols used on electronic drawings.

The template shown in Fig. 6-5 is typical of those used in electrical drafting, while that shown in Fig. 6-6 will prove useful in many electronic drawings. Together, they contain all the forms necessary for most electrical and electronic drafting. Their use is recommended to all electrical and electronic draftsmen, although a good draftsman can produce excellent drawings with only the basic tools.

Electrical Template 704. This template is one of the electrical draftsman's most useful tools in making drawings of power and lighting—especially on drawings that are drawn to a scale of 1/8″ = 1′0″. In order to realize just how valuable this tool can be, look at the plan view of an electrical system in Fig. 6-7. Every electrical symbol on this drawing can be made with the 704 template.

The room numbers are designated by the number enclosed in an ellipse. The ellipse is first drawn by using one of the openings in the 704 template as a guide, and the numbers are then lettered in the ellipse freehand by use of guidelines, as explained in Sec. 3.2.

All of the lighting fixtures on this drawing (Fig. 6-7) are then drawn by using some of the forms under "Lighting Fixtures" on the 704 template; that is, the type "9" fixture on the exterior of the building is drawn with the larger circle form—a straightedge being used to draw the lines indicating that it is a wall bracket fixture; the type "11" fixtures are drawn with the square form, and the small circle (indicating an outlet box) within this square is drawn with the appropriate circle form. The remaining lighting fixtures—type "1"—are again drawn with the rectangle form of the 704 template, and the circle (outlet box) is drawn with another circle form.

Notice that the "outlet" circles are already measured on the template for their respective fixture form; this again saves the draftsman much time. For example, once the outline of the fixture is drawn, merely slide the printed outline of the fixture on the template over the one drawn, and this automatically centers the circle opening within the fixture symbol. See Fig. 6-8.

Fixture designations on the electrical plan in Fig. 6-7 are shown by a triangle form with the fixture type number inside. Again, the triangle is drawn by means of one of the forms on the 704 template, and the numbers are then lettered freehand. The light-control switches (S and S_3) are also lettered freehand.

Circuit lines on the plan in Fig. 6-7 are curved in

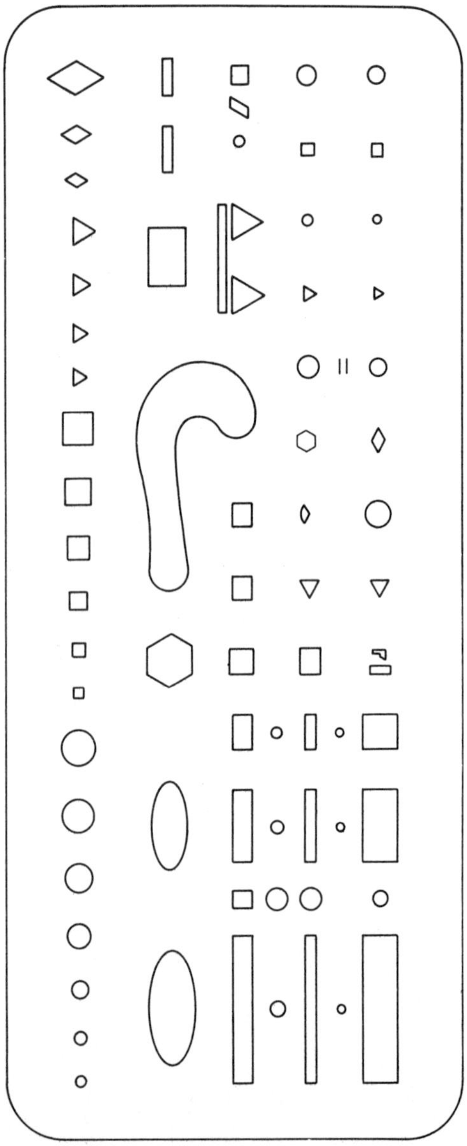

Figure 6-5. A drafting template typical of those used in electrical drafting.

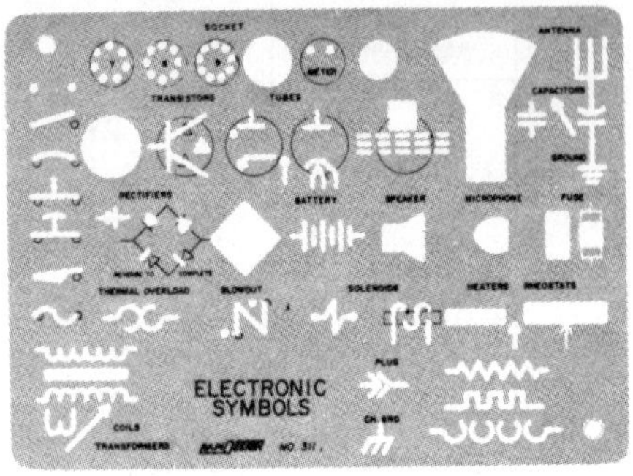

Figure 6-6. A template that will prove useful in electronic drafting.

order not to confuse them with building or equipment lines. These curved lines are also made with the 704 template by using the curved opening located approximately in the center of the template. Even the slash marks cutting these circuit lines (indicating the number of conductors or wires in the circuit) can be made with either the outside edge of the template or with one side of one of the square openings.

Type "14" fixtures on the drawing in question are recessed and are therefore shown on the drawings by the symbol ◎ . To make this symbol with the 704 template, use the group of circles in the lower right-hand corner of the template. To illustrate, use the circle on the far left of this group for the outside circle of the symbol, then use the third circle from the left to draw the smaller circle inside the first one.

The duplex receptacles are drawn using the symbol (⊜). The circle is first drawn with the 704 template and then the two straight lines (blades) are drawn with a straightedge as shown in Fig. 6-9.

Receptacle circuit lines are drawn with the curved opening in the center of the template as described previously. Notice that these receptacle circuits differ from the lighting circuits in that they have broken lines as opposed to the solid lines used for the lighting circuits. The solid lines indicate the circuit to be run in the ceiling or side wall, while the broken lines indicate the circuits to be run in the floor or ceiling below.

When one is using a template, the cutout or opening is placed in the desired position on the drawing and is then held firmly in place with the left hand (for right-handed draftsmen) against a T-square or parallel bar, as shown in Fig. 6-10. A conical-point lead is used to draw the figure on the tracing or drawing paper, the edges of the opening being used as a guide.

Practice is necessary in order to judge how much pressure to use when one is drawing various

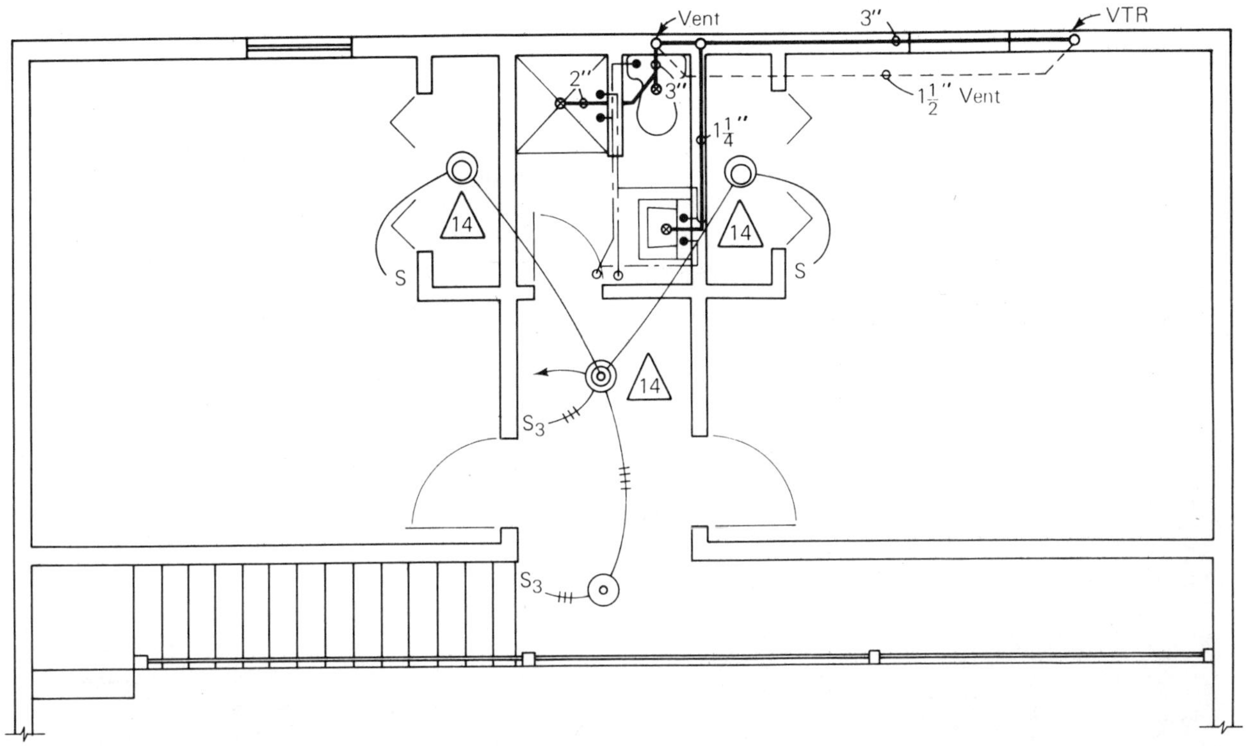

Figure 6-7. Plan view of an electrical system, where the 704 template was used to draw every electrical symbol shown.

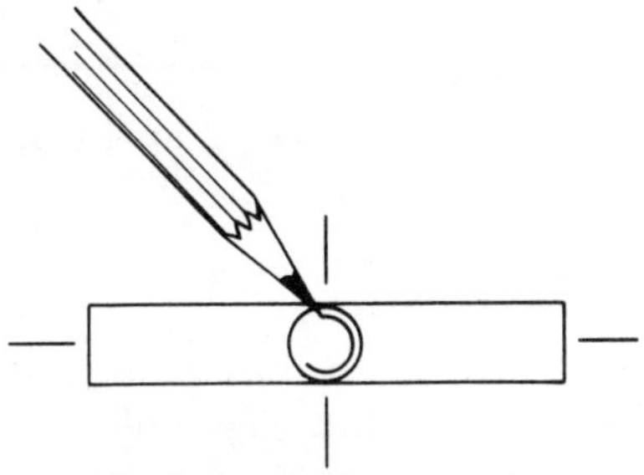

Figure 6-8. Method of centering a circle within a lighting fixture symbol.

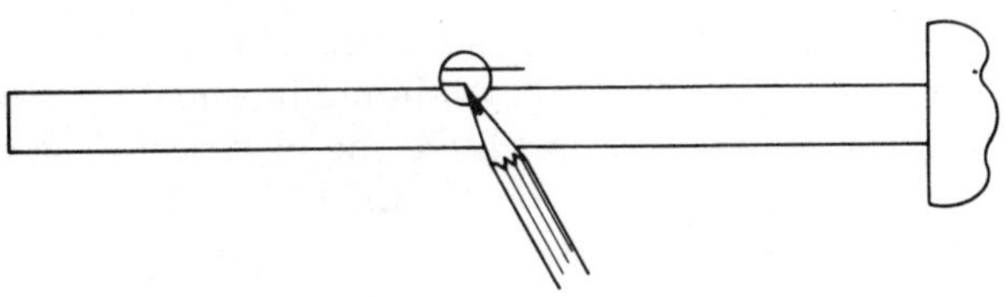

Figure 6-9. Finishing a duplex receptacle symbol with a straightedge.

Figure 6-10. Positioning a template for drawing.

objects. For example, a medium-heavy pressure is applied when a circle opening is used, in order to obtain clean lines that will show up. However, when one is drawing items such as the water closet in Fig. 6-11, care must be taken when the lead is moved around the sharp corners. Otherwise, the lead may not readily follow the edge of the cutout, resulting in an error like the one shown in Fig. 6-12. Another case where caution must be used is when the openings of the template cross, as in the letter X in a lettering template, or the cross of an exit light symbol (⊗).

Electrical draftsmen use various schematic symbols in designing power and control diagrams. A

Figure 6-11. (*Left*) Symbol used for a water closet.

Figure 6-12. (*Right*) Care must be taken when the lead is moved around sharp corners; otherwise, an error will result, as shown here.

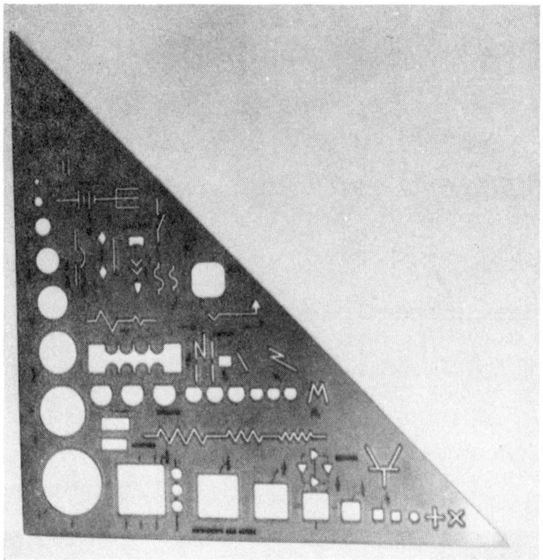

Figure 6-13. The template shown here is useful for making various electrical symbols used in schematic diagrams.

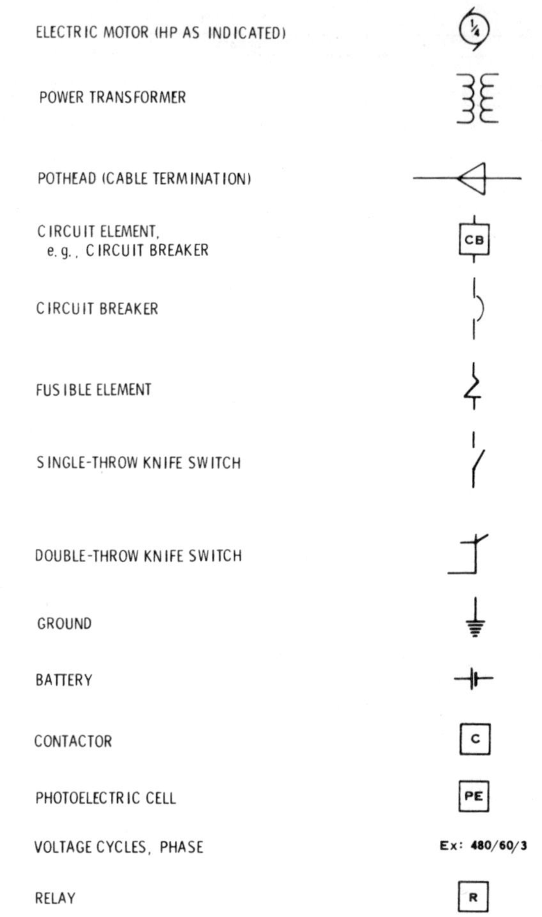

Figure 6-14. List of symbols used in electrical schematic diagrams.

template such as the one shown in Fig. 6-13 is very helpful in preparing such drawings; the chart in Fig. 6-14 shows the various symbols that can be made with this template—along with the respective meaning of each.

A set of electrical/electronic templates manufactured by RapiDesign is shown in Fig. 6-15. The set number is R-300 and consists of one 8″ × 5″ template R-301—Standard Electrical/Electronic Symbols; template R-302—Contacts, Switches, Contactor, Relays; and template R-303, titled Terminals & Connectors, Transmission Path, Etc. The set is one of the most complete on the market, and its openings are taken from USASI Y32.2 symbols. These symbols are grouped into sections for easy use, and are in correct relative size so that any size grid can be obtained through photographic reduction. Therefore, nearly all electrical and electronic symbols can be drawn with this set of templates. The electronic draftsman, especially, should obtain such a set.

Electrical draftsmen who work for architects, consulting engineers, and electrical contractors will be required to draw outlines of buildings in order to locate the various items of the electrical system on the drawings. In other words, the outline of the building must be shown in order to show reference to the electrical equipment. In doing so, items other than those related to the electrical work should be shown on the electrical drawings to facilitate the electrical design. For example, it is a good idea to show toilet or bath fixtures on electrical plans in order to show lighting fixtures over a lavatory mirror, a recessed light centered in a shower stall, etc. In showing plumbing fixtures, a plumbing fixture template is a great time-saving device for drawing these fixtures.

Figure 6-16 shows a typical plumbing fixture template, and Fig. 6-17 shows a water closet drawn with this template. This same plan view of a water closet could be drawn with the use of a straightedge and a French curve. However, the process would take

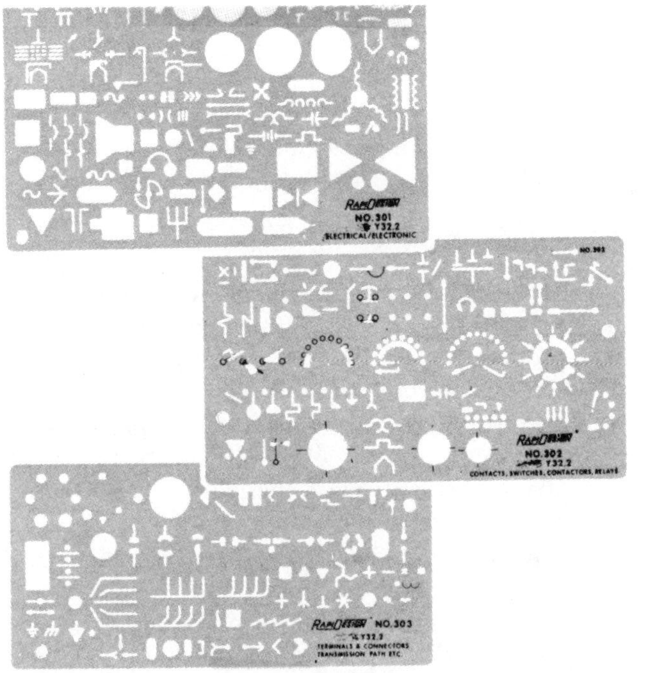

Figure 6-15. A good set of electrical/electronic drafting templates manufactured by RapiDesign.

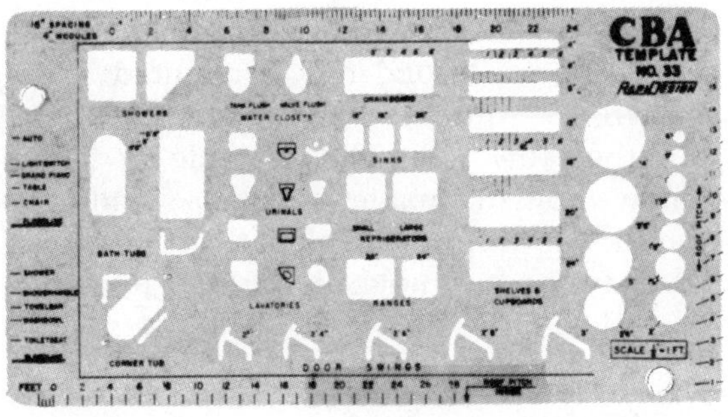

Figure 6-16. Typical plumbing fixture template.

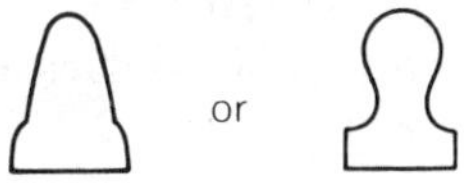

Figure 6-17. Water closet symbol drawn with the template in Fig. 6-16.

about five minutes for the average draftsman. On the other hand, the same view can be drawn with the template in less than five seconds. From this evaluation, we can see how templates save the draftsman much valuable time.

Lettering templates. Most lettering on electrical drawings is done freehand by methods described in Sec. 3.2. Sometimes, however, it is desirable to have drawing titles and title blocks stand out and to be extremely neat (as though the lettering had been printed on a press); this is especially true for many electronic drawings that must be reproduced. The mechanical lettering devices that were described in Sec. 2.3 will obtain these results, but the use of such devices is not always practical. The use of such mechanical lettering devices is limited because of the amount of time involved in their use and also because of the cost of such devices.

Another device, known as the lettering template, is a good compromise between freehand lettering and lettering with a mechanical lettering device like the Vari Typer, for example. Examples of various types of templates are shown in Fig. 6-18. Figure 6-19(a) shows a drawing title which has been lettered freehand; 6-19(b) shows this same title lettered with the use of a lettering template; 6-19(c) shows how the same title appears using a Leroy lettering set.

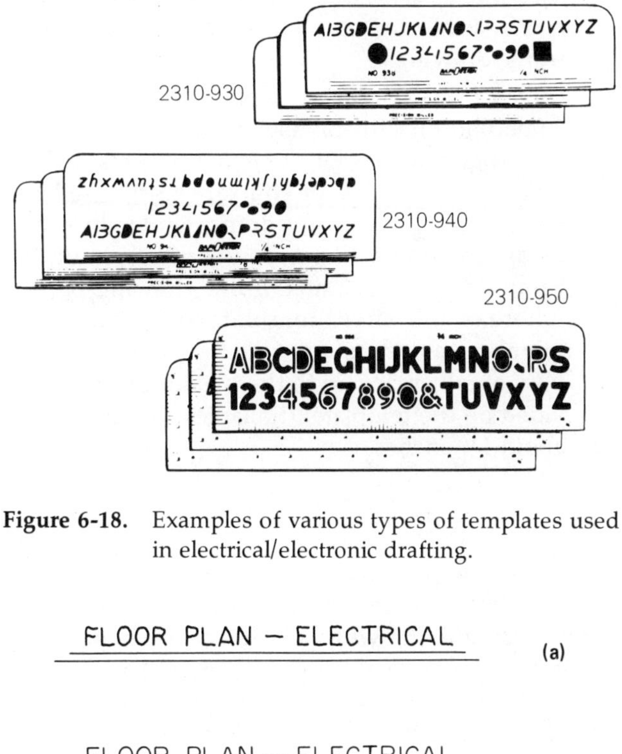

Figure 6-18. Examples of various types of templates used in electrical/electronic drafting.

FLOOR PLAN — ELECTRICAL (a)

FLOOR PLAN — ELECTRICAL (b)

FLOOR PLAN — ELECTRICAL (c)

Figure 6-19. Examples of various types of lettering.

When one is using the lettering template, it is best to use a hard lead to trace the outlines of the various letters on the drawing, and then to use a soft lead (such as H or F) to go over the original letters. The latter is usually done with a straightedge for all lines, and the curved parts of letters are drawn freehand with care.

When using any template, always retrace the original lines of the object at least once. Be careful not to move the template when doing this, as it is difficult to align the template exactly to its original position if it slips.

Other templates may be useful to the electrical and electronic draftsman, but the ones previously mentioned should suffice for most applications. Additional templates should be purchased only as the individual need arises. Some additional ones may include:

1. *Isometric Ellipse Template*. For drawing circular objects on isometric or pictorial drawings.
2. *Square Template*. For drawing panel boards, circuit boards, chassis, etc., on electrical and electronic drawings.
3. *Large Circle Template*. A large circle template is sometimes useful on electrical and electronic drawings where a lot of circular objects are required. A compass, of course, can be used for these objects, but many draftsmen try to avoid using the compass because the pivot pin of the compass punches a hole in the drawing paper, whereas the circle template does not. The template can, therefore, take the place of a compass on all circles up to the maximum size provided on the template, the largest usually being about 1¼ inches.

SUMMARY OF 6.2

Templates are powerful time-saving devices, and since the draftsman is expected to turn out neat and highly accurate drawings in the shortest possible time, templates are highly recommended for all work.

The objects drawn with templates should be redrawn at least twice; three or four times may be necessary, depending upon the desired darkness of the line.

In some cases, it may be necessary to draw the outline of the object very lightly at first, and then darken the lines with regular drafting tools or perhaps freehand. A good example of this is the use of lettering templates.

Exercise care when drawing around openings in templates with sharp turns or corners; also when template lines cross each other, as in the letter "X."

Before continuing with the text in this chapter, perform the following assignment.

ASSIGNMENT

1. Shown below are 20 symbols commonly found on electrical working drawings. In the space provided, place the letter corresponding to the correct answer found in the list. (Fig. 6-20)
2. The 20 electronic symbols shown in Fig. 6-20 are frequently used on electronic drawings. In the space provided, place the letter corresponding to the correct answer found in the list.
3. Prepare a sheet of 11″ × 17″ tracing paper with border lines as described previously. Leave room in the lower right-hand corner for a title block, as also explained before.
4. Beginning 1″ in from the left-hand border and 1″ down from the top border, draw the first electrical symbol of Fig. 6-3 using templates where applicable. Now measure to the right 1″ and draw the second symbol in the list, and so on until you have drawn all of the symbols in the list. When a second row is required, measure down 1″ from the row above it and begin another row, etc. The example in Fig. 6-21 shows how the first few symbols should appear on the drawing. When all of the symbols in Fig. 6-3 have been completed, continue (on the same sheet) with the ones in Fig. 6-4.

QUESTIONS

1. The purpose of an electrical or electronic drawing is to show how a piece of equipment, object, or system is to be ___, ___, ___, or ___.
2. Why are symbols used in place of actual pictures of a device on most electrical and electronic drawings?
3. What three drafting tools are used to make the majority of electrical and electronic symbols?
4. What are the three basic requirements for selecting a set of electrical/electronic symbols?
5. How do drafting templates save the draftsman much valuable time?
6. Would an electrical draftsman—working for an architectural/engineering firm—ever have need of a plumbing fixture template?

ASSIGNMENT

Shown below are 20 symbols commonly found on *electrical* working drawings. In the space provided, place the letter corresponding to the correct answer in the list.

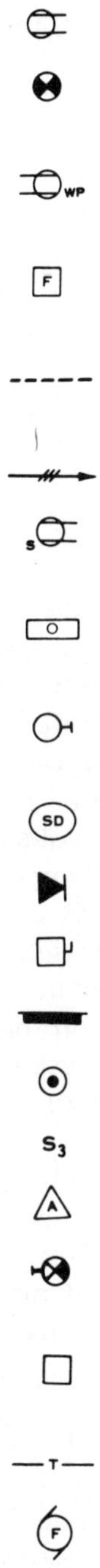

1.	______	A.	Fluorescent fixture
2.	______	B.	Incandescent fixture, recessed
3.	______	C.	Incandescent fixture, wall-mounted
4.	______	D.	Exit light, surface- or pendant-mounted
5.	______	E.	Exit light, wall-mounted
6.	______	F.	Indicates fixture type
7.	______	G.	Receptacle, duplex-grounded
8.	______	H.	Receptacle, weatherproof
9.	______	I.	Combination switch and receptacle
10.	______	J.	Receptacle, floor-type
11.	______	K.	Switch, three-way
12.	______	L.	Light or power panel
13.	______	M.	Disconnect switch
14.	______	N.	Conduit, exposed
15.	______	O.	Home run to panel
16.	______	P.	Telephone conduit
17.	______	Q.	Fan coil-unit connection
18.	______	R.	Fire-alarm striking station
19.	______	S.	Smoke detector
20.	______	T.	Telephone outlet, wall

Shown below are 20 symbols commonly found on *electronic* working drawings. In the space provided, place the letter corresponding to the correct answer in the list.

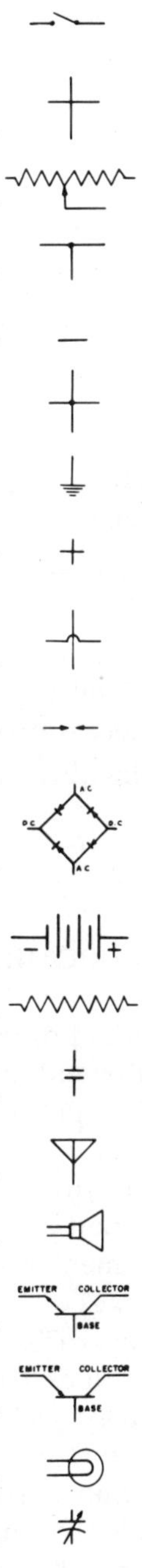

1.	______	A.	Variable or adjustable resistor
2.	______	B.	Contacts
3.	______	C.	NPN transistor
4.	______	D.	Switch
5.	______	E.	Positive sign
6.	______	F.	General antenna
7.	______	G.	Conductor crossover
8.	______	H.	Old symbol for conductor crossover
9.	______	I.	Loudspeaker
10.	______	J.	Adjustable capacitor
11.	______	K.	Capacitor
12.	______	L.	Earth ground
13.	______	M.	Resistor
14.	______	N.	Battery
15.	______	O.	Rectifier
16.	______	P.	Tap
17.	______	Q.	Crossover connection
18.	______	R.	Negative sign
19.	______	S.	PNP Transistor
20.	______	T.	Incandescent lamp

Figure 6-20. List of electronic symbols used for assignment.

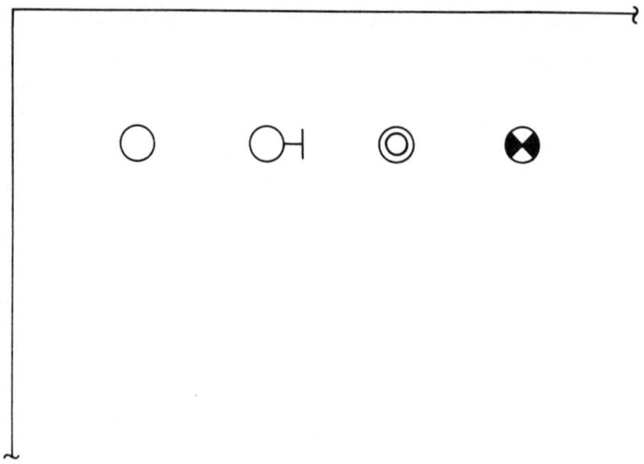

Figure 6-21. Example showing how the first few symbols should appear on the drawing.

7. What device is used for lettering which is thought of as a compromise between freehand lettering and a mechanical lettering device?
8. When using any template, always ____ the original lines of the object at least once.
9. Why should the draftsman exercise care when tracing around sharp corners of templates?
10. Will the draftsman ever find it necessary to draw over lines made with a template to make them darker?

6.3 Electrical and Electronic Schedules

In general, a schedule is a systematic method of presenting notes or lists of components, equipment, and the like on a drawing in tabular form. When properly organized and thoroughly understood, not only are they powerful time-saving methods for the draftsman, but they also save the specification writer, the engineer, the contractor, and the workmen on the job much valuable time.

The lighting-fixture schedule in Fig. 6-22, for example, lists the fixture type and identifies each fixture type on the drawings for a given project by number. The manufacturer and identification number of each type are given, along with the number, size, and type of the lamps for each. The "Volts" column gives the correct voltage to which the fixture is to be connected, and the "Mounting" column tells whether the fixture is wall-mounted, surface-mounted on the ceiling, or recessed. The remaining column may give such information as the mounting height above the finished floor, in the case of a wall-mounted lighting fixture, or any other pertinent data for the proper installation of the fixtures.

If this same data were to be specified in the written specifications, not only would it take the designer longer to describe the various fixtures and how to identify them, but it would also require that the estimator and the workmen on the job comb through page after page of the written specifications to obtain the same information. Many times, workmen do not always have access to the written specifications on the job, whereas they almost always have one or more copies of the working drawings.

From the previous paragraph, we can see that the schedule offers an excellent means of providing essential information in a clear and accurate manner, allowing the workmen to carry out their assignments in the least amount of time.

The following schedules are typical of those used on electrical and electronic drawings; each should be thoroughly studied.

Panelboard schedules. The panelboard schedule in Fig. 6-23 is typical of those used on electrical working drawings to give pertinent data on the service panel boards within a building. This schedule provides sufficient data to identify the panel number (as shown on the drawings), the type of cabinet (either surface-mounted or flush), the panel main bus bars and/or circuit breaker (in amperes, volts, and phase), the number and type of circuit breakers contained in the panel board, and the items fed by each. This type of schedule, however, does not give detailed information concerning the individual circuits, such as the wire size or the number of outlets on the circuit; this latter information must be given elsewhere on the drawing—usually in the plan view or in power-riser diagrams.

The draftsman draws the schedule on working drawings, using the straightedge and triangles—measuring off the various dimensions with the architect's scale. Guidelines are then made with the Ames lettering guide and the information is finally lettered (freehand or with a mechanical lettering device).

A practical application of this form is shown in Fig. 6-24. Here, the panelboard type is identified by the letter A. The location of this panel board within the building will be shown on the floor plans by the appropriate symbol and identified again as panel A. The schedule indicates that the panel cabinet is to be surface-mounted and is to contain a 200-ampere, 120/240-volt, single-phase (3-wire) main circuit

LIGHTING FIXTURE SCHEDULE						
FIXT. TYPE	MANUFACTURER'S DESCRIPTION	LAMPS		VOLTS	MOUNTING	REMARKS
		NO.	TYPE			

Figure 6-22. Typical lighting-fixture schedule.

PANEL BOARD SCHEDULE										
Panel No.	Type Cabinet	Panel Mains			Branches					Item Fed or Remarks
		Amps	Volts	Phase	1P	2P	3P	Prot	Frame	

Figure 6-23. Typical panelboard schedule.

PANEL BOARD SCHEDULE										
Pnl No	Type Cabinet	Panel Mains			Branches					
		Amps	Volts	Phase	1P	2P	3P	Prot	Frame	Items Fed
"A"	Flush	200A	120/240V	3φ 4WΔ	—	1	—	20A	70A	Time Clock
Square "D" NQOB W/Main Breaker					—	—	1	20A	70A	A.H.U.
					—	1	—	30A	70A	Water Htr.
					—	—	1	30A	70A	Cond. Unit
					5	—	—	20A	70A	Lights
					10	—	—	20A	70A	Receipts
					5	—	—	20A	70A	Spares
					12	—	—	—	—	Provisions Only

Figure 6-24. Practical application of a panelboard schedule.

breaker. The panel board is to be manufactured by ITE and is to be a type EQ.

The columns under the heading "Branches" give data pertaining to the overcurrent protection devices for the individual branch circuits. This schedule indicates that the panel will contain fourteen 1-pole circuit breakers (for 120-volt loads) with a trip rating of 20 amperes to furnish overcurrent protection for the lighting and receptacle circuits. The schedule also calls for six 2-pole circuit breakers rated at 20 amperes to furnish overcurrent protection for the air conditioning units and motors; one 2-pole, 30-ampere circuit breaker for the water heater; one 2-pole, 50-ampere circuit breaker for the kitchen unit (combination cooktop and refrigerator); one 2-pole, 40-ampere circuit breaker for the time switch controlling the outside lights; six 1-pole circuit breakers as spares for later electrical additions; and two blank spaces for future provisions also.

The wire size is not indicated on this schedule, but the electrician installing the system will know that a 20-ampere circuit requires no. 12 AWG wire, a 30-ampere circuit requires no. 10 AWG wire, etc. The wire sizes usually are indicated on the plan drawing also.

Figure 6-25 shows another type of panelboard

PANEL SCHEDULE

Panel No.	Designer			Checked			Date
MAINS	Ø	W		Volts	Ampere		
Circuit No.	Switch or Breaker			SERVES	Connected KVA	Demand Factor	Demand KVA
	Pole	Frame	Trip or Fuse				
1							
2							
3							
4							
5							
6							
7							
8							
9							
10							
11							
12							
13							
14							
15							
16							
17							
18							
19							
20							
21							
22							
23							
24							
25							
26							
27							
28							
29							
30							
31							
32							
33							
34							
35							
36							
37							
38							
39							
40							
41							
42							
				TOTALS			

DEMAND I
DESIGN I PERMISSIBLE Ed. FEEDER SIZE Ed/1000 A.F.
FEEDER LENGTH AMP. FT. M. ACTUAL Ed.

Figure 6-25. A different type of panelboard schedule.

schedule used on another project. This type not only has spaces for the data on the former schedule, but also has provisions so that the circuits may be numbered so that the designer can balance the load. This schedule gives information that includes: the total load on each circuit (in voltamperes or watts), the size and type of overcurrent protection, and the like. The space at the bottom of the panel is for the manufacturer and type of panel board to be installed as well as for providing other necessary information.

Electric-heat schedule. The electric-heat schedule is an excellent means of conveying necessary data for the installation of electric heating—baseboard, fan-forced wall heaters, ceiling heaters, etc. An identification mark is used with the location of the heaters on the floor plan. Then, adjacent to the mark on the schedule, manufacturers' data are entered and the physical dimensions, the rated voltage, the type of mounting, and remarks concerning the connection or mounting of these units are given. The floor plan corresponding to this schedule will show the physical location of the heaters, the circuits feeding them, and the location of the thermostat controlling them.

Electronic drawings will require similar schedules for nearly every project. The one in Fig. 6-26 for example, is used on a how-to project to identify the various components given in a schematic diagram. The identification number is listed on the schematic view, and then the schedule tells what the part is (by catalog number) and how many of each part listed are required to complete the project. Most of the schedules found on electronic drawings will be of this type; that is, material lists and where they are to be used in a particular circuit.

SEMICONDUCTOR CHART

COMPONENT NUMBER	HEATH PART NUMBER	MAY BE REPLACED BY	DESCRIPTION	LEAD CONFIGURATION
Q103 Q201 Q205	417-118	2N3393	NPN TRANSISTOR	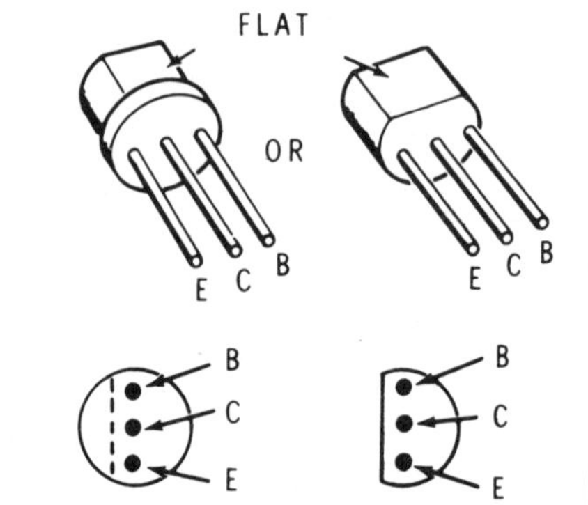
Q102	417-94	2N3416	NPN TRANSISTOR	
Q101, Q104 Q202, Q203 Q204, Q206	417-201	X29A829	PNP TRANSISTOR	
D101 D201 D202	56-56	1N4149	DIODE	HEATH PART NUMBERS ARE STAMPED ON MOST DIODES.
ZD203	56-63	MZ500-10	ZENER DIODE	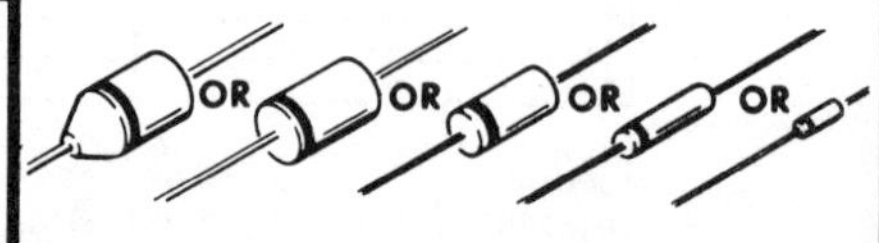

Figure 6-26. Schedule used on an electronic drawing. (*Courtesy of Heath Co.*)

SUMMARY OF 6.3

A schedule is a useful time-saving method of presenting notes or lists of equipment on drawings in tabular form.

The proper description of an item could require pages of written specifications, whereas a schedule can be used to describe the same item with only a minimum of notes, providing an excellent means of conveying essential information in a clear and accurate manner so that the workmen on the job can carry out their assignments in the least amount of time.

Only a few of the many electrical and electronic schedules were mentioned in this section, but the examples given should be quite sufficient to give the reader the proper knowledge to understand any schedule that he may encounter.

6.4 Electrical and Electronic Diagrams

Chapter 1 briefly covered the several types of drawings that the electrical and electronic draftsman is likely to encounter. Of the types mentioned, the diagrams are probably the type most used in all electrical and electronic drawings, mainly because a large part of all electrical and electronic drawings deal with circuits and their connections.

Schematic wiring diagrams, like the ones shown in Figs. 6-27 and 6-28, represent components in the system by symbols (described previously), the wiring and connection to each, and other detailed information. Sometimes the wires are shown in an assembly of several wires, which appear as one line on the drawing. When this method is used, each wire should be numbered where it enters the assembly and should keep the same number when it comes out of the assembly to be connected to some component in the system.

Schematic symbols are best drawn with templates, as described in Sec. 6.2. The connections are usually made with straightedges, and the components are listed in schedules or else are noted on the drawings. Such drawings are seldom if ever drawn to scale, but appear in diagrammatic form. The components, however, should be arranged in a neat and logical sequence so that they are easily traced and can be understood with the least amount of effort. Experience gained through study and practice drawing of schematic diagrams is the only way to consistently turn out schematic drawings of the highest quality, so much time should be spent in fulfilling these requirements.

Whereas schematic drawings are the type most frequently used on electronic drawings, block diagrams—known as power-riser diagrams—are the type most often drawn on electrical working drawings for building construction. Such diagrams give a picture of what components are to be used and how they are to be connected in relation to one another. This type of diagram is easily understood and requires much less time to draw or interpret than a schematic diagram—either detailed on one-line. As an example, compare the power-riser diagram in Fig. 6-29 with the one-line diagram in Fig. 6-30. Both diagrams are of an identical electrical system, but it is easy to see that the one in Fig. 6-29 is greatly simplified.

SUMMARY OF 6.4

Electrical and electronic diagrams usually fall into one of the following categories:

1. Complete schematic diagrams showing all details of the various components and every wire in the circuits connecting them.
2. Single-line diagrams, which are a simplified version of complete schematic diagrams.
3. Simplified block diagrams, often referred to as riser diagrams; that is, power-riser diagrams for service equipment, panel boards, and feeder circuits; telephone riser diagrams for telephone systems; fire alarm riser diagrams for fire alarm systems, etc.

IDENTIFICATION OF RESISTORS

In schematic drawings, resistors are shown by the symbol-/\/\/\/-with an identifying note such as R_1, R_2, etc., which gives the value in an accompanying schedule; or the type and value of the resistors are called out directly on the drawing. For example, a symbol for a resistor may have the value .01 μF lettered beside it. However, in pictorial assembly drawings, the electronic draftsman may be called upon to identify the various resistors as the builders of the circuit would have to identify them—by color codes.

Most one-half through 2-watt resistors are identified by colored bands on the resistor, and any electronic draftsman should be able to identify the value of these resistors.

To read the color code, the resistor should be viewed with the colored bands towards the left, like the drawing in Fig. 6-31. On the resistor in this drawing, the colored bands are nearer to one end than the

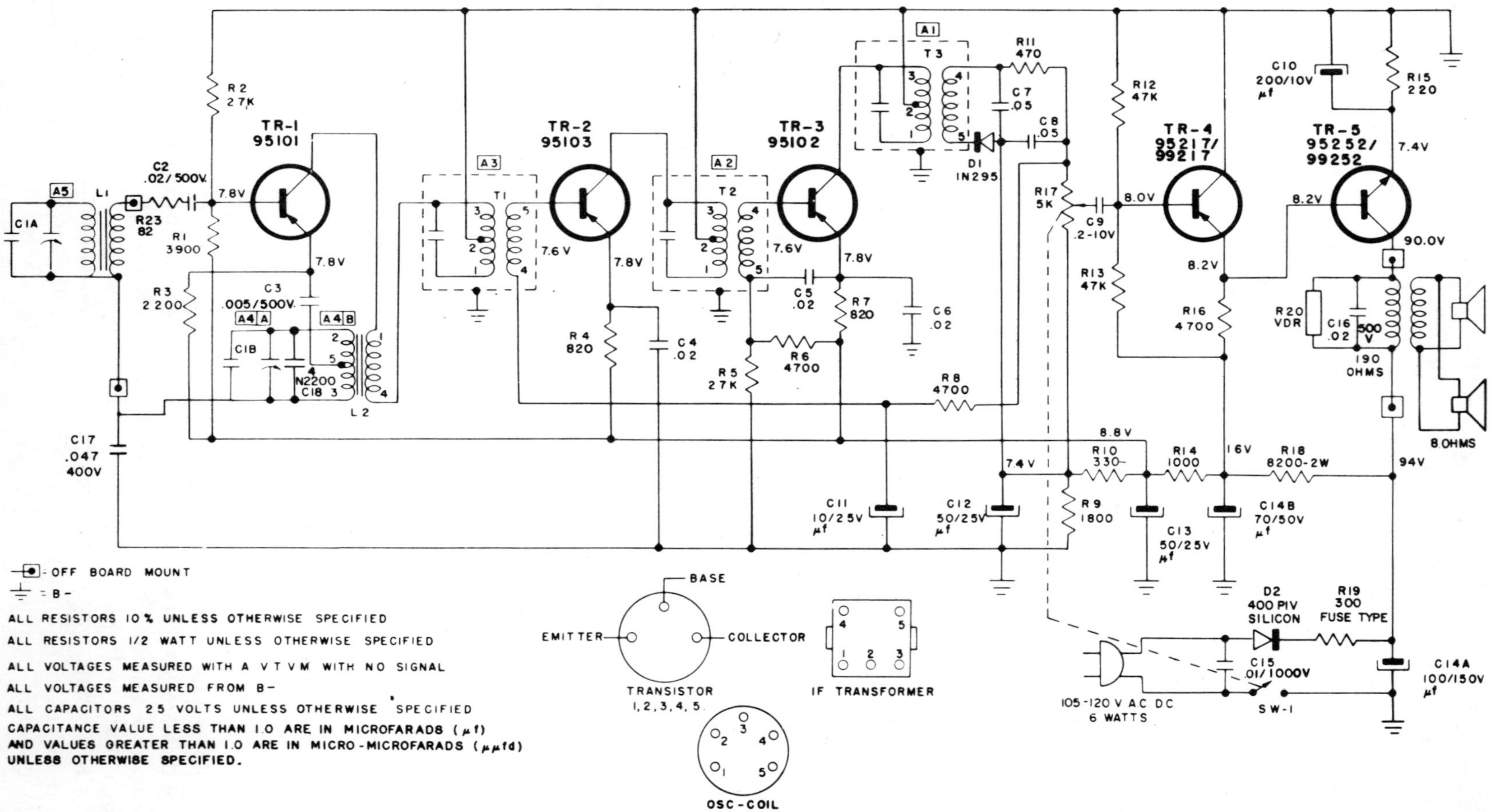

Figure 6-27. Schematic wiring diagram representing components in the system by symbols, the wiring and connection to each, and other detailed information.

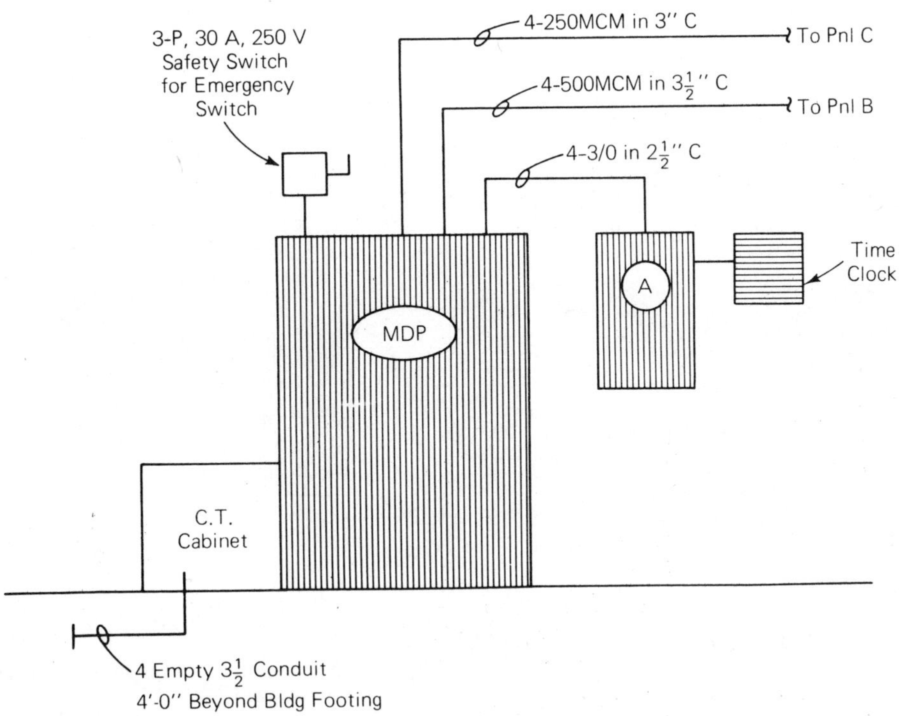

Figure 6-28. A typical power-riser diagram.

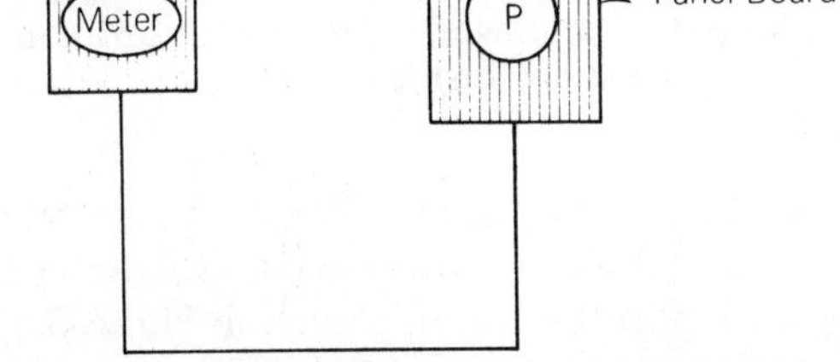

Figure 6-29. Power-riser diagram.

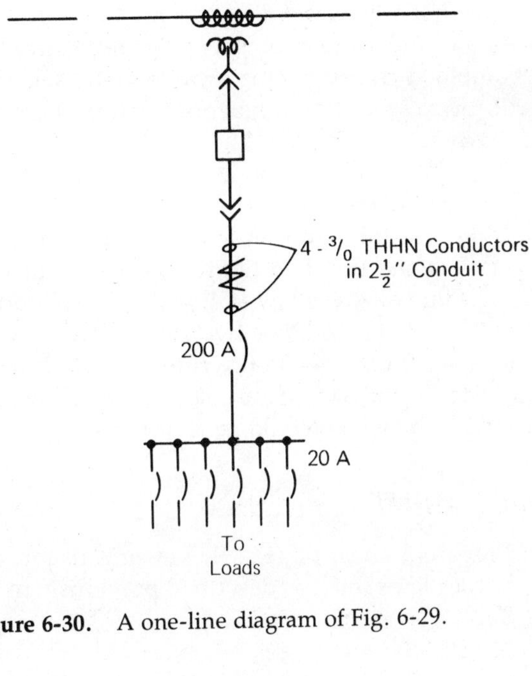

Figure 6-30. A one-line diagram of Fig. 6-29.

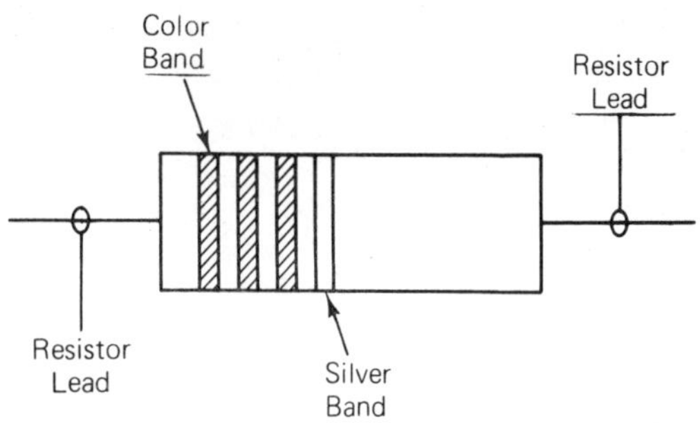

Figure 6-31. Resistor.

other. On other resistors, the colored bands may be centered; in this case, the gold or silver band should be on the right-hand side as it is viewed.

Each color represents a number, as follows:

Black	0
Brown	1
Red	2
Orange	3
Yellow	4
Green	5
Blue	6
Violet	7
Gray	8
White	9

Therefore, the resistor in Fig. 6-31 with the colors (from left to right) yellow (4), violet (7), and brown (1) is read as follows:

The first yellow band represents the number 4, and the second band represents the number 7; the two combined colors give us a value of 47. The third color (brown) gives the number of zeros in the value. Since brown represents 1, the third band indicates that the resistor has 1 zero in its value. So the value of the resistor is 470 ohms.

If the resistor had been color-coded yellow, violet, and red, the first two figures would again have the total value of 47 (yellow = 4, violet = 7). Red represents 2, so there would be two zeros, and the value would be 4700 ohms—often written 4.7 K ohms, as the "K" stands for 1000. If the colors were yellow, violet, orange, the value would be 47,000 ohms, etc.

ASSIGNMENT

1. Prepare a sheet of 11" × 17" tracing paper with border lines, etc., as described previously in this book.

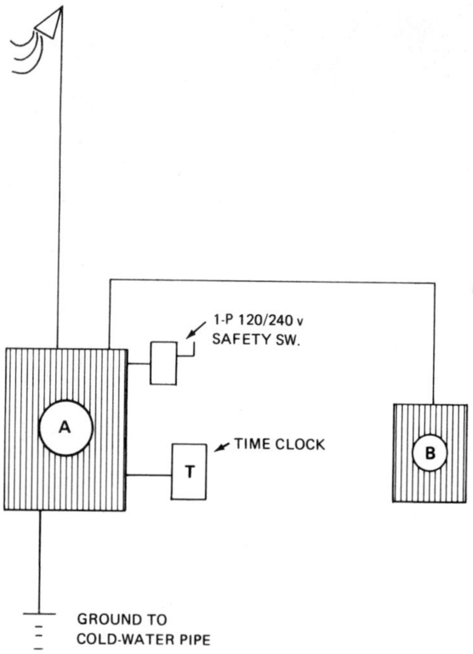

Figure 6-32. Sketch of power-riser diagram for assignment.

2. Using the architect's scale, scale the panelboard schedule that was shown in Fig. 6-23 and transfer this schedule to the left-hand side of the drawing paper on your board, enlarging it to twice the size. Use guidelines set at 3 on the Ames lettering guide for the smaller letters, and at 5 for the larger letters.
3. Using the completed schedule that was shown in Fig. 6-24 as a guide, transfer the following specification description of a panel board to the schedule which you have just drawn:

Provide and install a panel board manufactured by Sq. "D" of the type NQO. This panel board is to have a 120/208-volt main circuit breaker rated at 100 amperes and is to be designed for use on a 3-phase, 4-wire system. The branch circuit shall be protected by six 1-pole 20-ampere circuit breakers for the lighting and duplex receptacle circuits, one 3-pole, 40-ampere circuit breaker for the air conditioning unit, and one 30-ampere, 2-pole circuit breaker for the domestic water heater. The panelboard housing shall be designed for flush mounting.

4. On the right-hand side of the tracing paper, neatly draw a power-riser diagram exactly as shown in the rough sketch in Fig. 6-33, but use templates and straightedge to make good sharp, neat lines. Letter in all notes shown in the sketch also. Draw about twice the size.

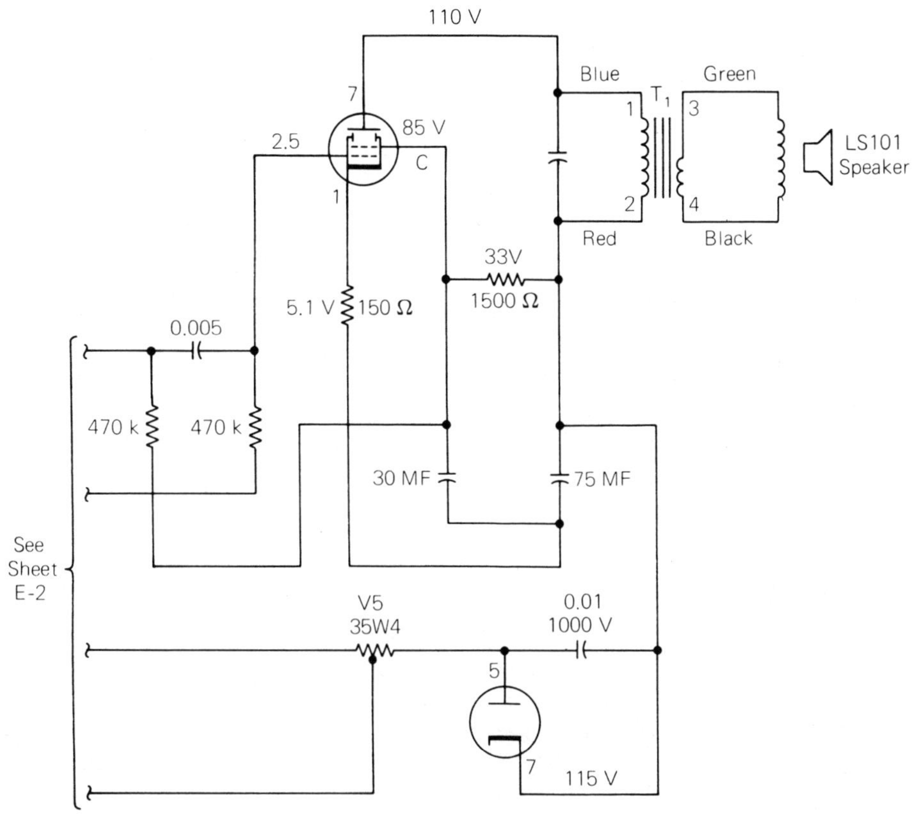

Figure 6-33. Schematic diagram for use in assignment.

5. Prepare another sheet of 11″ × 17″ tracing paper as done previously, and draw the schematic diagram shown in Fig. 6-30 to approximately twice the size and in the center of your sheet. Use templates for the symbols, straightedge for the lines, and an Ames lettering guide for guidelines (set at 3) for all the notes, numbers, etc.
6. Prepare a title block for each sheet, as described in Chapter 4.

Chapter 7

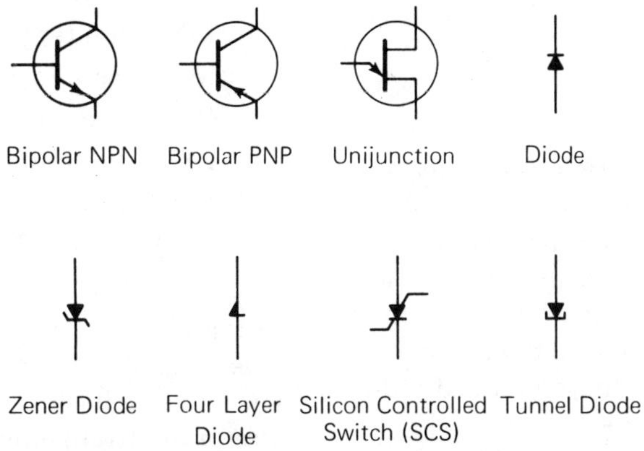

ELECTRONIC ESSENTIALS

Before one begins any actual electronic working drawings, there are several techniques that must be fully understood. This chapter, therefore, is designed to acquaint the reader with many essentials that will be extremely helpful to the electronic draftsman. Once learned, these techniques will help the draftsman prepare working drawings with more accuracy and with more speed.

7.1 Electronic Symbols

A large portion of all electronic drawings usually consists of wiring diagrams utilizing electronic symbols. These symbols are simplified pictorial drawings to represent the various components utilized in electronic equipment. A brief discussion of these symbols was given in Chapter 6, but because of the complexity of modern electronic circuits, a more detailed study will be presented here.

One of the simplest symbols is that of a single-cell battery, shown in Fig. 7-1. The longer of the two lines making up this symbol always represents the positive terminal of the battery, while the short line represents the negative terminal. When this symbol is drawn, the short line should be about half the length of the long one.

Multicell batteries can be shown with four or more lines, as shown in Fig. 7-2. However, it is common practice to use only four lines in multicell batteries—regardless of how many cells are involved. The exact number of cells and other characteristics of the battery can be described by notes on the drawings or can be listed in a component schedule.

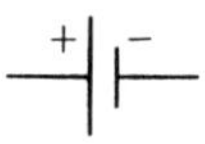

Figure 7-1. (*Left*) Symbol for single-cell battery.

Figure 7-2. (*Right*) Symbol for multicell battery.

Other electronic symbols are shown in Fig. 7-3; an explanation of each follows:

Amplifier. Several different basic types of amplifiers are in use, but the basic symbol for all of them is a triangle, as shown in Fig. 7-3. In general, an amplifier is used to produce an amplified reproduction of its input signal by drawing power from a source other than the input signal. Some amplifier letter combinations that may be used with amplifier symbols if needed for explanation are

BDG	bridging
BST	booster
CMP	compression
DC	direct current
EXP	expansion
LIM	limiting
MON	monitoring
PGM	program
PRE	preliminary
PWR	power
TRQ	torque

Antenna. Antennas are used on radio, TV, and other signal receivers for picking up signal waves. The antenna itself usually consists of a metallic structure or an arrangement of conducting wires or rods. Note in the symbol list that several types of antenna symbols are shown; that is, general antennas, dipole, loop, and counterpoise.

Arrestor, lightning. Figure 7-3 shows several symbols for lightning arrestors, including carbon block, electrolytic, horn gap, protective gap, sphere gap, etc. All are protective devices used to sidetrack or divert a discharge of lightning directly to ground without damaging an electronic device or its transmitting or receiving antenna.

Attenuator. An attenuator is a device that permits reducing the strength of a radio signal any desired amount without appreciably distorting the signal. Such a device usually consists of an arrangement of variable resistances.

Audible signaling device. Audible signaling devices consist of bells, buzzers, horns, loudspeakers, and the like, all of which are capable of being heard by the human ear. This means that any audible signaling device must be within 20 to 20,000 cycles—the range of human hearing.

Loudspeaker letter combinations used with symbols on electronic drawings include:

HN	electrical horn
HW	howler
LS	loudspeaker

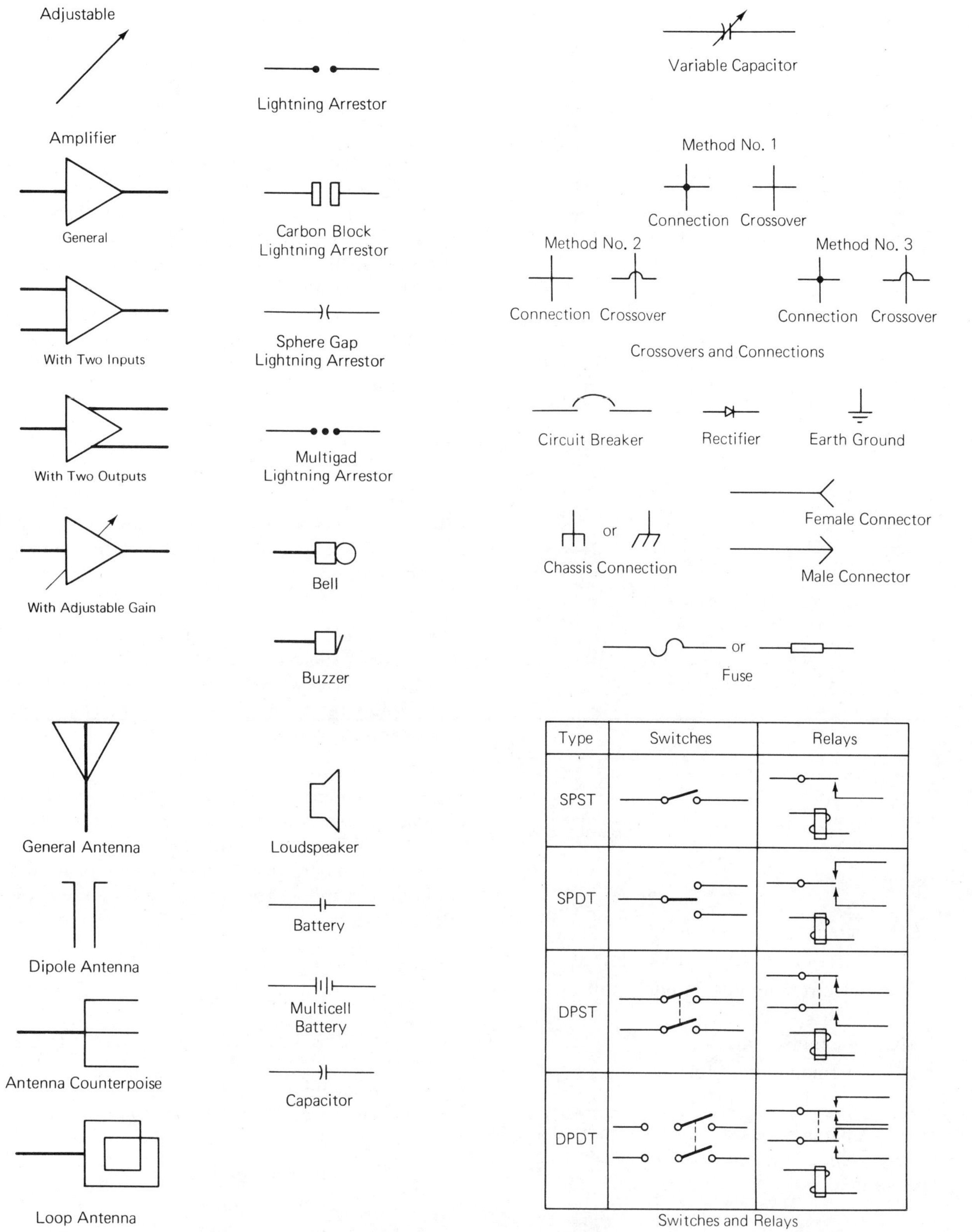

Figure 7-3. Electronic symbols.

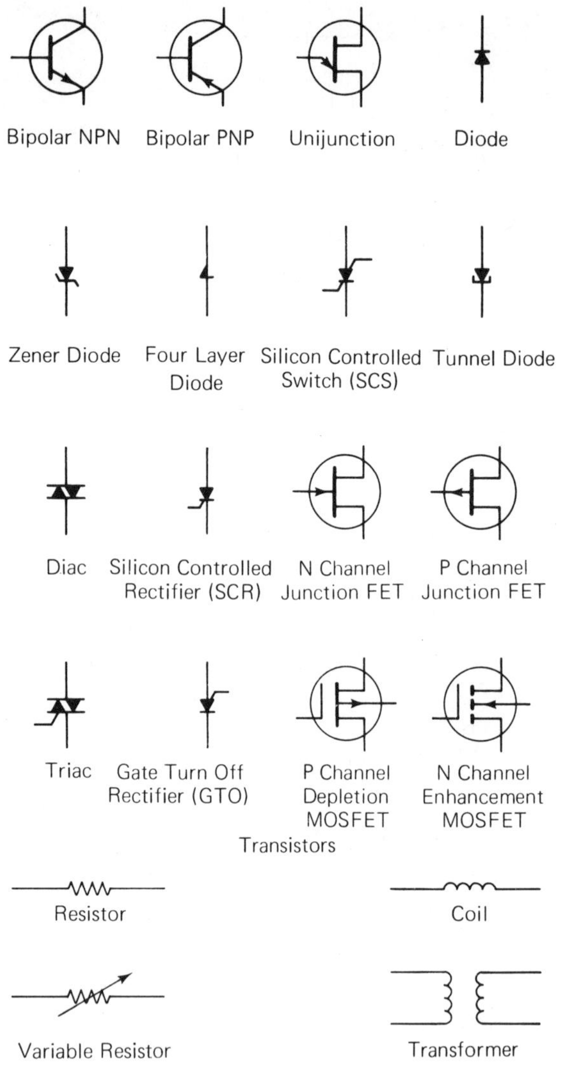

Figure 7-3. (cont.)

SN	siren
EM	electromagnetic with moving coil
EMN	electromagnetic moving coil and neutralized winding
MG	magnetic armature
PM	permanent magnet

Battery. A battery consists of one or more dry cells or storage cells connected together to serve as a dc voltage source.

Capacitor. A capacitor is used in electronic circuits to block the flow of direct current while allowing alternating current and pulsating current to pass. The device consists of two conducting surfaces separated from each other by an insulating material such as air, oil, paper, glass, or mica. A capacitor is capable of storing electrical energy. The curved line used in the symbol represents the outside electrode in fixed-paper and ceramic-dielectric capacitors, the moving element in variable and adjustable types, and the low-potential element in feed-through capacitors.

There are other symbols used to indicate capacitors that do not appear in some lists of standards. The electronic draftsman should be on the alert for such symbols. Other symbols related to the general capacitor symbol would include the variable capacitor, the adjustable capacitor with mechanical linkage, the continuously adjustable differential, the phase shifter, the split stator, and the feed-through capacitor.

Photoelectric cell. A type of light-sensitive cell in which a treated cathode, mounted in a glass envelope, emits electrons under the action of light, these electrons being collected by the anode. Another type varies its resistance with the change of light falling on the cell.

Circuit breaker. An electromagnetic device that opens or breaks a circuit automatically when the current rises above a predetermined value. Such devices are rapidly replacing fuses in almost all electronic devices and instruments.

Circuit element. A rectangle is often used on electronic drawings to represent any number of circuit elements. To indicate different components, letter combinations are placed either inside or adjacent to the basic symbol. The ones to follow are typical of those used.

CB	circuit breaker
DIAL	telephone dial
EQ	equalizer
FAX	facsimile set
FL	filter
FL-BE	filter, band elimination
NET	network
PS	power supply
RU	reproducing unit
RG	recording unit
TEL	telephone station
TPR	teleprinter
TTY	teletype

Ground. The earth, to which a connection is made by means of a buried conductor for electrical or electronic purposes. Also, the chassis of an electronic instrument when it is connected to ground and serves as the return path for signal circuits. Ground is usually considered to be at zero r.f. (reactive factor) potential.

Coil. A number of turns of wire wound on an iron core or on a coil form made of insulating material. A coil offers considerable opposition to the passage of alternating current but very little opposition to direct current.

Connector. Any device that joins or couples two or more parts. Note in the symbol lists that female and male connectors are shown separately.

Connection. Any wire or other conductor that joins another wire or component in an electronic circuit. Note in the symbol list that two systems are approved for showing connections. One is the dot system (this is the most accepted method). The other is using no other symbol than showing lines or devices touching; this system has some weaknesses and is not recommended for most electronic drawings.

When the dot system is used to show connections, crossovers are shown by merely crossing the wires, as would be shown for a no-dot connection. However, when the no-dot system is used to show connections, the crossovers must be shown as illustrated in Fig. 7-4.

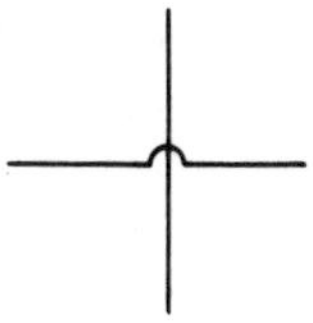

Figure 7-4. When the no-dot system is used to show connections, the crossovers must be as shown here.

Switch. The purpose of a switch is to open and close an electrical circuit, or to change the connections between components or circuits. The several types shown in Fig. 7-3 include single-pole; single-pole, double-throw; double-pole, single-throw; double-pole, double-throw, and a rotary selector switch. The last type is drawn with the viewer looking at the knob or at the rear and with the operational sequence in the clockwise direction.

Inductor. An inductor—used in electronic circuits—is usually in the form of a coil and is used for many purposes. One is to limit the flow of alternating current while allowing direct current to pass. When an inductor is used in this manner, it is usually referred to as a choke coil. Radio-frequency choke coils have air or pulverized iron cores, while AF choke coils and filter chokes have iron cores.

Resistor. An electronic component that offers resistance to the flow of electric current is called a resistor. Its resistance is specified in ohms or megohms, and its power rating is in watts. Since resistors offer a resistance to the flow of current, they are used to drop the voltage to the correct value for operation of transistors and also to isolate circuit components from each other, so that one will not interfere with the operation or action of the other.

The zigzag symbol—made with a 60-deg angle between adjacent lines—is probably the symbol that is used most on most electronic drawings. However, rectangles are also used in some drawings with the value of the resistor placed inside the rectangle or near it.

Resistors may be fixed, variable, or tapped, the symbols of which are shown in Fig. 7-3.

Relay. A relay may be used to control the current in one circuit by a much smaller current flowing in another circuit. It is an electromagnetic device with either normally open or normally closed contacts. These contacts act as switches and are turned on and off by action of the relay coils.

Transformer. A transformer is a device consisting of two or more coils constructed in such a way that magnetic lines of force produced by the flow of alternating or pulsating direct current through one coil will pass through the other coil and induce in it a corresponding alternating voltage. The symbols used for inductors are also used for drawing transformers on electronic drawings.

Loudspeaker. This device is used to radiate acoustic power effectively at a distance through the air. When incorporated into radios, TVs, etc., it usually is intended to reproduce and magnify the tone from the oscillator. A pictorial symbol is normally used to show a speaker on an electronic drawing. Templates will aid the draftsman in drawing this and other similar symbols.

PNP transistor. A transistor in which the base is made up of semiconductor material that has been doped with a donor material that gives an excess of electrons, while the emitter and collector are made of semiconductor materials that have been doped with an acceptor material that gives an excess of holes.

NPN transistor. A transistor in which the material arrangements are in reverse of a PNP transistor. Both types are best drawn on electronic drawings by the use of templates that are readily available.

Rectifier. A device that changes an alternating current into a pulsating direct current is called a rectifier.

Conductors. The common grouping of conductors may be indicated on electronic drawings in several different ways. For example, multiconductor cables are shown by encircling the conductors with a loop, as shown in Fig. 7-5. A broken line (Fig. 7-6) indicates that the cable is shielded. If this shield is connected to a ground, it may be shown as in Fig. 7-7. When the circuit paths are separated on the drawing, the common shielding is indicated as shown in Fig. 7-8. Other ways to show cable groupings on drawings are shown in Fig. 7-9.

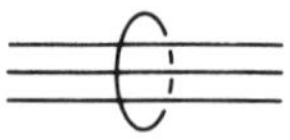

Figure 7-5. Multiconductor cables are shown on drawings by encircling the conductors with a loop.

Figure 7-6. A broken line loop indicates that the cable is shielded.

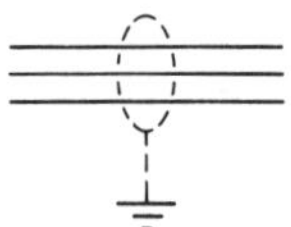

Figure 7-7. Symbol for a shielded cable connected to ground.

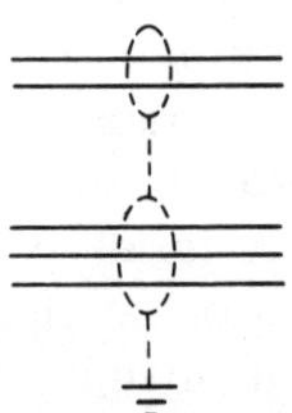

Figure 7-8. Symbol showing common shielding when the circuit paths are separated.

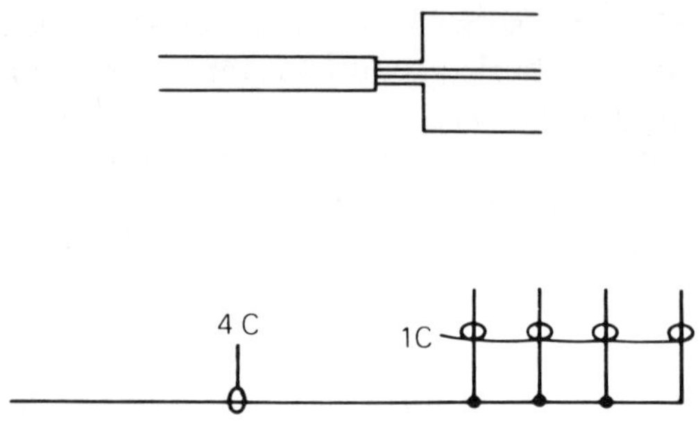

Figure 7-9. Two other ways of showing cable groupings on drawings.

7.2 Symbol Sizes and Proportions

As mentioned in Chapter 6, templates are available to aid the electronic draftsman in drawing symbols. There are also a variety of standard and special electronic symbols available on preprinted adhesive sheets. The size and detail of each symbol is governed by the size of the drawing, the available space, and the size of the template pattern or stick-on symbols at hand.

Theoretically, the size of a symbol is not important, since the drawings are diagrammatical. However, the relative sizes of symbols are an important matter. These relative sizes are shown in Table 7-1.

Table 7–1

ITEM	*MINIMUM (in inches)*	*MAXIMUM (in inches)*
Capacitor (height of straight line)	0.25	0.40
Capacitor (distance between gap)	0.06	0.10
Chassis ground	0.25	0.35
Connection dot	0.06	0.12
Coil (width of loops)	0.15	0.25
Resistor (height of zigzags)	0.15	0.30
Terminals (diameter of circle)	0.06	0.10
Transistor envelope (diameter)	0.60	0.80

Nearly every electronic device has a standard symbol that can be used to represent the device in a diagrammatic drawing of an electronic circuit. These symbols should be drawn with the heavy solid line as described in an earlier chapter, to highlight the symbol and to make it stand out over the other items on the drawing.

Although there are no particular sizes to which symbols should be drawn, there are minimum and maximum dimensions to be followed, and once a size is selected, this same size should be used throughout the drawing.

7.3 Semiconductors

The electronic draftsman of the past dealt with electron tubes. Now the electronic draftsman deals mostly with semiconductors, known as transistors, which perform the same function as their tube counterparts.

Although semiconductors are noted for their relatively poor conductivity, they also have the ability to conduct electricity in two different modes.

Like their tube counterparts, transistors are used in certain basic circuits; there are three basic circuits in which a triode transistor can be used: (1) the common-base circuit, (2) the common-emitter circuit, and (3) the common-collector circuit. Of these three, the common-emitter circuit is found more frequently than the other two, mainly because this is the circuit that gives the greatest voltage and power gain. However, the fundamentals of all three will be briefly described in order to give the electronic draftsman a basic knowledge of transistors so that he may approach his work more intelligently.

A typical common-base circuit is shown in Fig. 7-10; this is for an NPN transistor. The solid arrows indicate the direction of useful electron flow. Therefore, electrons flow from the negative terminal of battery B1, through resistor R1, through the NPN transistor, into the emitter, across the emitter-base junction to the base, then across the base-collector junction into the collector, and from the collector through the collector resistor R2 to the positive terminal of battery B2.

Characteristics of the common-base transistor circuit include a very low input impedance and a high output impedance. This output voltage is in phase with the input voltage and is greater than the input voltage. The current gain of the stage is less than one. A common-base circuit is often used in applications where it is desired to match a low impedance to a high impedance.

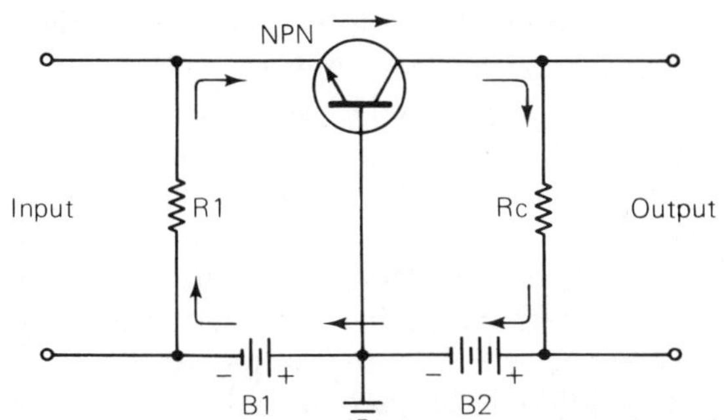

Figure 7-10. Typical common-base circuit for an NPN transistor.

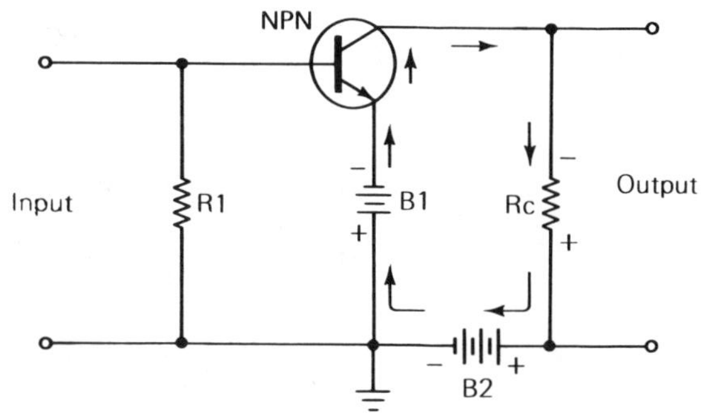

Figure 7-11. One of the most frequently used transistor circuits is the common-emitter circuit.

The most frequently used transistor circuit is the common-emitter circuit shown in Fig. 7-11. The circuit shown in this circuit is for an NPN transistor. The battery B1 provides the forward bias needed for the emitter-base junction, and battery B2 provides the reverse bias needed for the base-collector junction.

The emitter is made negative with respect to the base to provide forward bias for the emitter-base junction of the NPN transistor. The solid arrow on the diagram indicates the direction of electron flow in the circuit. Therefore, electrons leave the negative terminal of battery B1 and flow to the emitter. They cross the emitter-base junction, then flow through the base, across the base-collector junction, and through resistor R2 to the positive terminal of battery B2.

As stated previously, the common-emitter circuit is the most frequently used transistor circuit and has a voltage gain of from 80 to 100; there is also considerable current gain in this type of circuit. It has a medium input resistance and an output resistance of about 20,000 ohms. The output signal is 180 deg out of phase with the input signal.

The third possible circuit configuration using a triode transistor is the common-collector circuit. In this type of circuit the collector is operated at signal ground potential. Although this circuit is not found as often as the common-emitter circuit, it does have some characteristics that are useful in some special applications.

A common-collector circuit using an NPN transistor is shown in Fig. 7-12. Again, the arrows in the diagram show the direction of electron flow through the circuit. Therefore, electrons flow from the negative terminal of B1 through the emitter resistor R2 to the emitter. They flow across the emitter-base junction, through the base, across the base-collector junction, to the positive terminal of B2. Battery B1 biases the emitter-base junction in a forward direction, whereas B2 biases the base-collector junction in a reverse direction.

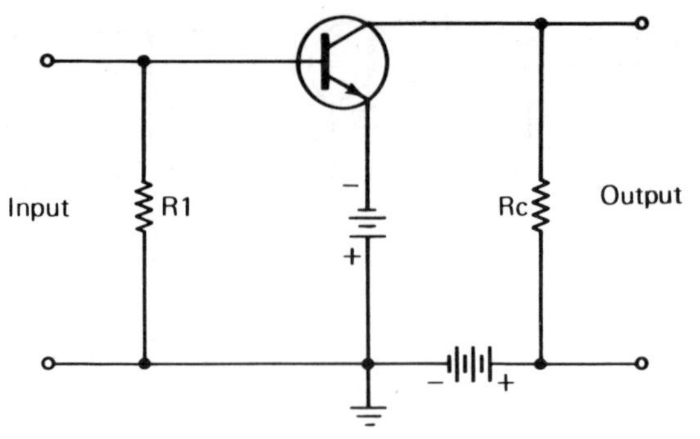

Figure 7-12. A common-collector circuit using an NPN transistor.

Electrons flowing through R2 set up a voltage drop across R2 with the polarity indicated on the diagram.

Other types of semiconductors include diodes that are used in rectifying applications. However, their inherent characteristics permit many uses other than simply as signal and power rectifiers. Table 7-2 lists special diode applications and the particular characteristic of the pn junction that led to this use.

7.4 Interpreting the Schematic

In order for the electronic draftsman to do his best work, he must learn how to interpret schematic drawings—not only from the viewpoint of being able to copy a designer's sketch, but he must also understand what is taking place. Understanding will only come with experience and by studying other books on basic electronics. However, some basic suggestions are in order. By following these suggestions, the reader should at least have a starting knowledge of interpreting electronic drawings and diagrams.

One of the best ways for a budding electronics draftsman to learn to read schematic diagrams is by doing—by actually building electronic circuits. There are many useful electronic gadgets available in kit form from Radio Shack, Heath Co., etc. These kits come with easy-to-follow directions, and most contain a pictorial drawing as well as a schematic drawing that you follow while constructing the kits. The beginner can refer to the pictorial drawing and then compare this with the schematic diagram. You can start off with an easy kit (costing around three to five dollars) and work up to the more sophisticated models. There is probably no better way to quickly learn how to read electronic schematic diagrams.

In general, electronic schematic diagrams indicate the scheme, or plan, according to which electronic components are connected for a specific purpose. Diagrams are not normally drawn to scale, and the symbol rarely looks exactly like the component. Lines joining the symbols representing electronic components indicate that the components are connected.

In order satisfactorily to serve all its intended purposes, the schematic diagram must be accurate, it must be understood by all qualified personnel, and it must provide definite information, without ambiguity.

The schematics for an electronic device must indicate all circuits in the device. Furthermore, it should be easy to read and follow an entire closed path in each circuit. If there are interconnections, they should be clearly indicated.

If at all possible, the conductors connecting the electronic symbols should be drawn either horizontally or vertically; rarely are they ever slanted.

A dot at the junction of two crossing wires means a connection between the two wires, while an absence of a dot—in most cases—indicates that the wires cross without connecting.

Schematic diagrams are, in effect, shorthand explanations of the manner in which an electronic circuit or group of circuits operates. As such, they make extensive use of symbols and abbreviations. The more commonly used symbols were explained earlier, and the reader should become thoroughly familiar with all of these symbols in order to correctly read electronic schematic drawings. The use of symbols presumes that the person looking at the diagram is reasonably familiar with the operation of the device and that he will be able to assign the correct meaning to the symbols.

Usually, every component on a complete schematic diagram has a number to identify the component. Supplementary data about such parts are supplied on the diagram or on an accompanying list in the form of a schedule which describes the component in detail.

To interpret schematic diagrams, remember that each circuit must be closed in itself, and each component should be in a closed loop connected by conductors to a source of electric current—such as battery, transformer, etc. There will always be a conducting path leading from the source to the component and a return path leading from the component to the source. The path may consist of one or more conductors. There may also be other components in the same loop, or additional loops branching off to other devices. But for each electronic component it must be possible to trace a completed conducting loop to the source.

Table 7-2
SEMICONDUCTOR DEVICES AND APPLICATIONS

Device	*Junction Configuration or Type*	*Semiconductor Material*	*Application*
Rectifiers (diodes)	pn junction	Silicon	Low fwd. voltage drop, signal and power rectification
	" "	Germanium	high rectification ratio, high inverse breakdown, high temperature, signal and power rectification
	Dry disc	Selenium	Power rectifier, low-frequency diode, self-healing
	" "	Copper oxide	Meter rectifier, low voltage power rectifier
	" "	Copper sulfide	Low voltage power rectifier
Transistors	pnp or npn	Germanium	low saturation voltage, general purpose to 75° C
	" " "	Silicon	High temp. use to 175° C, higher voltage
Field Effect Transistors (FET)	n or p channel types	Silicon	High input impedance, resistive, bidirectional output impedance
Unijunction Transistors	n-type bar with p-junction between ends of bar	Silicon	Relaxation oscillator, timing, trigger circuits, neg. res. device
Thyristors Silicon Controlled Rectifier (SCR)	pnpn	Silicon	Phase controlled rectifier, similar to thyratron tube
Triac	two SCRs in parallel and opposite orientation	Silicon	Bidirectional control of AC, light dimmers, power tool speed control
Tunnel Diode	heavily doped pn junction	Germanium Gallium arsenide	Negative resistance, microwave amplifier oscillators, converters

Table 7-2 (cont.)

Device	*Junction Configuration or Type*	*Semiconductor Material*	*Application*
Photo-diodes Photo-transistors	pn npn	Germanium Silicon	Photo-conduction and photo-voltaic for control applications
Photoelectric Cells	photoresistive	Lead sulfide	Infrared detector
	"	Lead teluride	Infrared detector
	photovoltaic	Cadmium sulfide	Light meter
	"	Selenium	Light meter
Varistors	Fired	Silicon carbide	Surge suppressors
	Dry-disc	Selenium	Contact protector
Thermistors	Fired	Mixed metal oxides	Temperature sensing, control, compensating
Zener Diodes	pn junction reverse biased to Zener breakdown	Silicon	Reverse biased, voltage regulator, voltage reference
Varactor Diode (varicap)	pn junction reverse biased, no current flow	Silicon	Voltage controlled capacitor, parametric amplifier, multipliers
Light Emitting Diode (LED)	pn junction	Gallium arsenide Phosphide	Visible, infrared, light emitting, displays

QUESTIONS

1. Explain in your own words for the meaning of the symbol 
2. What are the relative input and output impedances of the common-base amplifier circuit?
3. Is it possible to get a current gain by using the common-emitter circuit?
4. What is the relationship between the input signal voltage and the amplified output signal voltage of the common-emitter circuit?
5. In your own words, describe why constructing electronic kits can help the electronic draftsman learn to read schematic diagrams.
6. Why are electronic symbols used on schematic drawings rather than pictorial views of the various components?
7. Describe the basic operation of an amplifier as used in electronic circuits.
8. If an amplifier symbol is shown on an electronic drawing, and the letters "BST" appear near the symbol, what type of amplifier does this denote?
9. Name one purpose of a capacitor when used in an electronic circuit.
10. What purpose does a circuit breaker serve in an electronic circuit?
11. What other protective device is being replaced by the circuit breaker?
12. If a rectangle appears on an electronic drawing with the letters "EQ" in it, what device or component does this symbol denote?
13. A number of turns of insulated wire wound on an iron core is characteristic of what electronic components?
14. When the chassis of an electronic instrument is grounded, what purpose does this serve?
15. Name two purposes of a single-pole switch in an electronic circuit.
16. What are the three basic types of resistors?
17. What is the name of the device that is used to radiate acoustic power effectively at a distance through the air?
18. Give the definition of a rectifier.
19. From the information given in earlier chapters, what type of drawing line should be used for drawing electronic symbols?
20. What devices have all but replaced electron tubes in electronic circuits?

Chapter 8

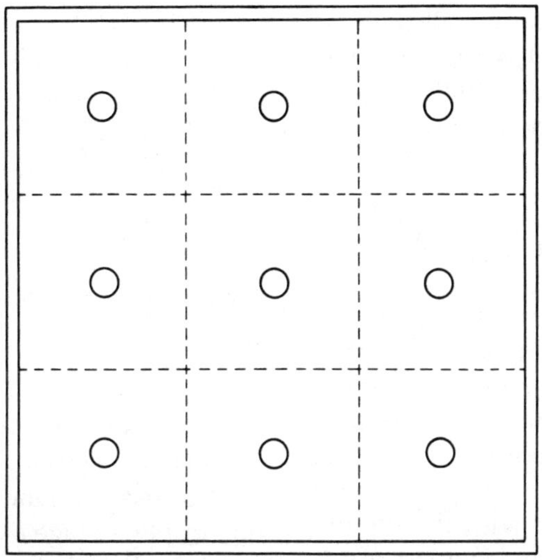

LIGHTING PLANS

The main objective of this chapter is to introduce the reader to basic lighting design. It further shows how to lay out lighting schemes, how to calculate the permitted loads per circuit, and, finally, how to actually draw the design in the form of working drawings so that the workmen on the job can install the systems. Lighting fixture details as well as schedules will also be covered, so that the draftsman will thoroughly understand the use of these time-saving devices.

8.1 Fundamentals of Lighting Design

The basic requirement for any lighting design for building construction is to determine the amount of light that should be provided (recommended illumination levels being used) and the best means of providing it; that is, lighting equipment should be selected that will provide the highest visual comfort and performance that is consistent with the type of area to be lighted and the budget provided. However, since individual tastes and opinions vary to a great extent, there naturally can be many solutions to any given lighting problem. Some of these solutions will be dull and commonplace, while others will show imagination and resourcefulness. The data presented in this chapter will get you started on the right foot, but only experience in the field of lighting design will enable you consistently to produce lighting designs of high quality.

In order to clearly show the basic steps of a lighting design, the floor plan in Fig. 8-1 will be used; it is a floor plan of a private office where regular office work will take place. Referring to the Table 8-1 and looking under "Offices," we see that 100 footcandles of illumination should be provided for offices where regular office work such as mail filing; reading good reproductions, etc., will take place. This recommended footcandle level should be noted, but in doing so, use a form such as the one in Fig. 8-2 to record this and other necessary data.

The next step is to record a complete description of the physical characteristics of the room on the form. This description should include the length, width, ceiling height, and total floor area of the room, and may be obtained from either the floor plans and elevations or by actual measurements. The floor plan in Fig. 8-1 is 8 feet wide by 12 feet long and the ceiling height is 8 feet; thus, the total area is (8 × 12) 96 square feet. All of these data are then entered in the proper spaces on the form in Fig. 8-2.

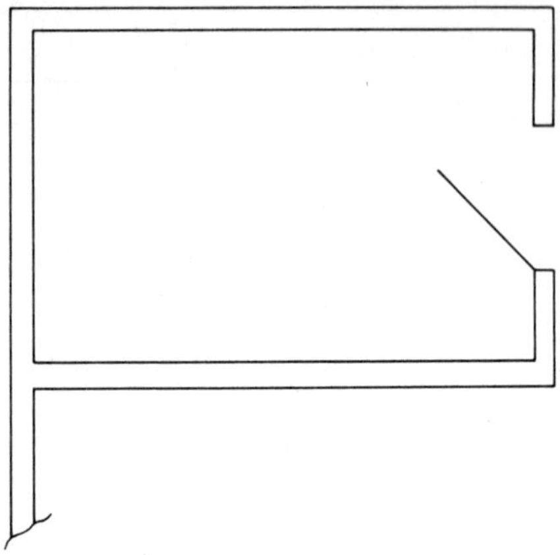

Figure 8-1. Table floor plan of a private office where regular office work will take place.

The amount of reflectance from the room surfaces is very important in determining the amount of light to use. In general, light-colored surfaces have a higher percentage of reflectance than those with dark finishes. We shall assume that the office walls are painted with a light-colored paint, and that the floor is a light-colored tile. By comparing color samples of known reflections with our office samples, we are able to determine the reflectances of the office surfaces as follows: ceiling, 80 percent; walls, 50 percent; and floor, 20 percent. Enter these percentages in the appropriate spaces on the form.

A modern method for calculating the correct amount of illumination to be used in a given area is called the *zonal-cavity method*. This system divides the room into three cavities, as shown in Fig. 8-3. Here we find that the desk top is approximately 2½ feet above the finished floor (the level of the desk top is known as the working plane), so the floor cavity in this example will be 2.5 feet. The ceiling cavity is the distance from the ceiling to the bottom of the lighting fixture, but since we have selected recessed lighting fixtures for this area, there will be no ceiling cavity. The remaining area between the ceiling and the floor cavity (the top of the desk) is, therefore, 5.5 feet.

The lighting fixture chosen for this office space is a conventional 2′ × 4′ recessed fixture containing four 40-watt fluorescent lamps. Some of the reasons for selecting this type of fixture are

1. The high efficiency and good color tones of the fluorescent lamps result in low operating cost.
2. The acrylic lens offers both good light distribution and diffused light for visual comfort.

	Recommended Footcandles (Minimum At Any Time)
Iron and Steel Manufacturing—	
Open hearth—	
Charging floor	20
Pouring slide—	
Slag pits	20
Control platforms	30
Hot top	30
Hot top storage, checker cellar	10
Mixer bulding	30
Calcining building, skull cracker	10
Rolling mills—	
Blooming, slabbing, hot strip, hot sheet	30
Cold strip, merchant, sheared plate	30
Pipe, rod, tube, wire drawing	50
Tin plate mills: tinning, galvanizing, cold strip rolling	50
Motor room, machine room	30
Inspection—	
Blackplate, bloom, billet chipping	100
Tinplate, other bright surfaces	200*
Laundries—	
Washing	30
Flatwork ironing, weighing, listing, and marking	50
Machine and press finishing, sorting	70
Fine hand ironing	100
Leather Manufacturing—	
Cleaning, tanning and stretching, vats	30
Cutting, fleshing, and stuffing	50
Finishing and scarfing	100
Leather Working—	
Pressing, winding, and glazing	200
Grading, matching, cutting, scarfing, sewing	300
Machine Shops—	
Rough bench and machine work	50
Medium bench and machine work, ordinary automatic machines, rough grinding, medium buffing and polishing	100
Fine bench and machine work, fine automatic machines, medium grinding, fine buffing and polishing	500
Extra fine bench and machine work, fine grinding	1000
Materials Handling—	
Wrapping, packing, labeling	50
Picking stock, classifying	30
Loading, trucking	20
Inside truck bodies and freight cars	10

* Specular materials or work surfaces may necessitate special consideration in selection and placement of lighting equipment, or orientation of the work.

	Recommended Footcandles (Minimum At Any Time)
Paint Manufacturing—	
General	30
Comparing mix and standard	200
Paint Shops—	
Dipping, spraying, rubbing, firing, ordinary hand painting and finishing art	50
Fine hand painting and finishing	100
Extra fine hand painting and finishing (automobile bodies, piano cases, etc.)	300
Paper Box Manufacturing: General Area	50
Paper Manufacturing—	
Beaters, grinding, calendaring	30
Finishing, cutting, trimming, paper-making machines	50
Hand counting, wef end of paper machine	70
Paper machine reel, paper inspection and laboratories	100
Rewinder	150
Plating	30
Polishing and Burnishing	100
Printing Industries—	
Type foundries—	
Machine and hand casting; font assenbly—sorting	50
Matrix-making, dressing type	100
Printing plants—	
Color inspection and appraisal	200
Machine composition, composing room	100
Presses	70
Imposing stones and proofreading	150
Electrotyping—	
Molding, finishing, leveling molds, routing, trimming	100
Blocking, tinning, electroplating, washing, backing	50
Photoengraving—	
Etching, staging, blocking	50
Routing, finishing, proofing, tint-laying, masking	100
Rubber Tire and Tube Manufacturing—	
Stock preparation—	
Bandury, plasticating, milling	30
Calendaring	50
Fabric preparation: stock cutting, bead building	50
Tube and thread tubing machines	50
Tire building—	
Solid tire	30
Pneumatic tire	50
Curing department: tube curing, casing curing	70
Final inspection: tube, casing	200

	Recommended Footcandles (Minimum At Any Time)
Chemical Works—	
Furnaces, driers, stills, evaporators, filtration,bleaching	30
Tanks, crystallizers, extractors, percolators, nitrators	30
Clay Products and Cements—	
Grinding, filter presses, kiln rooms	30
Molding, pressing, cleaning, and trimming	30
Color and glazing, rough work; enameling	100
Color and glazing, fine work	300
Cleaning and Pressing Industry—	
Checking and sorting	50
Dry and wet cleaning and steaming	50
Inspection and spotting	500
Pressing, machine and hand	150
Repair and alteration	200
Cloth Products—	
Cloth inspection	2000
Cutting and pressing	300
Sewing	500*
Coal Tipples and Cleaning Plants—	
Breaking and cleaning areas	10
Picking	300
Dairy Products: Fluid Milk Industry—	
Boiler room and bottle storage	30
Bottle sorting	50
Bottle washers	*
Can washing and cooling equipment	30
Filling: inspection	100
Gauges, meter panels, and thermometers (on face)	50
Laboratories	100
Pasteurizes, separators, and storage refrigerators	30
Tanks, vats—	
Light interiors	20
Dark interiors	100
Electrical Equipment Manufacturing—	
Impregnating	50
Insulating: coil winding	100
Testing	100
Flour Mills—	
Rolling, sifting, purifying	50
Packing	30
Product control	100
Cleaning screens, man lifts, walkways, bin checking	30

* Special lighting from sources large enough to cover the surface being inspected, and of sufficiently low brightness to provide comfortable contrast conditions.

INDUSTRIAL INTERIORS

	Recommended Footcandles (Minimum At Any Time)
Forge Shops	50
Foundries—	
Annealing, cleaning, shakeout	30
Core-making, medium	50
Core-making, fine	100
Grinding and chipping	100
Inspection, medium	100
Inspection, fine	500
Molding, large; pouring and sorting	50
Molding, medium	100
Cupola	20
Garages: Automobile and Truck—	
Service garages—	
Repairs	100
Active traffic areas	20
Parking garages—	
Entrance	50
Traffic lanes	10
Storage	5
Glass Works—	
Mix and furnace rooms, pressing and lehr, glass blowing machines	30
Grinding, cutting glass to size, silvering	50
Fine grinding, polishing, beveling	100
Inspection, etching and decoratoring	200
Glove Manufacturing—	
Pressing and cutting	300
Knitting and sorting	100
Sewing and inspection	500
Hat Manufacturing—	
Dyeing, stiffening, braiding, cleaning, and refining	100
Forming, sizing, pouncing, flanging, finishing, and ironing	200
Sewing	500
Inspection—	
Ordinary	50
Difficult	100
Highly difficult	200
Very difficult	500
Most difficult	1000

	Recommended Footcandles (Minimum At Any Time)
Airplane Manufacturing—	
Stock parts—	
Production	100
Inspection	200
Parts manufacturing—	
Drilling, riveting, and screw fastening	70
Spray booths	100
Sheet aluminum layout and template work; shaping and smoothing of small parts for fuselage; wing sections, cowling, etc.	100
Subassembly: landing gear, fuselage, wing sections, cowling, and other large units	100
Final assembly and inspection	100
Machine tool repairs	100
Airplane Hangars: Repair Service Only	100
Assembly—	
Rough easy seeing	30
Rough difficult seeing	50
Medium	100
Fine	500
Extra fine	1000
Automobile Manufacturing—	
Frame assembly	50
Chassis assembly line	100
Final assembly amd inspection line	200
Body manufacturing—	
Parts	70
Finishing and inspecting	200
Bakeries—	
Mixing room	50
Face of shelves (vertical illumination)	30
Inside of mixing bowl (vertical mixers)	50
Fermentation room	30
Make-up room—	
Bread	30
Sweet yeast raised products	50
Oven, proofing, and wrapping rooms	30
Fillings and other ingredients	50
Decorating and icing—	
Mechanical	50
Hand	100
Book Binding—	
Folding, assembling, pasting, etc.	70
Cutting, punching, stitching	70
Embossing and inspection	200

	Recommended Footcandles (Minimum At Any Time)
Candy Making—	
Chocolate department—	
Husking, winnowing, fat extraction, crushing and refining, feeding	50
Bean cleaning and sorting, dipping, packing, wrapping	50
Milling	100
Cream-making: mixing, cooking, and molding	50
Gum drops and jellied forms	50
Hand decorating	100
Hard candy—	
Mixing, cooking, and molding	50
Die cutting and sorting	100
Kiss-making and wrapping	100
Canning and Perserving—	
Initial grading raw material samples	50
Tomatoes	100
Color grading (cutting rooms)	200
Preparation—	
Preliminary sorting—	
Apricots and peaches	50
Tomatoes	100
Olives	150
Cutting and pitting, final sorting	100
Canning—	
Continuous belt and sink canning	100
Hand packing	50
Olives	100
Examination of canned samples	200
Central Station Indoor Locations—	
Auxiliaries, battery rooms, boiler feed pumps, tanks, compressors, and gauge area	20
Boiler platforms, cable room, and circulator or pump bay	10
Burner platform	20
Condensers; deaerator, evaporator, and heater floors	10
Control rooms—	
Vertical face of switchboards—	
Simplex or section of duplex facing operator—	
Type A—Large centralized control room at 66 inches above floor	50
Type B—Ordinary control room at 66 inches above floor	30
Section of duplex facing away from operator	30
Bench boards (horizontal level)	50
Area inside duplex switchboards	10
Rear of all switchboard panels (vertical)	10
Emergency lighting, all areas	3
Chemical laboratory	50
Screen house, switchgear power, telephone equipment room	20
Tunnels or galleries, piping	10
Turbine bay sub-basement	20
Turbine room	30

Table 8-1. Illumination tables.

Area	Recommended Footcandles (Minimum At Any Time)
Hospitals—(Continued)	
Fracture room—	
General	50
Fracture table	200
Laboratories—	
General	50
Close work	100
Libraries	70
Locker rooms	20
Lobby and lounge rooms (lobby daytime 50)	30
Medical records room	100
Nurses' station—	
General-Day	70
General-Night	30
Desk and charts	50
Medicine room counter	100
Nurses' workroom	30
Nurseries—	
General	30
Examination table	100
Play room, pediatric	30
Obstetrical—	
Clean-up and scrub-up rooms	30
Labor room	20
Delivery room, general	100
Delivery table	2500
Pharmacy—	
General, manufacturing	50
Work table	100
Active storage	30
Private rooms and wards—	
General	10
Reading	30
Psychiatric disturbed patients' areas	10
Radioisotope facilities—	
Radiochemical laboratory	30
Up-take measuring room	20
Examination table	50
Solariums	20
Storage, central—	
General	30
Office	70
Surgery—	
Instrument and sterile supply room	30
Clean-up room (instruments)	100
Operating room, general	100
Operating table	2500
Recovery room	30

Area	Recommended Footcandles (Minimum At Any Time)
Art Galleries—	
General	30
On paintings (supplementary)	30*
On statuary and other displays	100**
Auditoriums—	
Assembly	15
Exhibition	30
Banks—	
Lobby—	
General	50
Writing areas	70
Tellers' stations, posting, and keypunch	150
Depots, Terminals, and Stations—	
Waiting rooms, rest and smoking rooms	30
Ticket offices: general, ticket racks, counters	100
Baggage checking	50
Platforms and storage	20
Toilets and washrooms	30
Hospitals—	
Anesthetizing and preparation room	30
Autopsy and morgue—	
Autopsy room	100
Autopsy table	1000
Morgue, general	20
Central sterile supply—	
General	30
Needle sharpening	150
Dental suite—	
Operatory, general	70
Instrument cabinet	150
Dental chair	1000
Laboratory, bench	100
Recovery room	5
Emergency room—	
General	100
Local	2000
Examination and treatment room—	
General	50
Examining table	100
Exits, at floor	5
Eye, ear, nose and throat suite—	
Darkroom (variable)	0–10
Eye examination room, ear, nose and throat room	50

* Dark paintings with fine detail should have two to three times this footcandle level.
** In some cases much more illumination is necessary to bring out the beauty of the statuary.

Table 8-1. (cont.)

Area	Recommended Footcandles (Minimum At Any Time)
Museums (See Art Galleries)	
Offices—	
Reading high-contrast or well-printed material; tasks and areas not involving critical or prolonged seeing such as conferring, interviewing, inactive files, and washrooms	30
Reading or transcribing handwriting in ink or medium pencil on good-quality paper; intermittent filing	70
Regular office work; reading good reproductions, reading or transcribing handwriting in hard pencil or on poor paper; active filing; index references; mail sorting	100
Accounting, auditing, tabulation, bookkeeping, business machine operation; reading poor reproductions; rough layout drafting	150
Cartography, designing, detailed drafting	200
Corridors, elevators, escalators, stairways	20*
Post Offices—	
Lobby, on tables	30
Sorting, mailing, etc.	100
Residences—	
Specific visual tasks—	
Table games	30
Kitchen activities—	
Sink	70
Range and work surfaces	50
Laundry, trays, ironing board, ironer	50
Reading and writing, including studying—	
Books, magazines, newspapers	30
Handwriting, reproductions, poor copies	70
Study desks	70
Reading music scores—	
Simple scores	30
Advanced scores	70**
Sewing—	
Occasional periods, coarse thread, large stitches, high contrast thread to fabric	30
Occasional periods, light fabrics	50
Prolonged periods, light to medium fabrics	100
Dark fabrics, fine detail, low contrast	200
Shaving, make-up, grooming: on the face at mirror locations	50
Work shop, bench work	70
General lighting—	
Entrances, hallways, stairways, stair landings	10
Living room, dining room, bedroom, family room, sun room, library, game or recreation room	10
Kitchen, laundry, bathroom	30

* Or not less than 1/5 the level in adjacent areas.
** When score is substandard size, and notations are printed on the lines, 150 footcandles or more are needed.

Illumination level is only one characteristic of a lighting installation; many other important considerations enter into the design of a completely satisfactory visual environment (see Chapter Four). However, it is obvious that without the basic requirement of adequate illumination, in other words sufficient footcandles, no visual task can be performed correctly, rapidly, safely, and easily. The quantitative requirement for good illumination varies greatly with the nature of the activity, and is primarily a function of the difficulty of the visual task in terms of size of detail, brightness or color contrast, and speed demanded. Other factors such as the length of time for which the task is to be performed, the surrounding conditions, and the physiological state of the eyes that have to do the work are also involved.

Evidence providing a sound basis for definite footcandle recommendations is not easy to obtain. Much work in this field has been carried on over a period of many years, using various methods and various criteria of visual performance. Researches, reported in 1958 and later, have provided new basic data considered to be more directly applicable to the problems involved, as well as a definite method of evaluating real work situations in terms of the laboratory test results. On the basis of these studies the Illuminating Engineering Society has made footcandle recommendations for a wide variety of representative industrial operations and other visual activities. The material in this chapter is taken from these recommendations.

The following footcandle table is divided into two parts, general interior lighting and the lighting of industrial interiors. The illumination specified is to be provided *on the work surface*, whether this be horizontal, vertical, or oblique. Where there is no definite work area, it is assumed that the illumination is measured on a horizontal plane 30 inches above the floor. The values given are not to be construed as initial footcandles provided by a new installation; they are recommended *minimum footcandles at any point on the task at any time*. This means that the installation must be so designed that the collection of dirt on luminaires, lamps, walls, and ceilings, and the normal depreciation in light output of the lamps themselves will not at any time lower the illumination below the recommended level (see Chapter Six).

For many of the situations listed in the table, especially where there is a need for relatively high footcandles on the task, it would obviously be difficult and impractical to light the entire room to the recommended value. Under these circumstances the illumination on the work is obtained by a combination of general lighting and specialized supplementary lighting. Precautions must then be taken to ensure that luminance ratios throughout the visual field are kept within desirable limits, and the directional quality of the supplementary light must be carefully considered.

Area	Recommended Footcandles (Minimum At Any Time)
Restaurants, Lunch Rooms, Cafeterias—	
Dining areas—	
Intimate type—	
Subdued environment	3
Light environment	10
For cleaning	20
Leisure type—	
Subdued environment	15
Light environment	30
Quick-service type—	
Normal surroundings	50
Bright surroundings	100
Cashier	50
Food displays: twice the general levels, but not less than	50
Kitchen—	
Inspection, checking, and pricing	70
Other areas	30
Schools—	
Reading printed material	30
Reading pencil material	70
Reading spirit-duplicated material—	
Good	30
Poor	100
Drafting, bench work	100
Lip reading, chalkboards	150
Sewing	150
Stores—	
Show windows—	
Daytime lighting—	
General	200
Feature	1000
Nighttime lighting—	
Secondary business districts or small towns—	
General	100
Feature	500
Main business districts, highly competitive—	
General	200
Feature	1000
Store interiors—	
Circulation areas	30
Merchandising areas—	
Service	100
Self-service	200
Showcases and wall cases—	
Service	200
Self-service	500
Feature displays—	
Service	500
Self-service	1000

Area	Recommended Footcandles (Minimum At Any Time)
Rubber Goods, Mechanical—	
Stock preparation—	
Banbury, plasticating, milling	30
Calendaring	50
Fabric preparation: stock cutting, hose looms	50
Molded and extruded products, and curing	50
Inspection	200
Service Areas—	
Stairways, corridors, elevators	20
Toilets and washrooms	30
Sheet Metal Works—	
Presses, shears, stamps, punches, spinning, miscellaneous machines, medium bench work	50
Tin plate inspection, galvanized; scribing	200*
Shoe Manufacturing, Leather—	
Cutting tables, marking, buttonholing, skiving, sorting, vamping, counting, stitching on dark materials	300
Making and finishing: nailers, sole layers, welt beaters and scarfers, trimmers, welters, lasters, edge setters, sluggers, randers, wheelers, treers, cleaning, spraying, buffing, polishing, embossing	200
Shoe Manufacturing, Rubber—	
Washing, coating, mill run compounding	30
Varnishing, vulcanizing, calendaring, upper and sole cutting	50
Sole rolling, lining, making and finishing processes	100
Stone Crushing and Screening—	
Belt conveyor tubes, main line shafting spaces, chute rooms inside of bins	10
Primary breaker room, auxiliary breakers under bins	10
Screens	20
Storage Rooms and Warehouses—	
Inactive	5
Active—	
Rough bulky	10
Medium	20
Fine	50
Sugar Refining—	
Grading	50
Color inspection	200
Textile Mills, Cotton—	
Opening, mixing, picking	30
Carding, drawing, slubbing, roving, spinning, spooling	50
Beaming and slashing on comb—	
Grey goods	50
Denims	150

* Special lighting from sources large enough to cover the surface being inspected, and of sufficiently low brightness to provide comfortable contrast conditions.

Area	Recommended Footcandles (Minimum At Any Time)
Hospitals—(Continued)	
Therapy—	
Physical	20
Occupational	30
Toilets	30
Utility room	20
Waiting room—	
General	20
Reading	30
X-ray room and facilities—	
Radiography, fluoroscopy and darkroom	10
Deep and superficial therapy	10
Viewing room	30
Filing room, developed films	30
Storage, undeveloped films	10
Hotels—	
Bars and cocktail lounges (see Restaurants)	
Bathrooms—	
General	10
At mirror	30*
Bedrooms—	
General	10
Make-up	30*
Reading, ink writing	30
Dining areas (see Restaurants)	
Entrance foyer	30
Front office	50
Laundry—	
Washing	30
Flat work ironing	50
Machine and press finishing	70
Linen room—	
General	20
Sewing	100
Lobby—	
General	10
Reading and working areas	30
Marquee—	
Dark surroundings	30
Bright surroundings	50
Municipal Buildings; Fire and Police—	
Police—	
Identification records	150
Jail cells and interrogation rooms	30
Fire hall—	
Dormitory	20
Wagon room and recreation room	30

* For close inspection, 50 footcandles.

Area	Recommended Footcandles (Minimum At Any Time)
Textile Mills, Cotton—(continued)	
Inspection—	
Grey goods (hand turning)	100
Denims (rapidly moving)	500
Automatic tying-in	150
Drawing-in by hand	200
Weaving	100
Textile Mills, Silk and Synthetics—	
Manufacturing: soaking, fugitive tinting, and conditioning or setting of twist	30
Winding, twisting, rewinding, coning, quilling, slashing—	
Light thread	50
Dark thread	200
Warping (silk or cotton system) on creel, on running ends, on reel, on beam, on warp at beaming	100
Drawing-in, on heddles or reed	200
Weaving	100
Textile Mills, Woolen and Worsted—	
Grading	100
Drawing or spinning (frame or mule): white	50
Drawing or spinning (frame or mule): colored	100
Twisting: white	50
Warping: white, at reed: white	100
Warping: colored	100
Warping at reed: colored	300
Winding: white	30
Winding: colored	50
Weaving: white	100
Weaving: colored	200
Grey goods room—	
Burling	150
Sewing	300
Folding	70
Wet finishing: fulling, scouring, crabbing, drying	50
Dyeing	100
Dry finishing—	
Napping, conditioning, pressing, folding	70
Shearing	100
Inspecting (perching)	2000
Tobacco Products—	
Drying, stripping, general	30
Grading and sorting	200
Welding—	
General illumination	50
Precision manual arc welding	1000
Woodworking—	
Rough sawing and bench work	30
Sizing, planing, rough sanding, medium machine and bench work, glueing, veneering, cooperage	50
Fine bench and machine work, fine sanding and finishing	100

ILLUMINATION CALCULATION SHEET

FOR USE WITH THE IES ZONAL CAVITY METHOD

GENERAL INFORMATION

PROJECT IDENTIFICATION: ______________________________
(GIVE NAME OF AREA AND/OR BUILDING AND ROOM NUMBER)

AVERAGE MAINTAINED ILLUMINATION FOR DESIGN: ____________ FOOTCANDLES.

LUMINAIRE DATA:

MANUFACTURER: ____________ CATALOG NUMBER: ____________

LAMPS (TYPE & COLOR): ____________ NUMBER PER LUMINAIRE: ____________

TOTAL LUMENS PER LUMINAIRE: ____________ MAINTENANCE FACTOR: ____________

SELECTION OF COEFFICIENT OF UTILIZATION

STEP 1: FILL IN SKETCH AT RIGHT.

STEP 2: DETERMINE CAVITY RATIOS BY THE FORMULA:

$$\frac{5 \times h \begin{bmatrix} CC \\ RC \\ FC \end{bmatrix} \times (L + W)}{L \times W}$$

ROOM CAVITY RATIO, RCR = ______

CEILING CAVITY RATIO, CCR = ______

FLOOR CAVITY RATIO, FCR = ______

L= ______

W= ______

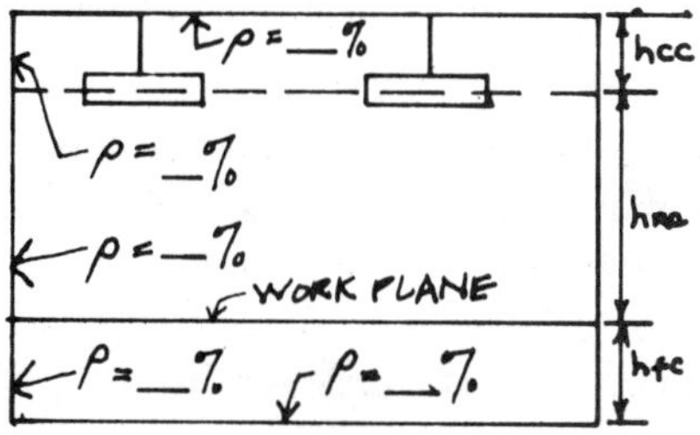

STEP 3: OBTAIN EFFECTIVE CEILING CAVITY REFLECTANCE (ρCC). ρCC = ______

STEP 4: OBTAIN EFFECTIVE FLOOR CAVITY REFLECTANCE (ρFC). ρFC = ______

STEP 5: OBTAIN COEFFICIENT OF UTILIZATION (CU) FROM MANUFACTURER'S DATA. CU = ______

CALCULATIONS

STEP 6:

AVERAGE MAINTAINED ILLUMINATION LEVEL

$$\text{FOOTCANDLES} = \frac{\text{(TOTAL LAMP LUMENS PER LUMINAIRE)} \times \text{(CU)} \times \text{MAINTENANCE FACTOR}}{\text{AREA PER LUMINAIRE}}$$

= ______________________________

= ______________ FOOTCANDLES ON WORK AREA

AREA PER LUMINAIRE: (THIS AREA DIVIDED BY THE LUMINAIRE LENGTH GIVES THE APPROXIMATE SPACING BETWEEN CONTINUOUS ROWS, OR IT MAY BE DIVIDED INTO THE TOTAL ROOM AREA TO DETERMINE THE NUMBER OF LUMINAIRES REQUIRED.)

$$\text{AREA PER LUMINAIRE} = \frac{\text{(TOTAL LAMP LUMENS PER LUMINAIRE)} \times \text{(CU)} \times \text{MAINTENANCE FACTOR}}{\text{FOOTCANDLES}}$$

= ______________________________

= ______________ SQUARE FEET.

CALCULATED BY: ______________________ DATE: ______________

Figure 8-2. A handy form for making lighting calculations.

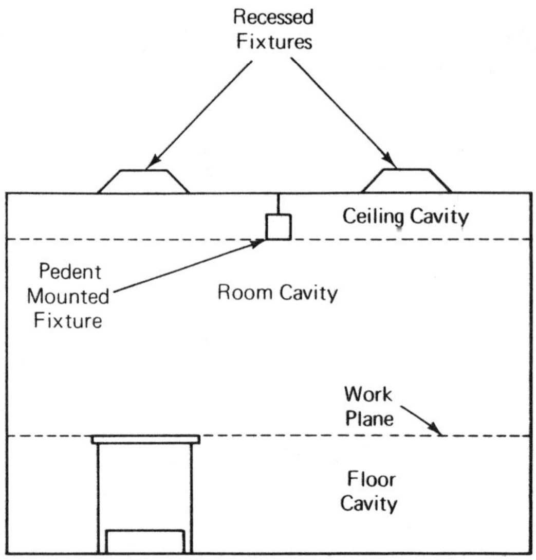

Figure 8-3. The zonal-cavity method divides the room into three cavities, as shown here.

3. The physical size of the fixture permits easy installation, as the fixtures will lie in the 2′ × 4′ frames of the T-bar ceiling.
4. The appearance of the lighting fixtures is suited for office areas, and they are reasonably priced.

Each lighting fixture contains four 40-watt cool white fluorescent lamps and by referring to a lamp data catalog, we find that each lamp has approximately 3200 initial lumens. These data are also entered in the form to be used later in the lighting calculations. Figure 8-4 shows a typical page from a light data catalog and shows the column from which the lumens of the lamp were obtained. A different lamp would have a different amount of initial lumens. With these data accounted for, we are now ready to begin the actual calculations to determine how many lighting fixtures will be required in this area to produce the desired 100 footcandles of illumination.

Rooms are classified according to shape by cavity-ratio numbers. The basic formula for obtaining cavity ratios in rectangular-shaped rooms is

$$\text{Cavity ratio} = \frac{5 \times \text{height (length + width)}}{\text{length} \times \text{width}}$$

where height is the height of the cavity (either floor, ceiling, or room) under consideration.

To illustrate the use of the formula, the room in Fig. 8-1 is 8 feet wide by 12 feet in length. We previously determined that the floor cavity is 2.5 feet and the room cavity is 5.5 feet. Therefore, the floor cavity is found by substituting the known values in the formula, and we have

$$\text{Floor cavity ratio} = \frac{5 \times 2.5(12 + 8)}{12 \times 8} = 2.6, \text{ or } 3$$

The room cavity is found in a similar way; that is

$$\text{Room cavity ratio} = \frac{5 \times 5.5(12 + 8)}{12 \times 8} = 5.7, \text{ or } 6$$

Notice in the form in Fig. 8-2 that a space is provided for the coefficient of ultilization; this term is used in the formula for determining the amount of required lighting fixtures necessary to produce a given amount of illumination in an area. Manufacturers usually supply photometric data for their own individual lighting fixtures; and the table in Fig. 8-5 is typical of such table. This table supplies the photometric data for the lighting fixtures that we intend to use in the office space in Fig. 8-1. It contains coefficients of ultilization for various surface reflections and room cavity ratios. To use this table, look under the heading of 80 percent effective ceiling reflectance (determined previously) and under this heading, look in the column under 50 percent wall reflectance (also found earlier). Since we have calculated the room cavity ratio to be 6, the coefficient of utilization adjacent to this column is 45 percent or 0.45. Since the floor cavity reflectance in our example is 20 percent, no further adjustment is necessary.

One other factor must be considered prior to the actual calculation—the "maintenance factor." This factor takes into account the reduction in light output because of lamp aging and because of dirt accumulation. Determining an appropriate maintenance factor can be a study all to itself, so for our purposes, we shall assume a maintenance factor of 0.70, which is average for office areas.

The number of lighting fixtures and lamps can be calculated from the following formula or equation:

$$NF = \frac{FA \times DF}{LPF \times LPL \times CU \times MF}$$

where

NF is number of fixtures required for desired illumination.
FA is floor area.
DF is desired footcandles.
LPF is lamps per fixture.
LPL is lumens per lamp.
CU is coefficient of utilization.
MF is maintenance factor.

GENERAL ELECTRIC LARGE LAMPS

FLUORESCENT LAMPS

40-WATT RAPID START—PREHEAT LAMPS (Continued)

MAINLIGHTER™ LAMPS

Life ratings published below for Mainlighter Lamps are for rapid start circuits only. On preheat circuits, rated average life at 3 hours per start is 12,000 hours; at 12 hours per start—15,000 hours. All F40 Mainlighter Lamps are excellent alternates for F40T12/LT lamps.

T-12 Approx. 1½" Diameter

To determine approximate lamp length (pin ends to pin ends), deduct 1/4" from the nominal lamp length shown below.

Nominal Lamp Watts	Bulb	Nominal Length (Inches)	Base	Lamp Ordering Code	List Price	DESCRIPTION (See Fluorescent Lamp Footnotes—Page 63)	Std. Pkg. Qty.	Approx. Hours Life	Approx. Initial Lumens	Approx. Lumens at 40% Rated Avg. Life
40	T-12	48	Medium Bipin 	**F40CW**	**$1.29**	Cool White—MAINLIGHTER.	24	*18000+	3150	2805
				F40CW 6PK	**1.45**	Cool White—MAINLIGHTER. 6-Pack	24	*18000+	3150	2805
				F40CW 48PK	**1.29**	Cool White—MAINLIGHTER. 48-Pack	48	*18000+	3150	2805
				F40CW 540-PALLET	**1.29**	Cool White—MAINLIGHTER. Palletized—540 Lamps	540	*18000+	3150	2805
				F40CWX	**1.50**	DeLuxe Cool White—MAINLIGHTER	24	*18000+	2200	1850
				F40CWX 6PK PM	**1.65**	DeLuxe Cool White—MAINLIGHTER. 6-Pack **PRICE MARKED**	24	*18000+	2200	1850
				F40WWX	**1.50**	DeLuxe Warm White—Improved Color. MAINLIGHTER	24	*18000+	2150	1795
				F40WWX 6PK PM	**1.65**	DeLuxe Warm White—Improved Color. MAINLIGHTER. 6-Pack **PRICE MARKED**	24	*18000+	2150	1795
				F40D	**1.50**	Daylight—MAINLIGHTER	24	*18000+	2600	2315
				F40D 6PK	**1.65**	Daylight—MAINLIGHTER. 6-Pack	24	*18000+	2600	2315
				F40W	**1.50**	White—MAINLIGHTER	24	*18000+	3200	2850
				F40WW	**1.50**	Warm White—MAINLIGHTER	24	*18000+	3200	2850
				F40WW 6PK	**1.65**	Warm White—MAINLIGHTER. 6-Pack	24	*18000+	3200	2850
				F40WW 540-PALLET	**1.50**	Warm White—MAINLIGHTER. Palletized—540 Lamps	540	*18000+	3200	2850
				F40SGN	**1.95**	Sign White—MAINLIGHTER	24	*18000+	2440	2050
				F40/C50	**1.95**	Chroma 50—Chromaline™—5000°K. MAINLIGHTER	24	*18000+	2150	1805
				F40/C75	**1.95**	Chroma 75—Chromaline™—7500°K. MAINLIGHTER	24	*18000+	1990	1630
				F40N	**1.90**	Natural (formerly Soft White—Natural) MAINLIGHTER	24	*18000+	2100	1765
				F40N 6PK	**2.00**	Natural (formerly Soft White—Natural) MAINLIGHTER. 6-Pack	24	*18000+	2100	1765
				F40CG 6PK	**2.00**	Cool Green—MAINLIGHTER. 6-Pack	24	*18000+	2850	2395
				F40PL 6PK	**4.60**	Plant Light—MAINLIGHTER 6-Pack	24	*18000+	900	675
				F40B 6PK	**3.10**	Blue—MAINLIGHTER. 6-Pack	24	*18000+	1160	740
				F40G 6PK	**3.10**	Green—MAINLIGHTER. 6-Pack	24	*18000+	4500	2340
				F40GO 6PK	**3.10**	Gold—MAINLIGHTER. 6-Pack	24	*18000+	2400	1800
				F40PK 6PK	**3.10**	Pink—MAINLIGHTER. 6-Pack	24	*18000+	1160	440
				F40R 6PK	**3.10**	Red—MAINLIGHTER. 6-Pack	24	*18000	200	60
				F40BL 6PK	**3.45**	Black Light. 6-Pack	24	*18000+	...	...
				F40BLB 6PK	**18.25**	Black Light Blue—Integral Filter. 6-Pack	24	*18000+	...	...
				FR40CW	**3.70**	Cool White—Reflector. MAINLIGHTER. 135°window (308)	24	*18000+	2600	...

Figure 8-4. Typical page from a lamp data catalog.

ATLAS – K-12 ACRYLIC **AG-7244-4**

Average Luminance in Footlamberts

Angle From Nadir	End-wise	Cross-wise
45	1350	1520
55	1035	995
65	705	650
75	440	440
85	525	395

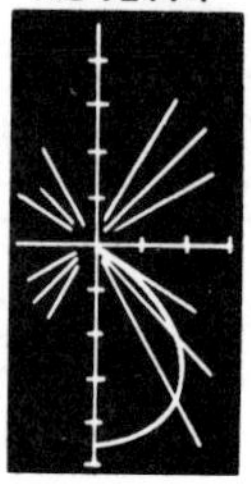

Distribution and Efficiency	Spacing Ratios and Reflectance
CIE % Luminaire 0.0 ↕ 100.0	Spacing not to Exceed 1.4 x Mounting Height
	Maintenance Factors Good: .75 Med.: .70 Poor: .65

Room Cavity Ratio	Effective Ceiling Cavity Reflectance, R_{cc} 80%			50%			10%		
	% Wall Reflectance, R_w								
	50%	30%	10%	50%	30%	10%	50%	30%	10%
	COEFFICIENTS OF UTILIZATION for 20% Floor Cavity Reflectance, R_{fc}								
1	75	73	71	71	69	67	66	64	63
2	68	64	61	64	61	59	60	58	56
3	61	57	53	58	55	51	55	52	50
4	55	50	46	53	49	45	50	47	44
5	50	44	40	48	43	40	45	42	39
6	45	40	36	43	39	35	41	37	35
7	41	35	32	39	34	31	37	34	31
8	37	31	27	35	31	27	34	30	27
9	33	28	24	32	27	24	30	26	23
10	30	25	21	29	24	21	28	24	21

Figure 8-5. Photometric data as supplied by manufacturer.

In our example (Fig. 8-1) we have previously determined the floor area to be 96 ft^2, the desired footcandles to be 100, 4 lamps per fixture, 3200 lumens per lamp, the coefficient of utilization to be 0.45, and the maintenance factor to be 0.70. By inserting these values in the equation, we have

$$NF = \frac{96 \times 100}{4 \times 3200 \times 0.45 \times 0.70} = 2.37 \text{ fixtures}$$

This resulting figure can be rounded off to two lighting fixtures to obtain approximately 100 footcandles of illumination for our example.

Locating the lighting fixtures depends on the general architecture, the size of bays, the type of lighting fixture under consideration, and other factors. In our example we have decided to space the fixtures as shown in the floor plan in Fig. 8-6, since the desk is located in the approximate center of the room.

It may be desirable to calculate the amount of illumination in a given area from the existing or proposed conditions. Our example called for 2.37 lighting fixtures, but two were used. We then know that we will have slightly less than the desired 100 footcandles. The amount of actual illumination then can be recomputed by the formula

$$FC = \frac{TL \times LPL \times CU \times ME}{A}$$

where

FC is footcandles.
TL is total lamps.
LPL is lumens per lamp.
CU is coefficient of utilization.
MF is maintenance factor.
A is area.

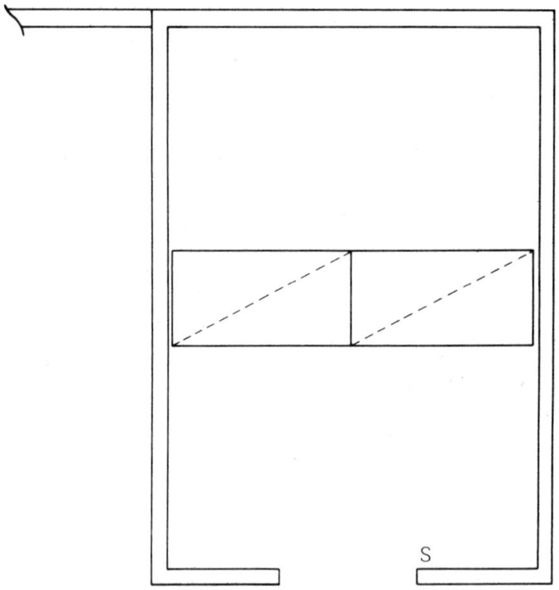

Figure 8-6. Layout of lighting fixtures for the office space mentioned in the text.

Thus,

$$FC = \frac{8 \times 3200 \times 0.45 \times 0.70}{96} = 84 \text{ footcandles}$$

8.2 Lighting Layouts

In most cases lighting fixtures are laid out to provide even distribution of illumination for an area when the fixtures are used for general illumination. Other fixtures may be located at various positions for special effects or for highlighting certain objects. Manufacturers normally supply spacing recommendations for each of their lighting fixtures and, for general illumination, these spacing recommendations should not be exceeded. These recommendation ratios are supplied in terms of maximum spacing to mounting height. In many cases, however, it is necessary to locate fixtures closer together than these maximums, in order to obtain required illumination levels.

Other than the spacing recommendations given previously (ratios in terms of maximum spacing to mounting height), no general rule can be given for locating the fixtures for general illumination. However, it is usually desirable to lay out the fixtures in a uniform pattern. This is accomplished by dividing the room or area into equal squares, the number of which should equal the number of lighting fixtures to be used. If this method is used, the lighting fixtures should be placed in the centers of these squares, as shown in Fig. 8-7, rather than at the corners, as shown in Fig. 8-8. The latter arrangement will, of course, provide a rather low illumination level at the walls, as compared with that at the center of the room.

Once the general lighting is laid out, supplemental lighting fixtures on various seeing tasks may be added, or perhaps decorative lighting may be desired.

Figure 8-9 shows a typical lighting layout for a department store. Conventional 2′ × 4′ fluorescent lighting fixtures (drawn to scale) provide the general lighting in the sales area. Note that most of these are installed in tandem (in pairs) in order to obtain the desired illumination level. This type of fixture is designated on the drawings as fixture type "A," and the word "all" under the identifying mark indicates that all fixtures in this area are type "A" unless otherwise noted.

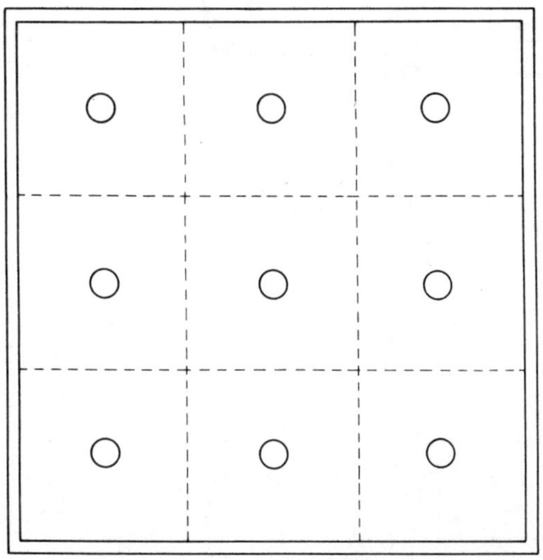

Figure 8-7. Correct method of spacing lighting fixtures.

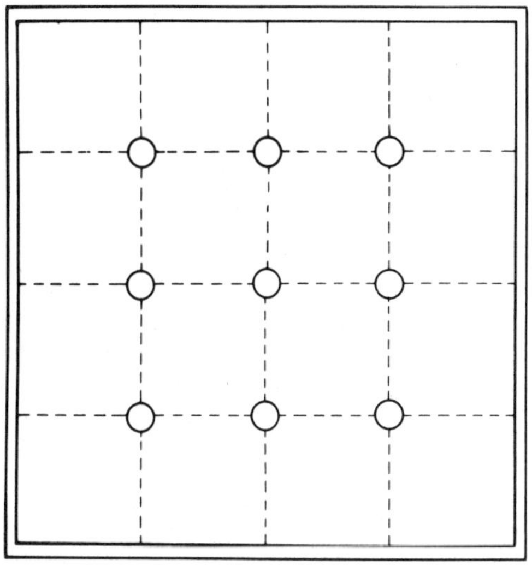

Figure 8-8. Incorrect method of spacing lighting fixtures.

The fixtures near the front entrance shown by the symbol ◎ are directional recessed spotlights to highlight the jewelry display in this area. The two different types used in this area are again designated by letters within a triangle—"H" and "J."

Other fixtures in the building are combinations of incandescent and fluorescent lighting fixtures of different types and sizes; all are identified by letters enclosed in a triangle and are then completely described in a lighting-fixture schedule appearing on another drawing sheet or else are described as to catalog number, manufacturer, etc., in the written specifications that accompany most construction documents. The type of mounting used for all lighting fixtures (surface, recessed, wall, etc.) is also usually indicated in a lighting-fixture schedule.

The mounting heights of wall-mounted fixtures are usually indicated in the symbol lists or legend, especially where most of the wall-mounted fixtures are to be mounted at one height. For example, an apartment project with, say, 50 units may have a wall-mounted fluorescent fixture over the lavatory mirror in each bathroom. A note adjacent to the symbol in the legend might read, ". . . wall outlet with fluorescent fixture mounted 4 inches above top of mirror to centerline of box unless otherwise indicated." If some of these fixtures need to be mounted at a different height, they could be so noted, as shown in the example in Fig. 8-10.

The lighting symbols used on the drawing in Fig. 8-9 consist mainly of a round circle to indicate an outlet or an incandescent lighting fixture or else a pictorial drawing (drawn to scale) to indicate a fluorescent fixture. However, for special lighting fixtures,

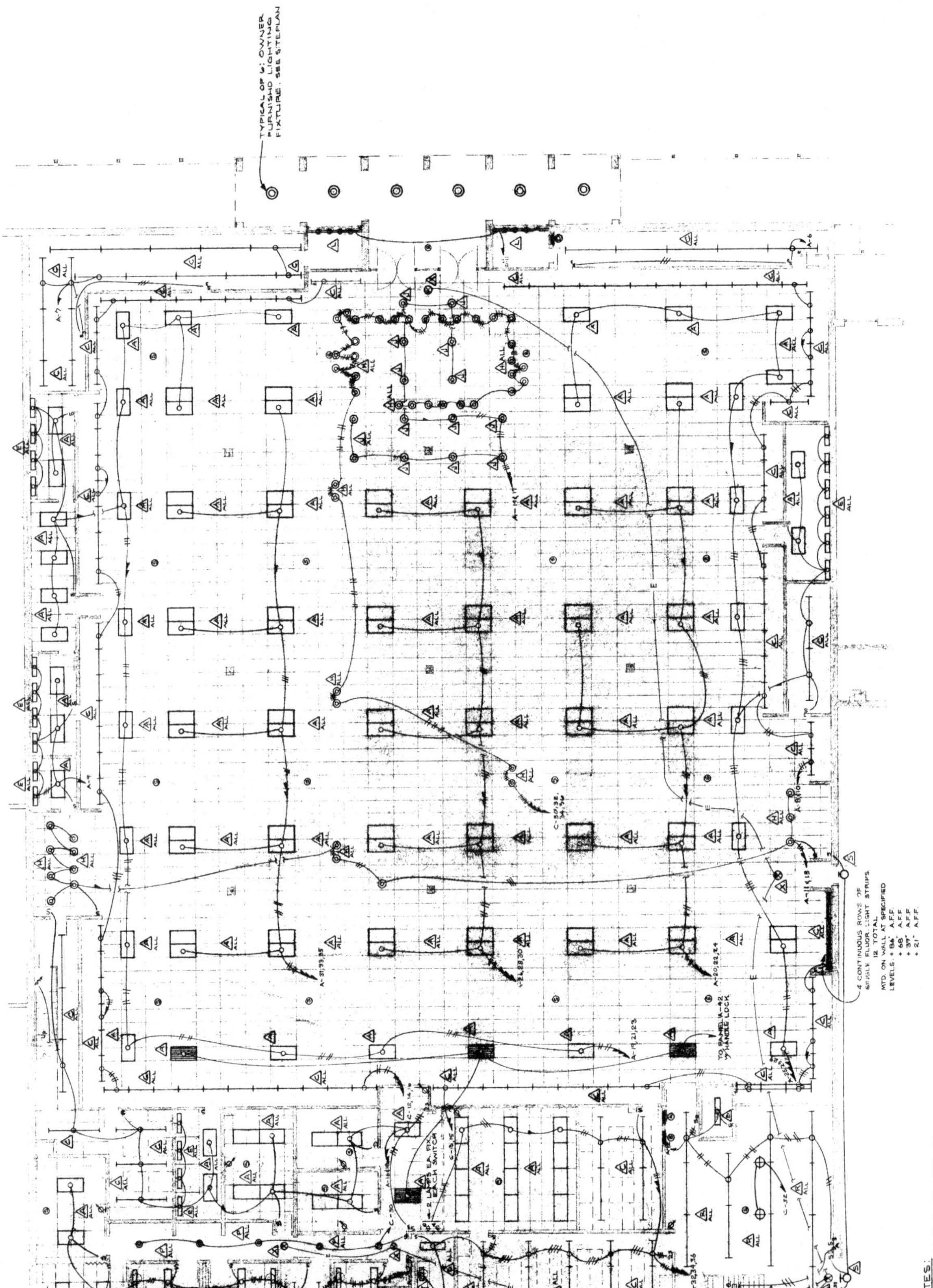

Figure 8-9. Typical lighting layout for a department store.

Figure 8-10. One method of noting height of lighting fixtures on working drawings.

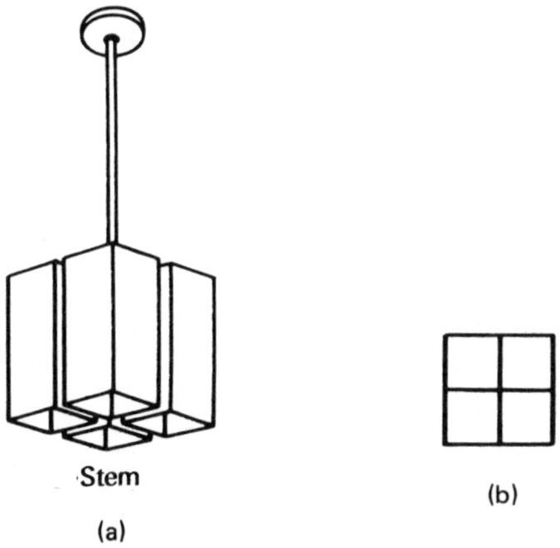

Figure 8-11. (a) Pictorial drawing of a lighting fixture consisting of four cubes. (b) Method of showing the lighting fixture in (a) on working drawings.

these symbols are sometimes modified to show the physical shape of a particular fixture. The lighting fixture shown in Fig. 8-11(a) consists of an arrangement of four 6" cubes and may be shown on the drawings as shown in Fig. 8-11(b)—drawn as closely as possible to scale.

Modification of the symbols used for fluorescent lighting is also common on electrical construction drawings. For example, cove lighting using bare-tube (no diffuser) fluorescent strips may be indicated on the drawings as shown, in Fig. 8-12.

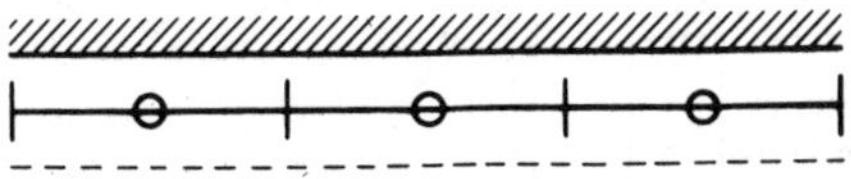

Figure 8-12. Method of showing cove lighting on working drawings.

8.3 Lighting Circuits

The wires or conductors that extend from the panel board (containing the fuses or circuit breakers) to the various lighting outlets are called branch circuits and are defined by the National Electrical Code as ". . .that portion of a wiring system extending beyond the final overcurrent device protecting the circuit . . ."

In general, the size of the branch-circuit conductors for lighting will be either number 14, 12, or 10 AWG sizes, with number 12 AWG size being the more common on most projects. A simple branch circuit requires two of such wires to provide a continuous path for the flow of electric current—operating at either 120 or 277 volts in most lighting applications.

On electrical drawings, the branch-circuit wiring is usually indicated by a single line drawn from outlet to outlet and then to the panel board, which provides overcurrent protection for the circuits. Sometimes the line from the last outlet on the circuit to the panel board is interrupted and "home-run" arrowheads are used near the last outlet to indicate that the circuit is to be run to the panel board.

The lines indicating lighting branch circuits can be solid lines to show that the conductors are to be run in the ceiling or wall, broken lines to show that the conductors are to be run in the floor or ceiling below, or evenly broken lines (dashes) to show that the wiring is to be run exposed. Figure 8-13 shows examples of these three types of branch-circuit lines.

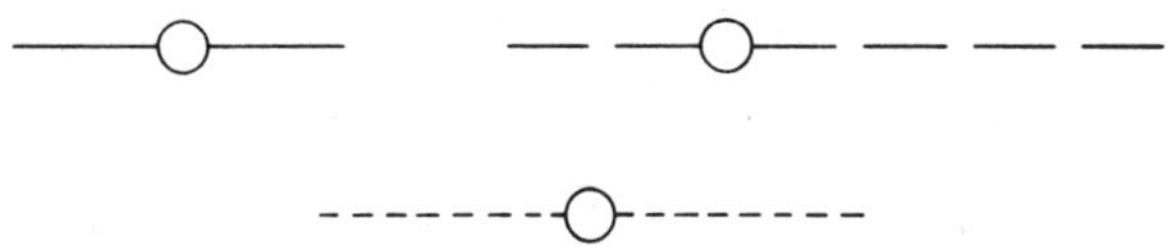

Figure 8-13. Examples of branch-circuit wiring as it should appear on working drawings.

The slash marks cutting the circuit lines indicate the number of conductors in the circuit, and the numerals above these slash marks indicate the size of the conductors. The number of "home-run" arrowheads indicates the number of circuits in the run. For example, the circuit in Fig. 8-14 tells us that the run consists of three number 12 AWG conductors installed in the ceiling, and the two arrowheads indicate that two circuits are involved—both terminating in panel board "D."

All modern lighting systems must be controlled to provide maximum convenience in buildings, and again standard symbols are used to show the location and type of control, while lines, slash marks, etc.,

Figure 8-14. A circuit homerun indicating three number 12 AWG conductors installed in the ceiling, the two circuits terminating in panel board "D."

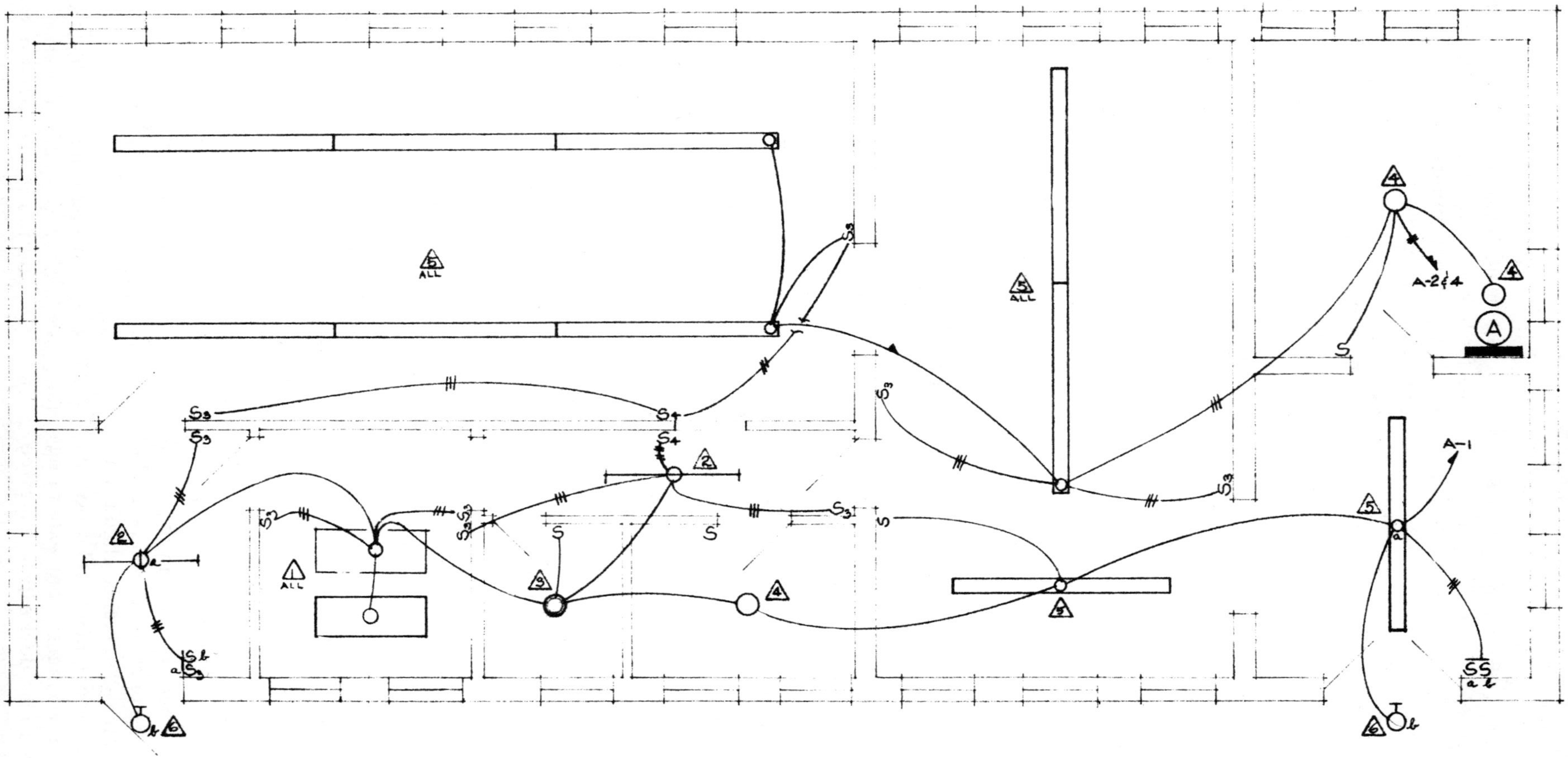

FLOOR PLAN · LIGHTING
SCALE : 1/4" = 1'- 0"

Figure 8-15. Practical applications of lighting controls used on electrical construction drawings.

show the type of circuits and the points of connection. Figure 8-15 illustrates some practical applications of lighting control used on electrical construction drawings. Notice that the single-pole switches (S) are used to control a light or a group of lights from one point, while the three- and four-way switches (S_3 and S_4, respectively) are used in combination to control a light or series of lights from two or more points.

A 2-pole switch (S_2) is normally used to control a series of lights on two separate circuits with only one motion of the switch handle. The switch and pilot light combination (S_p) is used where it is practical so that one can tell if a particular hidden light is energized from the switch location. A light in a closet, for example, more often is controlled by a wall switch on the outside of the closet door. If the light is accidentally left on when the door is closed, the lighted switch would reveal this. A light in a residential attic would be another good example where a switch/pilot light should be used.

Door switches in residential construction are also quite common for the control of closet lights. When the closet door is opened, the door-switch button is released and energizes the circuit to the closet light. When the door is closed, the light is automatically turned off. Review Chapter 6 for the symbol used for this and other types of light switches.

Other common lighting controls include magnetic relays to control several lighting circuits at one time and from one switching point, dimmers to change the intensity of illumination levels, time clock switches to automatically turn lights on or off at a predetermined time, and photoelectric switches to turn lights (usually outside lights) on at dusk and off at dawn.

8.4 Lighting in Electronics

Among the various lights used in electronic circuits is an annunciator used as a visual signalling device. The symbol commonly used is shown in Fig. 8-16; a shutter type annunciator is shown in Fig. 8-17. Indicator lamps are shown by the symbol in Fig. 8-18, and a letter indicator should be drawn adjacent to the symbol to show the color of the lamp. The following letters should be used to represent their respective colors:

A amber
B blue
C clear
G green
NE neon
O orange
OP opalescent
P purple
R red
W white
Y yellow

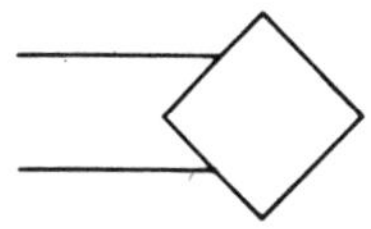

Figure 8-16. Symbol for annunciator.

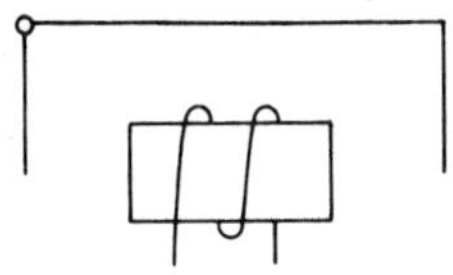

Figure 8-17. Shutter type annunciator.

Figure 8-18. Symbol for indicator lamp.

If confusion is likely to occur due to many other circular symbols used in a particular drawing—such as symbols for meters, tubes, etc.—the symbol in Fig. 8-19 can be used to represent an indicating lamp.

Lamp sockets are not usually identified on electronic drawings except in the cases of test lamps as shown in Fig. 8-20. Here the lamp is connected to the signal line in the normal position, but is directly operated by means of momentary contacts for test purposes.

Figure 8-19. Another symbol for an indicating lamp.

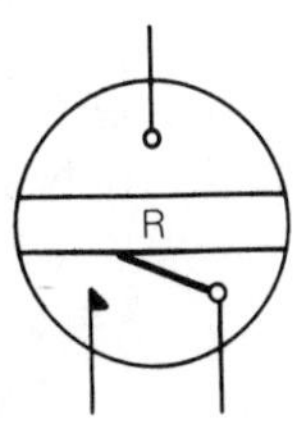

Figure 8-20. Symbol for test lamp.

Glow lamps of the neon variety are considered to be sturdier than incandescent lamps and are, therefore, preferred for most electronic instruments. Further advantages of the neon-type lamps are less consumption of power and the fact that they can be directly operated from ac or dc supplies, provided that the voltage is always above the firing potential. However, these lamps are not recommended when color is significant for indication, because of the poor color quality of the neon glow.

Fluorescent lamps are also used to some extent in electronic circuits. The symbol for a cold-cathode lamp is shown in Fig. 8-21, while the symbol in Fig. 8-22 represents a heated cathode or one with filaments. Light-emitting diodes are also used for light-emitting and infrared electronic displays, etc. These diodes will emit light in the infrared and visible ranges, depending upon the exact types of materials, treatment and dopants used during their construction. They are used in many types of indicators and alphanumerical displays because of their compatibility with solid-state devices and long life.

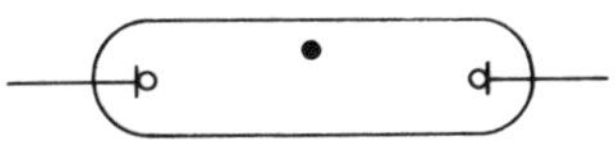

Figure 8-21. Symbol for cold-cathode lamp.

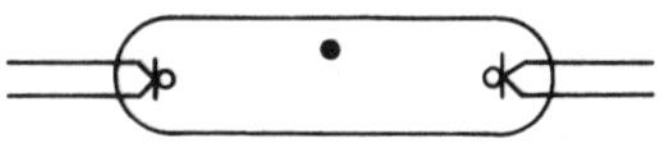

Figure 8-22. Symbol for heated-cathode lamp.

Light-sensitive photocells are used mainly in electronic circuits to translate light variations into electric signals. Therefore, they find application in industrial control systems, protection devices, street lighting, etc. The two basic types of photocells are the photovoltaic and the photoconductive cells. As the names imply, the first type (photovoltaic) produces a voltage and a resultant current flow in a closed circuit upon the application of incident light to the cell, whereas the latter type undergoes a resistive change upon the application of light. They come in various sizes and in a variety of types, which include vacuum or gas-filled tubes, solid-state diodes, and transistor units.

SUMMARY

The basic requirements for any lighting design for building construction are to determine the proper level of illumination required for a given area and then to use the best means of providing it.

In most cases, lighting fixtures are laid out to provide even light distribution over a given area.

Lighting fixtures and their controls are shown on electrical construction drawings by means of standard symbols; these symbols are connected by means of solid or broken lines (depending upon the case).

ASSIGNMENT

Secure a piece of 11″ × 17″ tracing paper to your drawing board and draw border lines and a title block as described previously.

The floor plan of a commercial store building is shown in Fig. 8-23 and is illustrated in this book at a scale of 1/8″ = 1′0″. Use your architect's scale to get the correct dimensions and then draw this building to a scale of 1/4″ = 1′0″ on the sheet of tracing paper.

The floor plan shows a partial layout of 2′ × 4′ fluorescent lighting fixtures; finish laying out the fixtures—using the same pattern. There should be a total of 24 such fixtures in the building.

Draw the curcuit lines on the drawing also, using a French curve as a guide. Six fixtures should be on each circuit, requiring a total of four circuits in all. Use "home-run" arrowheads to show the run from the last fixture on the circuit to the panel board "A." Figure 8-24 shows a slightly different layout similar to the assignment; this may be used as a guide.

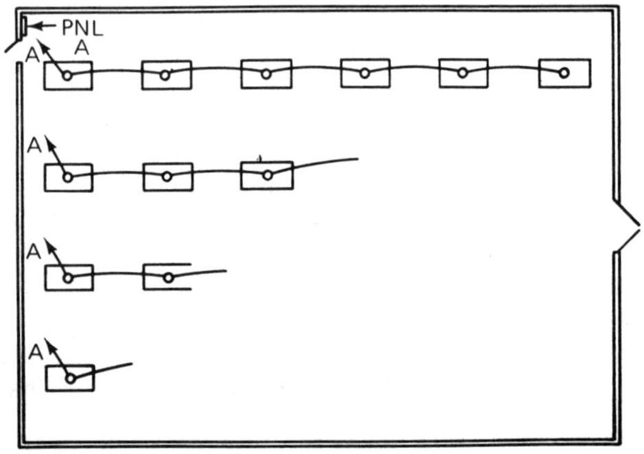

Figure 8-23. Drawing for chapter assignment.

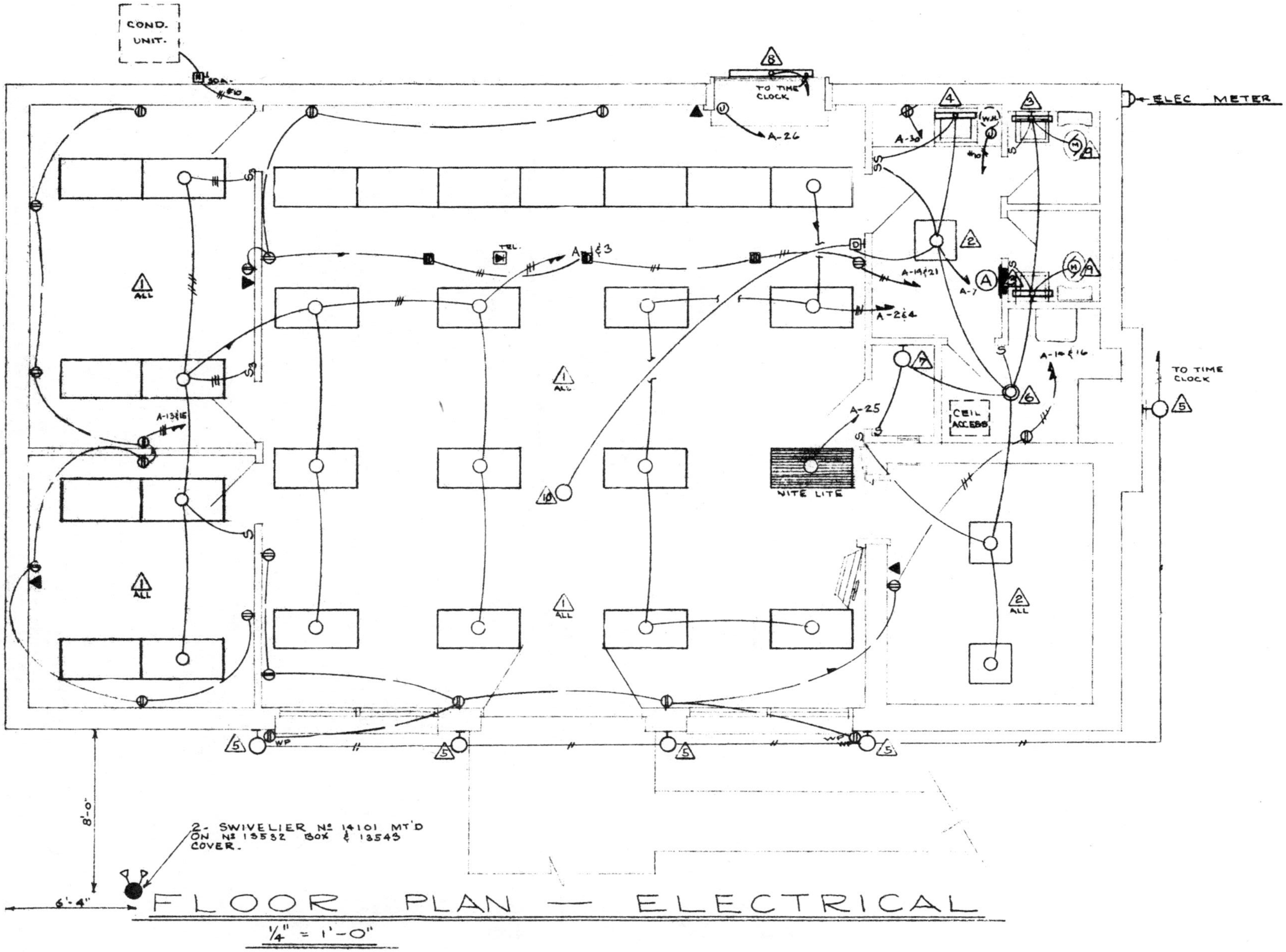

Figure 8-24. Example of a lighting layout showing placement of lighting fixtures, circuits, outlets, homeruns, etc.

Chapter 9

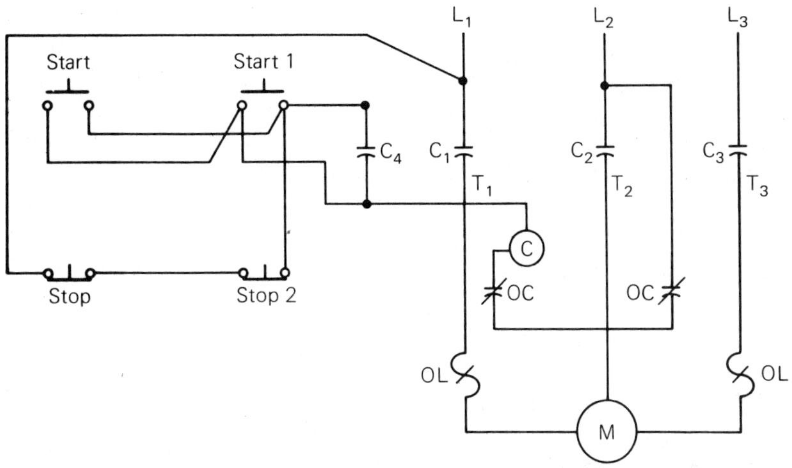

POWER PLANS

The point at which the electrical equipment is connected to an electrical wiring system is commonly called an electrical outlet. While there are many classifications of electrical outlets, this chapter will deal with those outlets known as *power outlets*—receptacle outlets, motor outlets, special-purpose outlets for appliances, etc.,—as opposed to *lighting outlets*, covered in Chapter 8.

This chapter will further cover the correct layout of power outlets for electrical working drawings and also the method of connecting them with circuit lines. Special controls—such as motor controls—will be included also.

9.1 Layout of Convenience Outlets

The ordinary wall plug (duplex receptacle or convenience outlet, as they are most often called) used in residential, commercial, and industrial buildings is the most common of all power outlets. This type of outlet is shown on electrical drawings by a symbol consisting of a circle with two lines drawn through it, and the connecting wiring is indicated by a single line drawn from the panel board (or by "home-run" arrowheads indicating that the circuit goes to the panel board) to the outlet, or from outlet to outlet when there is more than one outlet on the circuit.

The lines indicating branch circuits for convenience outlets can be solid to show that the conductors are to be run in the ceiling or wall, slightly broken to show that the conductors are to be run in the floor or ceiling below, or evenly broken to show that the wiring is to be run exposed on the building surfaces.

The slash marks, as previously described for lighting branch circuits, indicate the number of wires or conductors in the circuit, and a numeral placed near these slash marks gives the wire size in AWG (American Wire Gauge). When no slash marks or numerals are provided, this normally means that the circuit is to contain two number 12 AWG wires.

For residential wiring, the National Electrical Code specifically states the minimum requirement for the location of receptacles:

> In every kitchen, family room, dining room . . . receptacle outlets shall be installed so that no point along the floor line in any wall space is more than 6 feet, measured horizontally, from an outlet in that space, including any wall space 2 feet wide or greater and the wall space occupied by sliding panels in exterior walls . . . Receptacle outlets shall, insofar as practical, be spaced equal distances apart. Receptacle outlets in floors shall not be counted as part of the required number of receptacle outlets unless located close to the wall. . . .

With this NEC article in mind, the electrical draftsman would then lay out the various receptacles by measuring along the floor line of the wall and around corners of the room but not across doorways, fireplaces, passageways, or other spaces where a flexible cord extended across the space would be unsuitable.

Figure 9-1 shows the layout of electrical outlets for a typical residential bedroom. Note that at no point along the wall line is any receptacle more than 12 feet away from another or more than 6 feet from any door. The outlets are connected by a slightly broken line, indicating that the wiring is to be run under the floor; the two slash marks with the numeral "12" adjacent to them shows that the circuit contains two number 12 AWG conductors. In residential electrical construction, it is commonplace to connect both lighting and convenience outlets on a single circuit because of the relatively light loads on each.

Sometimes it is desirable to split-wire a duplex receptacle; that is, to have half the receptacle (one of the two plugs) "hot" all of the time and the other half switched by a conventional wall switch near the entrance of the room. Figure 9-2 shows a floor plan of a room with this arrangement. Note that the usual duplex receptacle symbol is used with half of it shaded in; the wiring for this type of arrangement also requires that three wires—in place of two—be used in the circuit wiring.

In residential hallways, vestibules, etc., it is recommended that at least one convenience outlet be located in the area regardless of its size. In larger hallways, one outlet should be provided for every 15 linear feet of hallway so that a vacuum cleaner, a floor polisher, and table lamps may be used in the area without extension cords.

The National Electrical Code also requires that every residence have two or more 20-ampere branch circuits for small appliance outlets in the kitchen and dining area of the home as well as one additional 20-ampere circuit for the laundry area. Fifteen hundred watts is the figure used when calculating residential appliance circuits and number 12 AWG wire must be used on these circuits to carry the required 20-ampere load.

A separate laundry circuit is provided to lessen the danger of overloads when the automatic washer is operating and also to accommodate the amount of current drawn from irons.

The floor plan of a typical residential kitchen is shown in Fig. 9-3, where six duplex receptacles are provided. To comply with the NEC, two 20-ampere circuits would have been sufficient to feed these receptacles, but to provide greater capacity, three 20-

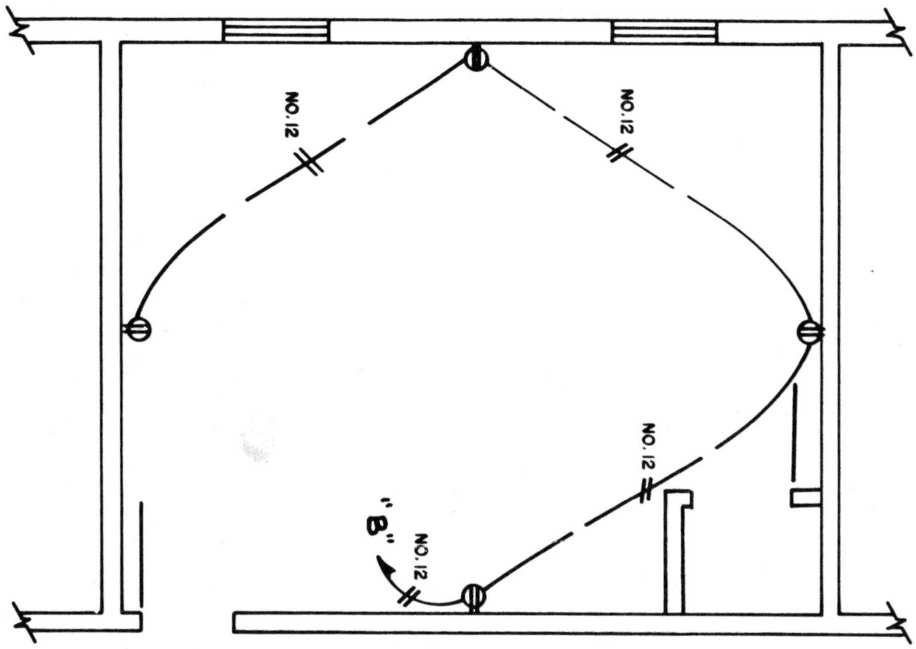

Figure 9-1. Layout of electrical outlets for a typical residential bedroom.

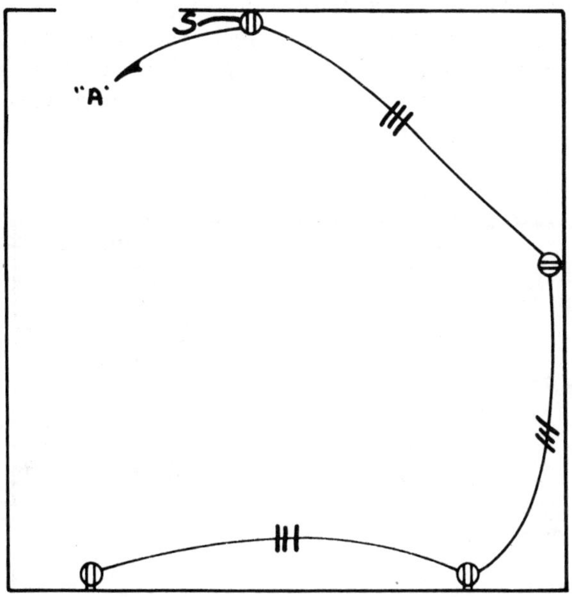

Figure 9-2. Floor plan of a room with duplex receptacles split-wired.

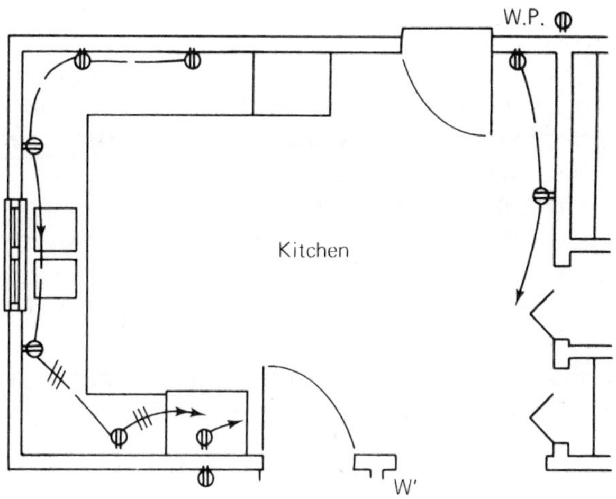

Figure 9-3. Floor plan of a typical residential kitchen.

ampere branch circuits were used. Note that one receptacle—for the portable dishwasher—is on a circuit by itself because of the large amount of current drawn by this appliance. This single receptacle must be rated at 20 amperes, since the NEC states, " . . . a single receptacle installed on an individual branch circuit shall have a rating of not less than the rating of the branch circuit."

In commercial and industrial buildings, convenience outlets are usually placed on separate circuits and are not made a part of the general lighting circuits, as is sometimes done in residential wiring. Convenience outlets used in commercial and industrial buildings must have a minimum of 180 watts allowed per outlet. Therefore, since a duplex receptacle has two plugs, a minimum of 360 watts shall be al-

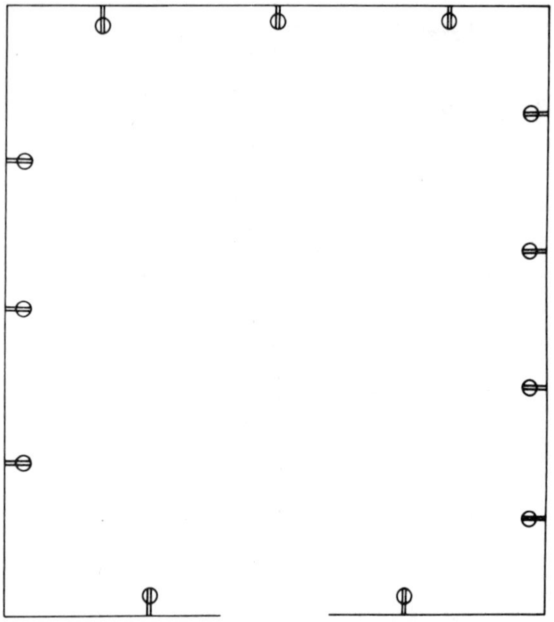

Figure 9-4. Floor plan of a room with several duplex receptacles located at various intervals.

lowed for each duplex receptacle. However, in actual practice, most electrical designers and draftsmen allow 400 watts for each duplex receptacle in order to have a slight margin to ensure that sufficient capacity is provided; it also makes it easier to keep the figures in one's head as the outlets are laid out.

The floor plan in Fig. 9-4 shows a room with several duplex receptacles located at various intervals within the room. Assuming that 20-ampere branch circuits are used, let us see how many circuits will be required to feed all of the receptacles.

First, find the number of receptacles that can be connected to a 20-ampere branch circuit.

$$\frac{120\text{ (volts)} \times 20\text{ (amperes)}}{400\text{ (watts per duplex receptacle)}}$$

400 (watts per duplex receptacle) = 6 duplex receptacles per circuit

Since there are twelve duplex receptacles in the area, two branch circuits will be required and will appear on the working drawing as shown in Fig. 9-5.

The electrical draftsman will also be required to show multi-outlet assemblies on electrical working drawings. Multi-outlet assemblies are a type of surface or flush raceway designed to house wiring and plug receptacles. They are usually assembled in the factory, but are sometimes assembled in the field or on the project. When such assemblies are used as a fixed piece of equipment, each five feet or fraction thereof of each separate and continuous length shall be considered as one outlet of not less than 180 watts capacity, except in locations where a number of appliances are likely to be used simultaneously. In this case, every one foot of length shall be considered as an outlet of not less than 180 watts.

The plan view in Fig. 9-6 gives an example of the use of multi-outlet assemblies on electrical working drawings. In this figure, the assembly was used to provide power for several television sets and small appliances that were to be demonstrated in this area. However, multi-outlet assemblies are not limited to such locations, but are also practical anywhere in residential or commercial applications where convenience receptacles are desired fairly close together.

Circuiting of the convenience outlets (or for any outlets, for that matter) should be done in a manner so that the homeruns are as short as possible; all connecting wiring between outlets should also be made as short as possible. On any given electrical installation, however, there will be many different combinations or groupings of outlets into circuits—all of which will be technically correct. The problem is to quickly determine which of the possible combinations will conserve the most wire and in turn keep the cost and voltage drop to a minimum.

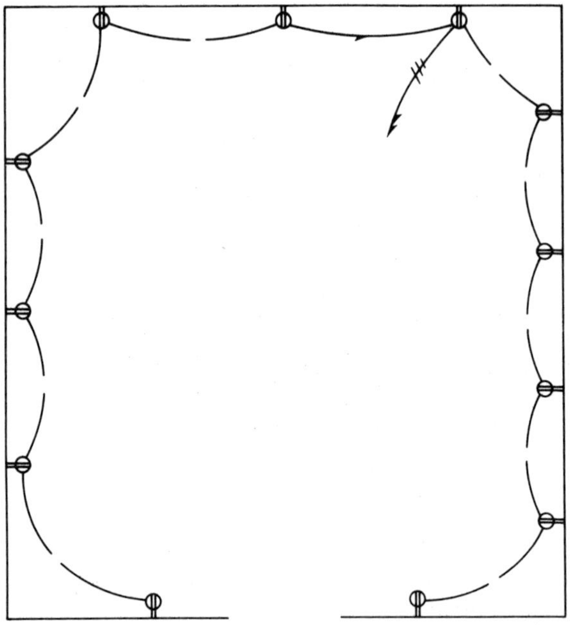

Figure 9-5. Layout of the twelve receptacles in Fig. 9-4 after circuit wiring has been drawn.

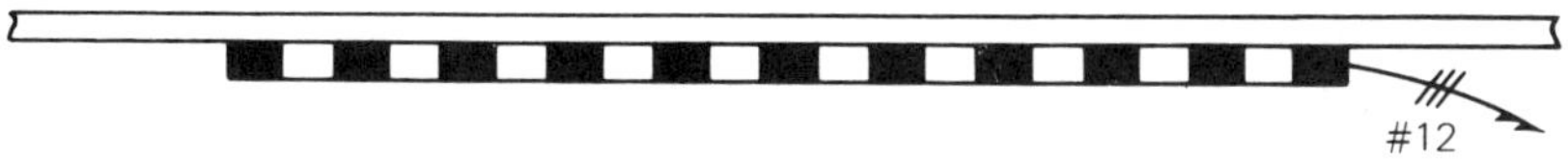

Figure 9-6. Example of multi-outlet assemblies used on electrical working drawings.

For most electrical projects—except jobs such as warehouses and similar projects where very little wiring will be required—separate floor plans should be used for the lighting and power. This enables the drawing to be more easily read by the workmen on the job as well as making it easier for the draftsman laying out the system. Notice the power layout in Fig. 9-7; this drawing is typical of a good layout that is easily read and understood.

In general, the duplex receptacles on the drawing in Fig. 9-7 are laid out five to a circuit, and circuits are combined into three or four separate circuits on the homeruns in order to conserve conduit and wire. For example, a common neutral wire may be used on three-phase circuits for three separate circuits, thus saving three additional runs to the panel board.

Receptacles are provided on the exterior of the building with weatherproof cover plates (marked WP). Duplex receptacles are also provided in the corridors for connecting vacuum cleaners, buffing machines, etc. All receptacles are mounted 15 inches above the finished floor, except where indicated otherwise—as in some cases where they are marked "48 inches up." Most of the circuits are drawn with the slightly broken line indicating that the circuits are to be run in the floor and side walls.

Some of the duplex receptacles over the countertops in the lavatory areas are marked with a standard receptacle symbol with a slash through it. This outlet is to be mounted 4 inches above the countertop, as is so indicated in either the legend (symbol list) or in the written specifications.

Most electrical designers and draftsmen lay out convenience outlets for both commercial and residential projects in the same manner; that is, all the outlets for obvious or known needs are laid out first, and then additional outlets are provided so that no point along any usable floor line is more than 6 ft from any outlet. Although this is not a NEC requirement for commercial buildings, it is still good practice to do so.

Common sense will normally indicate the amount and location of outlets in most commercial applications. Common sense and good judgment will come with experience, but in the meantime, the new designer or draftsman should study existing installations, ask questions concerning them, and then attempt to practice the procedures learned.

9.2 Layout of Special-purpose Outlets

Special-purpose outlets, as we shall use the term, are outlets for a specific piece of electrical equipment of known wattage and requiring electrical conductors of size number 10 AWG or smaller. Those circuits requiring wire size larger than number 10 AWG will be classified as *electrical feeders* and will be covered in Chapter 10.

Residential water heater circuits will serve as a typical example of a special-purpose outlet. Sizes of water heaters for home use will range in wattage from 1200 to 4500 watts; the smaller sizes are rated at 120 volts, and the larger ones are rated at 240 volts. In most cases, water heater circuits are connected directly to a junction box provided on the unit; one is shown in a floor plan view in Fig. 9-8.

Other special-purpose outlets would include an outlet for an electric water cooler (Fig. 9-9), a 240-volt outlet for an electric clothes dryer (Fig. 9-10), a 240-volt outlet for a through-wall heating/cooling unit (Fig. 9-11), and similar items.

When special-purpose outlets are shown on electrical working drawings, several methods are used. One is to use a universal symbol, , and then to indicate its function by subscript letters; that is, 20-ADW for dishwasher outlet, 30-ADC for clothes dryer, etc. Sometimes these outlets are used with code letters like a, b, c, etc., and then a schedule is provided on the drawings to indicate exactly what type of outlet is required, the number of conductors, and the wire size. See Fig. 9-12.

Another method is to indicate a 120-volt single receptacle by the symbol and then place the ampere rating near the symbol. For 240-volt receptacles, the symbol is used with the ampere rating adjacent to the symbol. If a receptacle is not required for the piece of equipment, a direct connection is made, and the symbol is used for a junction box or direct connection. Should a schedule be provided, the draftsman should indicate in the schedule what type of connection is required.

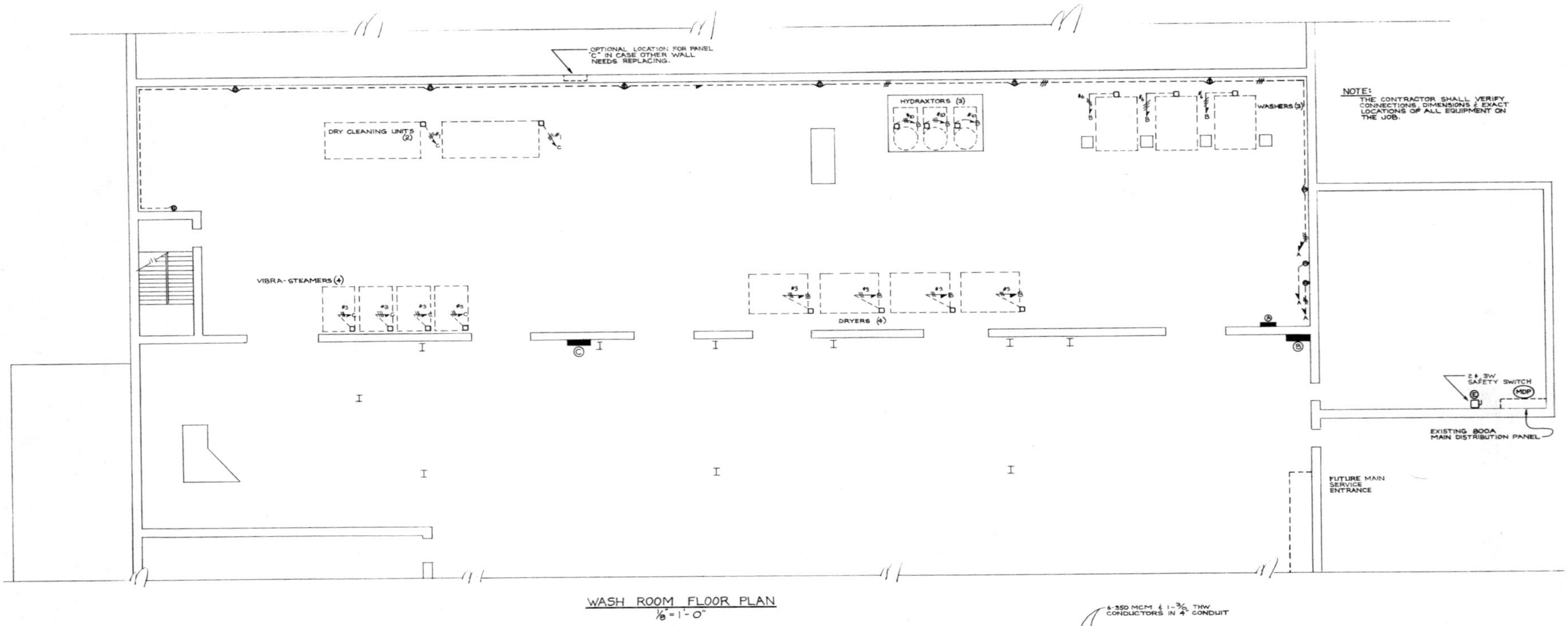

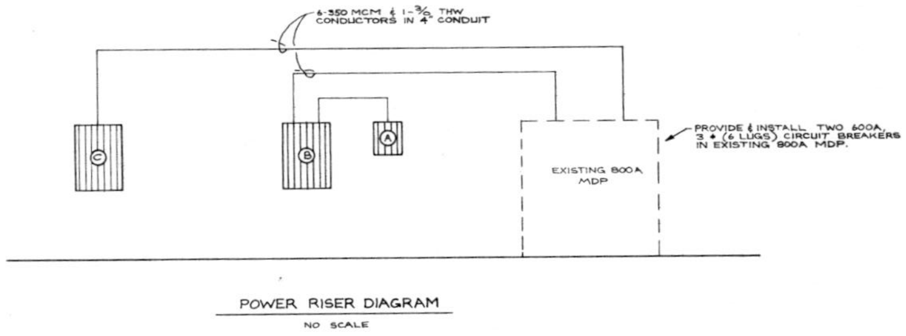

PANELBOARD SCHEDULE

PANEL Nº.	TYPE CABINET	PANEL MAINS AMPS	VOLTS	PHASE	BRANCHES 1P	2P	3P	PROT	FRAME	ITEM FED OR REMARKS
B	SURFACE	600 A	120/240 V	3φ 4W			3	25A	30A	HYDRAXTORS
SQ "D" TYPE I LINE W/ MAIN LUGS ONLY OR APPROVED EQUAL							3	60A	100A	WASHERS
							4	90A	100A	DRYERS
C	SURFACE	600 A	120/240 V	3φ, 4W			4	90A	100A	VIBRA-STEAMERS
SQ "D" TYPE I LINE W/ MAIN LUGS ONLY OR APPROVED EQUAL							2	150A	200A	DRY CLEANING UNITS

Figure 9-7. A good power layout that is easily read and understood.

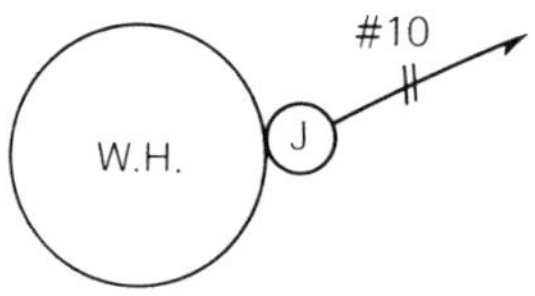

Figure 9-8. Method of showing water heater connection and circuit on a floor plan.

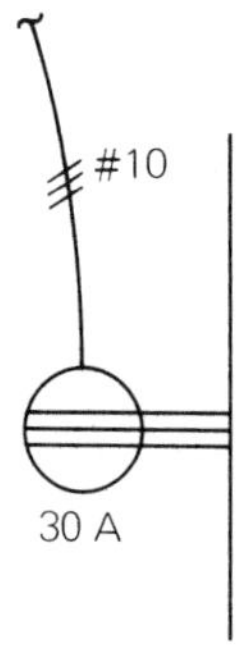

Figure 9-10. A 240-volt outlet for an electric clothes dryer.

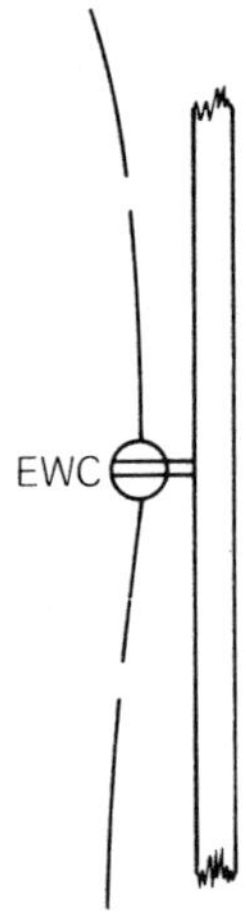

Figure 9-9. Method of indicating an outlet for an electric water cooler.

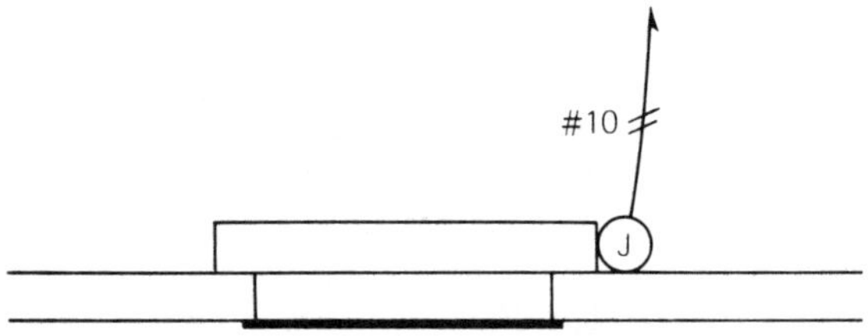

Figure 9-11. A 240-volt outlet for a through-wall heating/cooling unit.

SCHEDULE - RECEPTACLE TYPES —

Symbol Φ	Amp. Rating	Wire and Poles	Voltage Rating	Nema Type	Configuration	Catalog No *	Remarks
A Φ	15	3WG 2P	125	5-15R		5262	DUPLEX
B	20	3WG 2P	125	5-20R		5361 5362	SINGLE DUPLEX
C	30	3WG 2P	125	5-30R		9308	SINGLE
D	50	3WG 2P	125			9368	SINGLE
E	20	3WG 2P	250	6-20R		5461 5462	SINGLE DUPLEX
F	30	3WG 2P	250	6-30R		9330	SINGLE
G	50	3WG 2P	250	6-50R		9367	SINGLE
H	20	4WG 3P	250			7250	SINGLE 3 PHASE
J	30	4WG 3P	250			8340	SINGLE 3 PHASE
K	50	4WG 3P	250			8450	SINGLE 3 PHASE
R	50	3WG 2P	125/250	14-50R		9450	RANGE

• HUBBEL CATALOG NOS.- FOR EXAMPLE

NOTES

1) ALL RECEPTACLES SHALL BE GROUNDING TYPE.

2) SPECIAL RECEPTACLES AS CLOCK OUTLETS, WEATHERPROOF RECEPTACLES, ETC. SHALL BE AS DESCRIBED IN THE SPECIFICATION.

3) VERIFY LOCATION OF ALL EQUIPMENT OUTLETS, INCLUDING HEIGHT OF RECEPTACLES AND SWITCHES BEFORE ROUGHING IN.

Figure 9-12. A schedule of special-purpose outlets.

9.3 Motor Circuits

Electric motors range in size from small, fractional-horsepower, 120-volt motors to the very large, high-voltage, synchronous motors. They are designed for light or heavy duty and also for general-purpose, definite-purpose, or special-purpose types of application. Their characteristics include voltage, frequency, alternating or direct current, type of wiring, and phase in the case of alternating current motors.

Starting and speed-control equipment varies from simple, thermal tumbler switches to large switchgear-type controllers, the type of which depends basically upon the motor to be controlled, the type of the control circuit, and the location of the controller. Usually the controller is mounted adjacent to the motor or in the immediate vicinity. Sometimes, however, the controller may be located some distance away, such as in a motor control center where individual motor controllers or a factory-built self-contained unit with all the motor controllers are installed.

Motor controllers are available for either manual or automatic operation either at the controller or at a point remote from the controller and the motor. Any number of remote push-button stations may be used with automatic controllers.

When a disconnect means is required for the motor, it may be a separate externally operated switch, or it may be combined in the same enclosure as the controller. In some instances the motor circuit disconnecting means may serve as the motor disconnect.

The garbage disposal shown by symbol and notation on the floor plan in Fig. 9-13 is rated at 7.5 amperes at 120 volts. A separate 15-ampere, 120-volt branch circuit is shown going to panel A from the junction box provided for connection purposes on the disposal unit. Most garbage disposal units are powered by split-phase, 120-volt motors of between ⅓ and ½ horsepower rating; according to the National Electrical Code, running-overcurrent protection is required and must not exceed 125 percent of the full-load current rating of the motor.

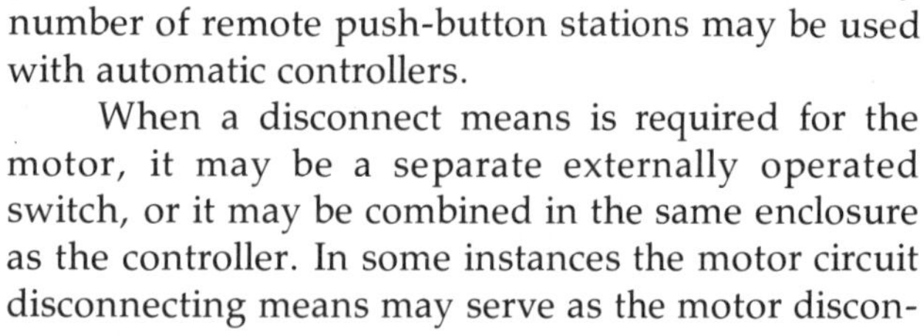

Figure 9-13. Method of indicating a connection and circuit for a garbage disposal on an electrical floor plan.

The garbage disposal unit shown in Fig. 9-13 was provided with a running-overcurrent protection by using a thermal protector ("heater") built into the unit. This thermal protector is of the manual reset type. In other units, however, automatic resetting of the thermal protector is featured.

Where running-overcurrent protection is an integral part of the disposal unit, the electrical draftsman merely shows a circuit feeding the unit, and a regular toggle switch if needed for control. However, where running-overcurrent protection is not an integral part of the disposal unit, the designer must specify one, as shown on the drawings in Fig. 9-14.

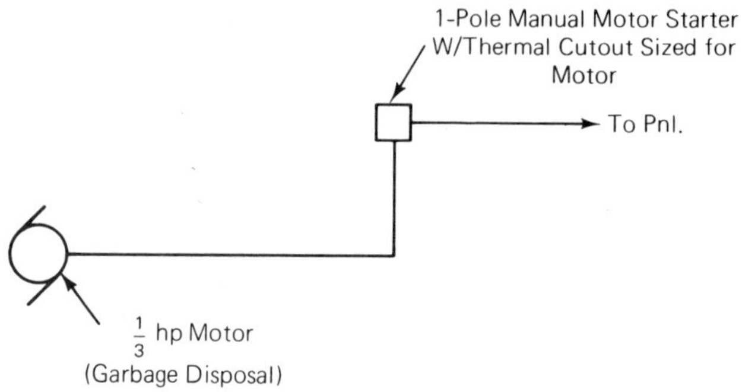

Figure 9-14. Method of specifying running-overcurrent protection when such is not an integral part of the disposal unit.

Exhaust fans are another type of motor-driven device that uses electric current for power. They may be equipped with an integral pull chain switch for starting and stopping, or a separate wall switch may be used for control. When an automatic system of humidity control is desired, these fans may be controlled with a humidistat.

Single- or multispeed exhaust fans used for ventilating residential kitchens, bathrooms, utility rooms, or laundry rooms have a very small power demand and are usually connected to lighting circuits. Larger fans for commercial kitchens of industrial applications, however, require motors of several horsepower and usually are controlled by an elaborate motor control system.

A wiring diagram for a magnetic motor started with two push-button stations is shown in Fig. 9-15. Electrical draftsmen working for industrial firms will be required to make numerous drawings of this type.

All symbols, wiring diagrams, and the like for power wiring are made with straightedges, French curves, and templates, the use of which has been discussed previously. The exercises at the end of this chapter will give the reader a sampling of actual applications of this type of drafting.

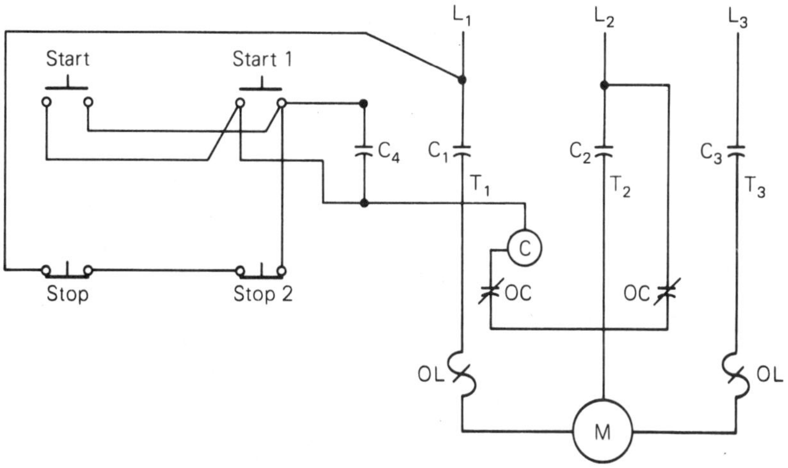

Figure 9-15. A wiring diagram for a magnetic motor starter controlled by two push-button stations.

9.4 Electronic Power Supplies

An electronic schematic diagram shows the method in which electricity flows through the circuit and the various components within the circuits that either use or change the electrical values. In all such diagrams, a proper sequence should be followed which shows where the electrical energy is provided (battery, transformer, etc.), the route in which it travels, and where it is utilized.

Sometimes these schematic diagrams become highly complex, but certain revisions can be made to simplify them somewhat. For example, substitutions can be made, certain lines can be omitted, portions of the circuit can be separated for simplicity, or the circuits can be divided into subcircuits.

In general, a line or a schematic diagram may be eliminated if it:

1. Connects to or feeds several common points.
2. Serves the same function for a number of separated symbols.
3. Can be replaced by an equivalent symbol.
4. Can be replaced by a suitable reference designation.

Figure 9-16 shows a relatively complex schematic diagram; a simplified version of the same diagram is shown in Fig. 9-17. Both drawings serve the same purpose, but the one in Fig. 9-17 is easier to read and also requires less drawing time.

ASSIGNMENT

1. Prepare a sheet of 11″ × 17″ drawing paper with border lines and title block. Then transfer the

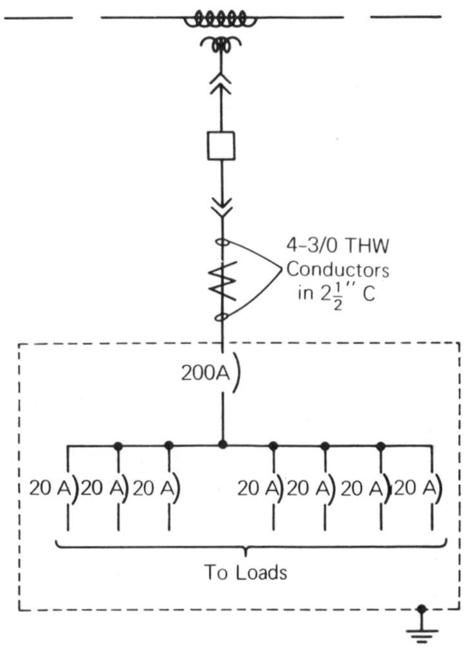

Figure 9-16. A relatively complex schematic diagram.

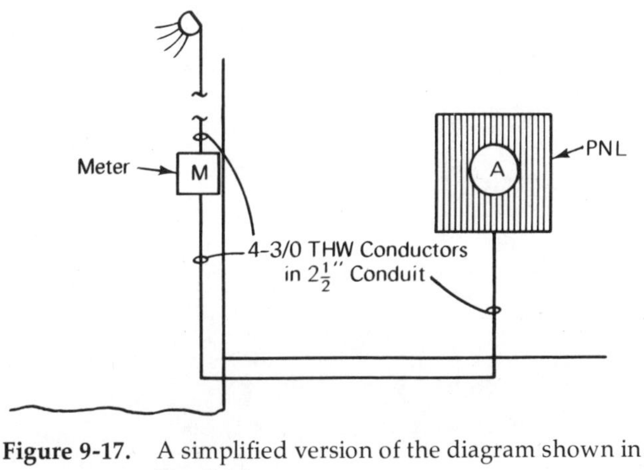

Figure 9-17. A simplified version of the diagram shown in Fig. 9-16.

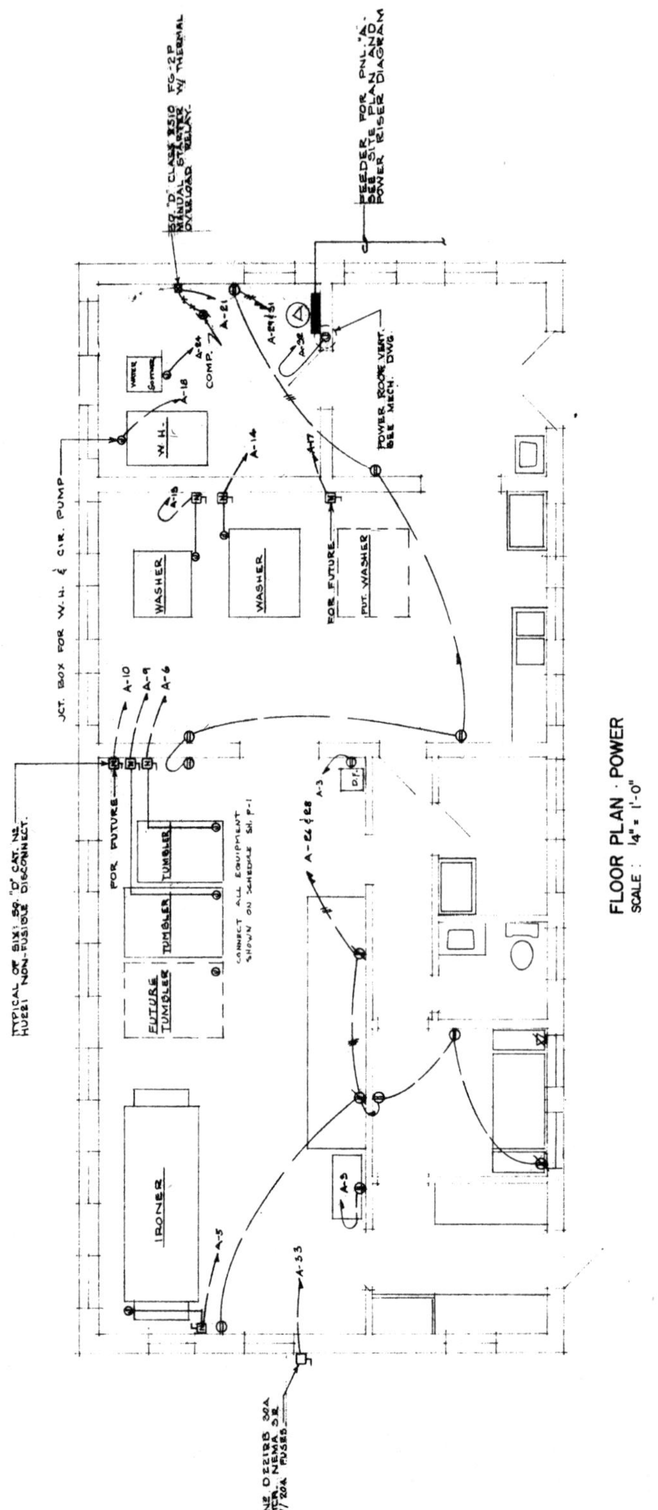

Figure 9-18. Floor plan for assignment 1.

floor plan in Fig. 9-18 (drawn at a scale of 1/8" = 1'0") to your drawing sheet. Draw this to a scale of 1/4" = 1'0". Include the duplex receptacles shown in your drawing and then circuit them, using your French curve, for the circuit lines. Use the methods described in this chapter to determine the number of outlets allowed on each circuit.

2. Prepare another sheet of tracing paper with border lines and title block. The schematic diagram in Fig. 9-19 is rather complex, whereas the sketch in Fig. 9-20 shows the same circuit somewhat simplified. Prepare a neat drawing using straightedge, templates, etc., of the sketch in Fig. 9-20; letter all notes, component identification, etc.

9.5 Power Supplies for Electronic Equipment

One of the most important sections in electronic equipment is the power supply, since this is the section that furnishes the operating voltages and currents required by the various stages.

The power supply in transistorized equipment, for example, will have to reduce the voltage to some value less than the 120-volt supply, whereas in tube-operated equipment, the power supply will often have to raise the ac line voltage to a much higher dc voltage. On the other hand, the current requirements of the power supply in a transistorized piece of equipment may be considerably higher than those of a power supply in a similar tube-operated device.

Tubes and transistors used in electronic equipment require dc operating voltages. Therefore, in a power supply designed to operate from a conventional 120-volt ac outlet, the power supply must provide some means of filtering or smoothing the pulsating dc to get pure dc. This is normally done by using a filter network in the circuit; this separates the ac and dc components of the pulsating dc at the rectifier output, so that only the dc appears at the output of the filter network.

Many power supplies also have some type of voltage-divider network designed to provide several different operating voltages from one power supply, because all tubes and transistors used in electronic equipment may not operate at the same voltage. It is more economical to use a single power supply and a voltage divider than to use a separate power supply for each voltage needed.

The power supply in most modern electronic equipment uses solid-state rectifiers in the majority of applications. These devices are designed to allow current to pass in one direction but not in the other. Any device capable of performing this operation can be used as a rectifier.

A selenium rectifier, for example, is made up of a series of selenium discs with a coating of selenium oxide on the surface of one side of each disc. Electrons can flow from the selenium to the selenium oxide quite readily, but they cannot readily flow in the opposite direction. This device can then be used as a rectifier, because it permits current to flow through it in one direction, but offers a high resistance to current flow through it in the other direction.

The selenium rectifier has been replaced recently by the silicon rectifier, which is much smaller than the selenium rectifier and is, therefore, more suited for use with solid-state equipment. In addition to their small size and high current-handling capabilities, silicon rectifiers are relatively inexpensive to manufacture, and their life is almost indefinite.

The bridge-rectifier circuit in Fig. 9-21 uses four rectifiers to function properly. The operation of this type of rectifier is simple. Follow the diagram as you read the following:

When terminal A is positive and terminal B is negative, current will flow from terminal B through the rectifier marked 2 on the diagram and then through the load to the junction of rectifiers 3 and 4. It will then flow through rectifier 4 back to terminal A on the transformer. During the next half-cycle, when terminal A is negative and terminal B is positive, current will flow from terminal A on the transformer through the rectifier marked 1 on the diagram and then through the load back to the junction of rectifiers 3 and 4. This time, the current will flow through rectifier 3 back to terminal B of the power transformer.

During each half-cycle, current flows through two of the rectifiers; that is, during one half-cycle it will flow through rectifiers 2 and 4, and during the other half-cycle it will flow through rectifiers 1 and 3. Also notice that current flows during both half-cycles and that this is therefore classified as a full-wave rectifier.

Bridge rectifiers are used in electronic equipment where comparatively high operating voltages and high currents are required.

Since the output from rectifiers is not pure dc, some means of purifying the pulsating dc is necessary. This is normally accomplished by a filter circuit. One of the simplest filters is the single capacitor filter

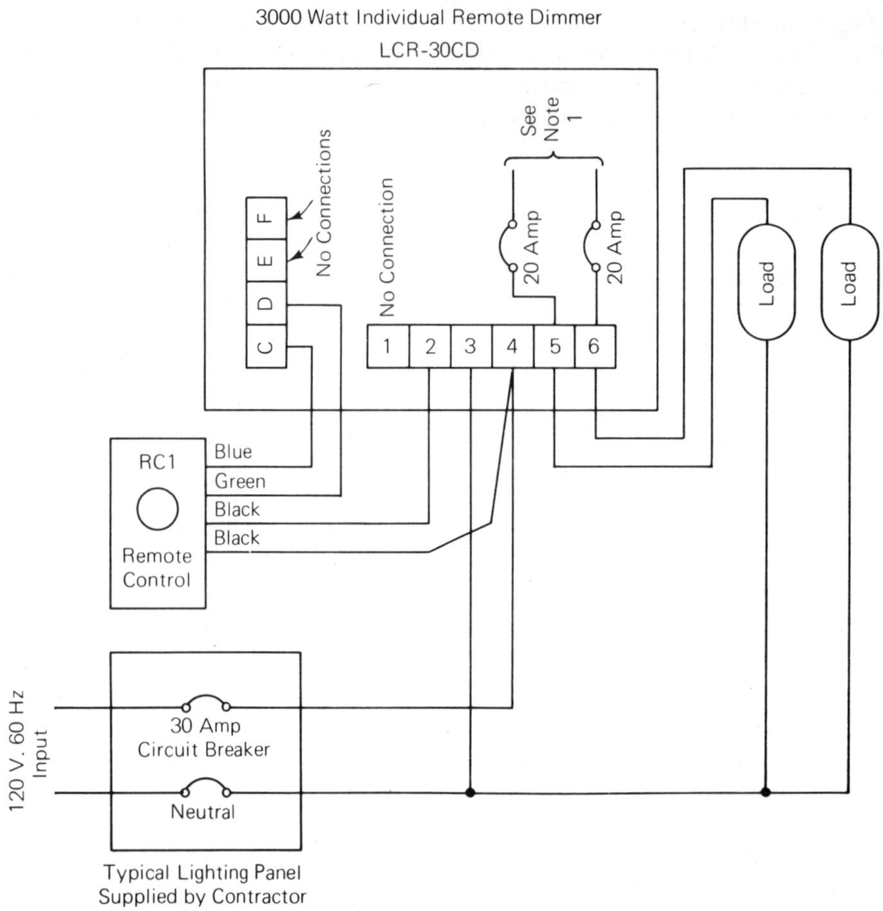

Figure 9-19. A rather complex schematic diagram.

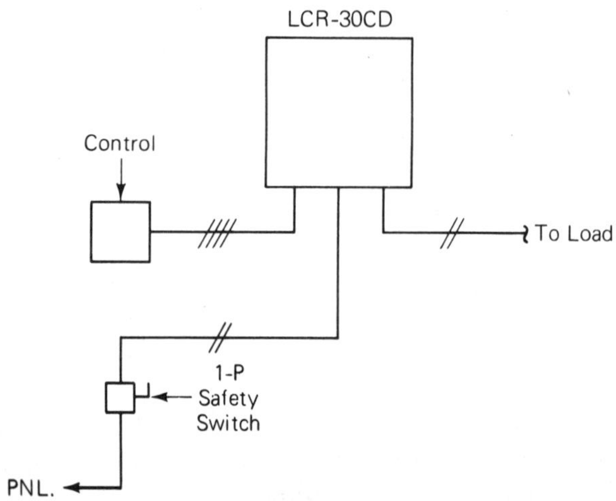

Figure 9-20. Simplified sketch of the circuit in Fig. 9-19 for use in assignment 2.

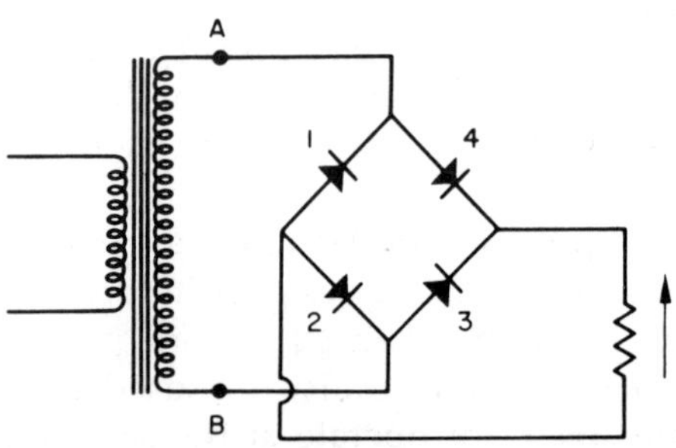

Figure 9-21. Schematic of bridge-rectifier circuit.

shown in Fig. 9-22. This type of filter is sometimes used in circuits where the current drained from the power supply is low. If the rectifier must supply high current to the circuit, this type of filter is generally unsatisfactory, because there will be too much ripple or hum present across the load.

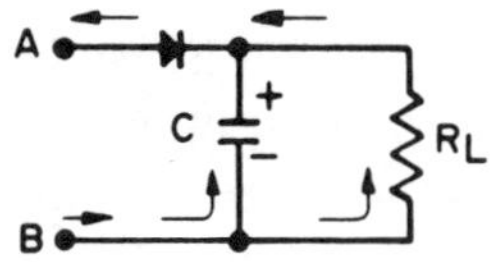

Figure 9-22. Single capacitor filter circuit.

An improved filter is shown in Fig. 9-23. This filter consists of two capacitors and a filter resistor. However, this circuit also has some disadvantages—for example, the large size of resistor and capacitor needed to obtain effective filtering in this type of circuit.

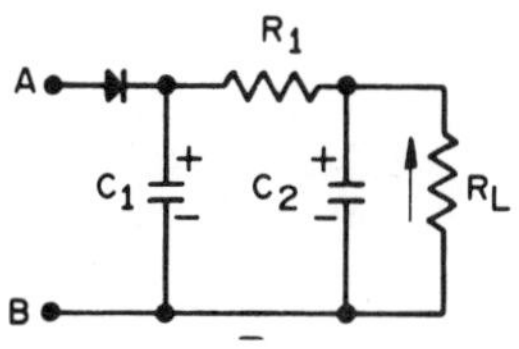

Figure 9-23. An improved filter circuit.

A choke-input filter is shown in Fig. 9-24. Here the rectifier tube is connected to a filter choke rather than to a capacitor. Power supplies of this type will be found in electronic equipment where comparatively high voltages are encountered.

In any filter network containing a filter choke or a filter resistor, the dc current flowing through the load must also flow through the filter choke or filter resistor. There will then be a voltage drop across this choke or resistor, the exact voltage of which will depend upon the dc resistance and the current flowing.

Filter networks vary from comparatively simple filters consisting only of a capacitor up to networks

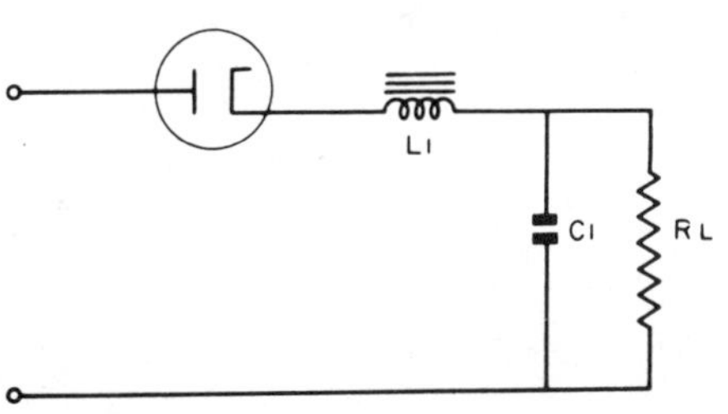

Figure 9-24. Choke-input filter circuit.

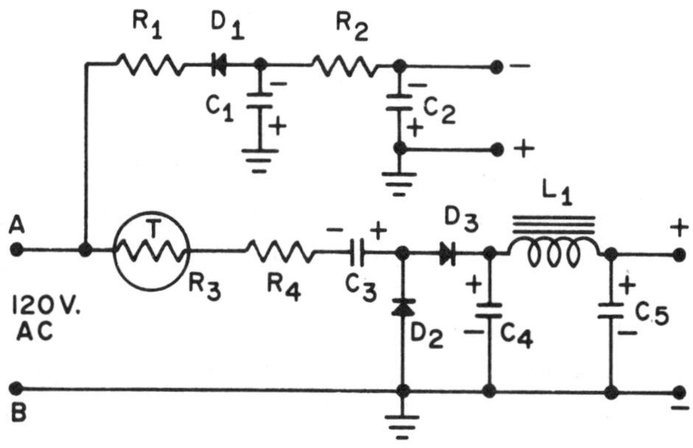

Figure 9-25. Diagram of typical dual-voltage power supply.

containing two chokes and two filter capacitors. The simpler types can be used where the current drain is low and where the filtering does not have to be very good. The more elaborate filters are used in power supplies having a high current drain and in cases where good filtering is required to supply pure dc at the power supply output.

The universal ac-dc power supply is so called because it can be operated from either an ac or a dc voltage source. A diagram of a typical dual-voltage power supply is shown in Fig. 9-25 while a regulated power supply employing transistor voltage regulator is shown in Fig. 9-26. These power supplies are typical of those that you are likely to encounter in your work as an electronic draftsman. You will, of course, find many variations of these circuits, but these are the basic circuits.

QUESTIONS

1. What are the advantages of a bridge-rectifier circuit?
2. In what type of application may a simple capacitor-type filter be used?

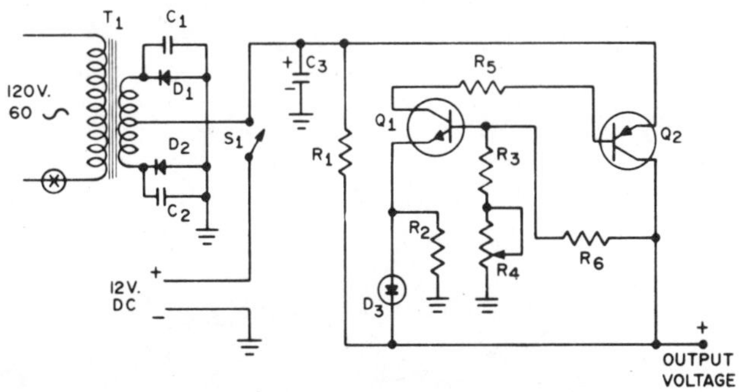

Figure 9-26. Diagram of typical regulated power supply.

3. Why are filters used in electronic circuits?
4. What advantage does a silicon rectifier have over a selenium rectifier?
5. Name one characteristic of the universal ac-dc power supply.
6. The power supply in electronic equipment is very important, since it furnishes the operating _____ and _____ required by the various stages.
7. In transistorized equipment, the power supply will normally have to _____ the voltage when connected to 120 volts.
8. In tube-operated equipment, the power supply will normally _____ the voltage when connected to a 120-volt source.
9. When the tubes or transistors that are used in electronic equipment operate on different voltages, what device is normally used to produce the various voltages?
10. What type of rectifier is used in the majority of modern electronic equipment?
11. In what type of electronic equipment will filter chokes normally be used?

Chapter 10

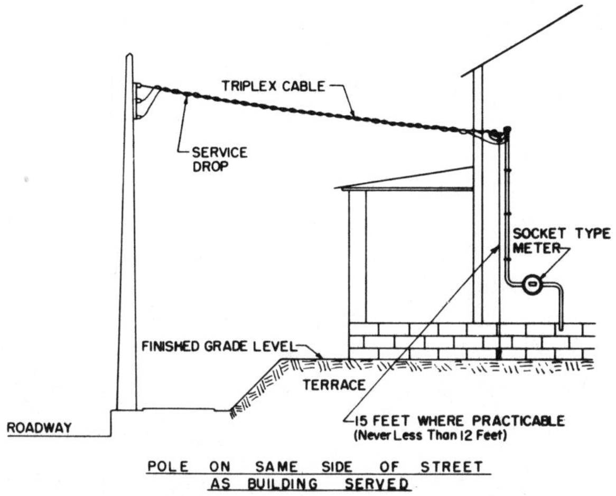

SERVICE AND FEEDER DRAWINGS

All buildings containing an electrical system require an electrical service. This chapter, therefore, is designed to show the electrical draftsman how such services are shown on electrical working drawings. This chapter further covers feeder circuits (the larger wire sizes that feed panel boards and other electrical equipment); it also describes how certain electrical details are shown on working drawings.

10.1 Electric Service Details

The overhead conductors—through which electrical service is supplied between the last utility pole and the point of their connection to the service facilities located at the building, power pole, transformer, or other support used for the purpose—are called the *service drop* (Fig. 10-1).

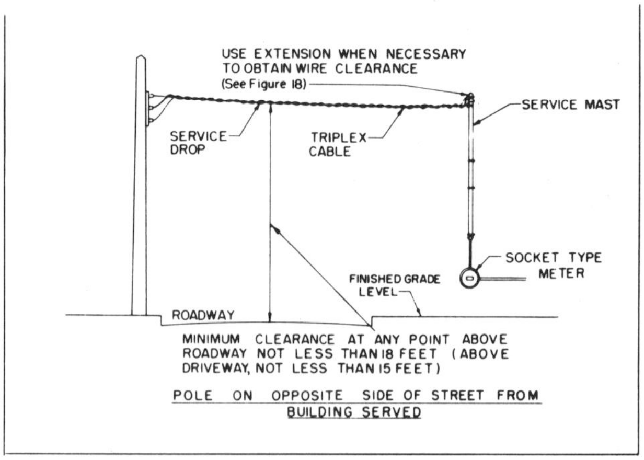

Figure 10-1. The overhead conductors, through which electrical service is supplied, are called the *service drop*.

All components between the point of termination of the overhead service drop and the building main switch (or disconnect) are known as the *service entrance* (see Fig. 10-2). The wires or cable used in this portion of the electric service are called the *service-entrance conductors*.

Service-entrance equipment (like the main overcurrent protection provided by a main switch or circuit breaker) provides a means of protecting the service conductors and feeders, as well as a means of disconnecting the feeders from energized service conductors. The electric metering equipment provides a means of measuring the amount of energy used so that the customer can be billed by the power company.

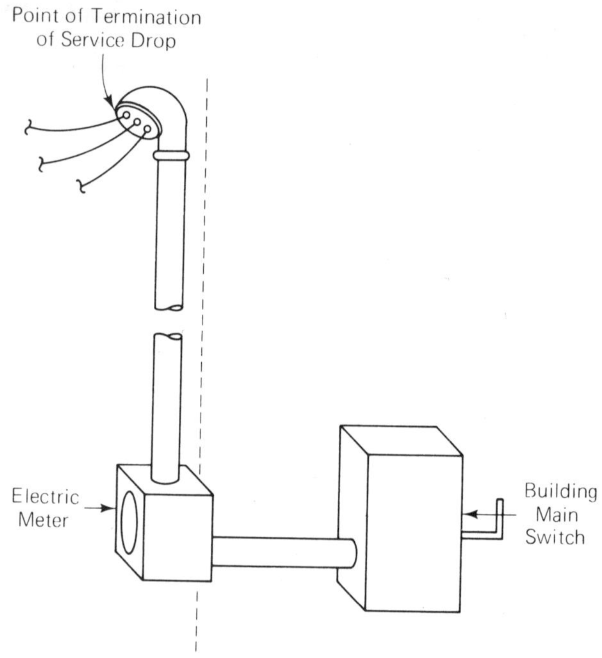

Figure 10-2. All components between the point of termination of the overhead service drop and the building main switch (or disconnect) are known as the service entrance.

The voltage supplied to the building depends upon the power company's available facilities, the method of servicing the customer, and the size and type of the customer's load. However, for most residences and other buildings requiring only relatively small electrical services, the power company will supply a single-phase, 120/240-volt, three-wire electrical service. When three-phase services are required, 120/208-volt "Y" or 277/480-volt wye is usually preferred over delta-connected services, because the "red-leg" or "high-leg" phase on the delta system prevents proper balancing of 120-volt loads.

Where a 277/480-volt service is used, most of the larger electrical equipment, such as heating and cooling equipment and motors, is connected to the 480-volt lines. Most of the electric lighting is connected to 277 volts (one phase to ground), and convenience outlets and some lighting circuits are connected to 120 volts, utilizing a dry type of transformer to reduce the higher voltage to the proper level.

All services require a safe substantial support, and requirements set forth in the National Electrical Code must be followed. Figures 10-3 through 10-6 show various overhead and underground service and meter installations that comply with the NEC and are acceptable to most power companies.

Service-entrance equipment usually consists of either a factory-built switchboard with all switches,

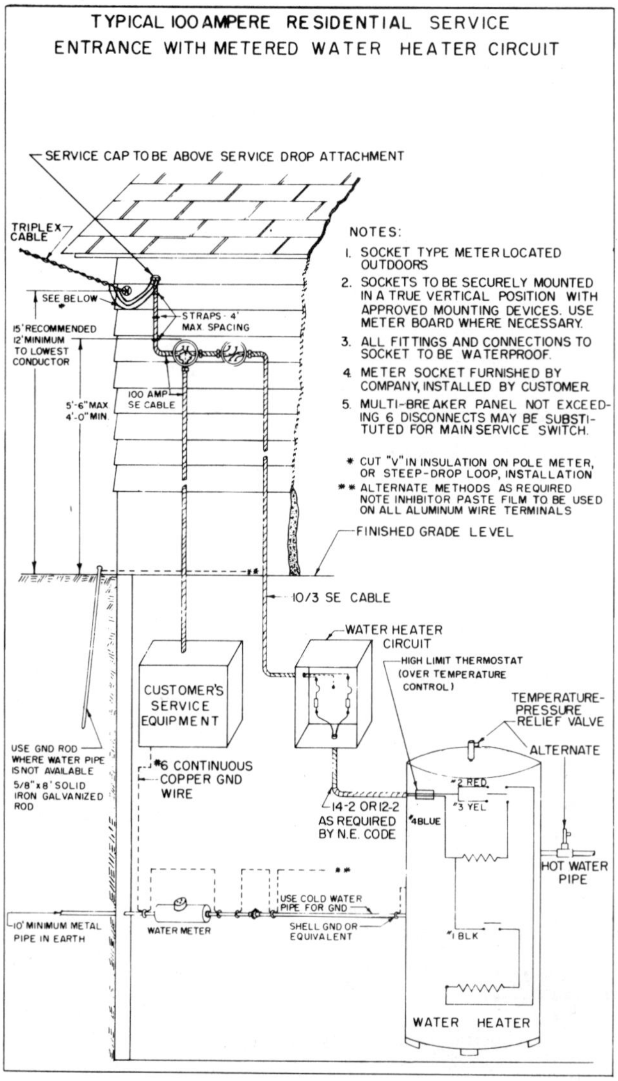

Figure 10-3. Typical 100-ampere residential service entrance with metered water heater circuit.

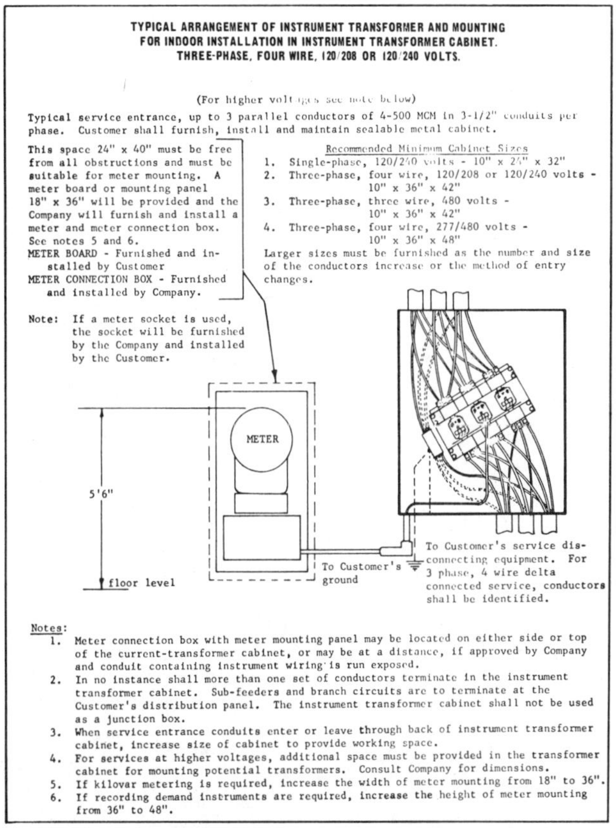

Figure 10-4. Typical arrangement of instrument transformer and mounting for indoor installation in instrument transformer cabinet. Three-phase, four-wire, 120/208 or 120/240 volts.

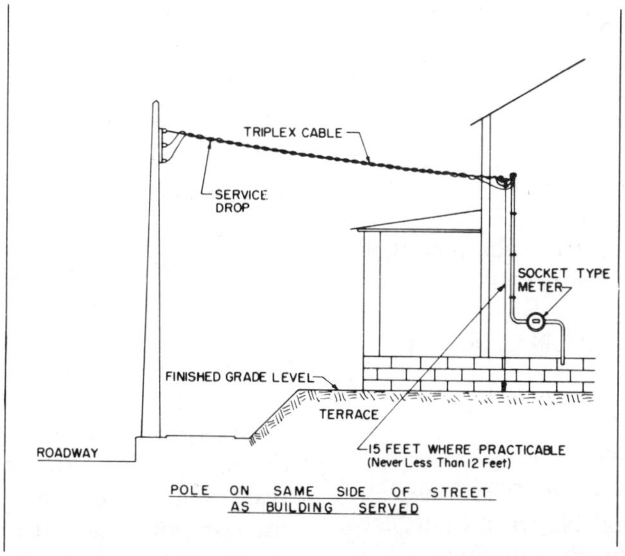

Figure 10-5. Overhead service installation with pole on same side of street as the building served.

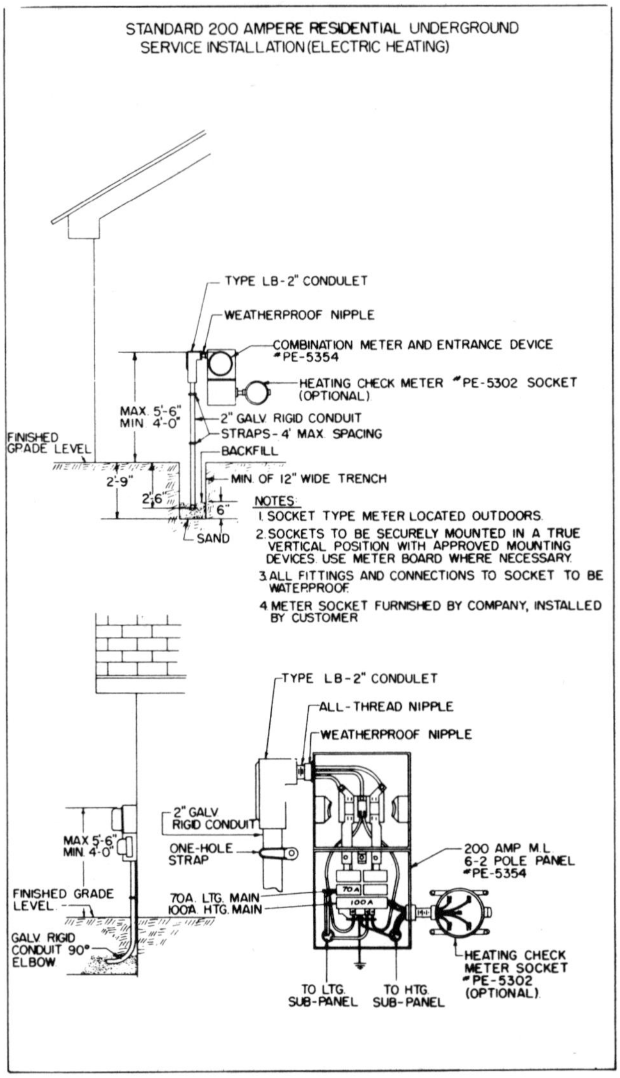

Figure 10-6. Standard 200-ampere residential service installation.

fuses or circuit breakers, and in some instances the metering equipment built in, or an assembly of individual units of switches or circuit breakers, metering equipment, and auxiliary gutters.

In the design of service-entrance equipment, the following factors must be taken into consideration:

1. Service characteristics available from the power company.
2. Total connected electrical load.
3. Total demand of electrical load.
4. Capacity for present and future requirements.
5. Requirements of National Electrical Code, local ordinances, and local power company.
6. Type and cost of equipment.
7. Delivery date of equipment.
8. Physical dimensions of equipment and space available in building.

The floor plan of a small commercial building is shown in Fig. 10-7. Note that the shaded rectangle depicts panel board "A," which, on this project, is the only panel board or overcurrent equipment used. The plan in itself does not tell too much about the characteristics of the panel board except to show its location in relationship to the rest of the building. A power-riser diagram (Fig. 10-8), however, begins to clarify the components of the service entrance as it shows the various conduit sizes, the type, size, and number of conductors, etc. When a panelboard schedule (Fig

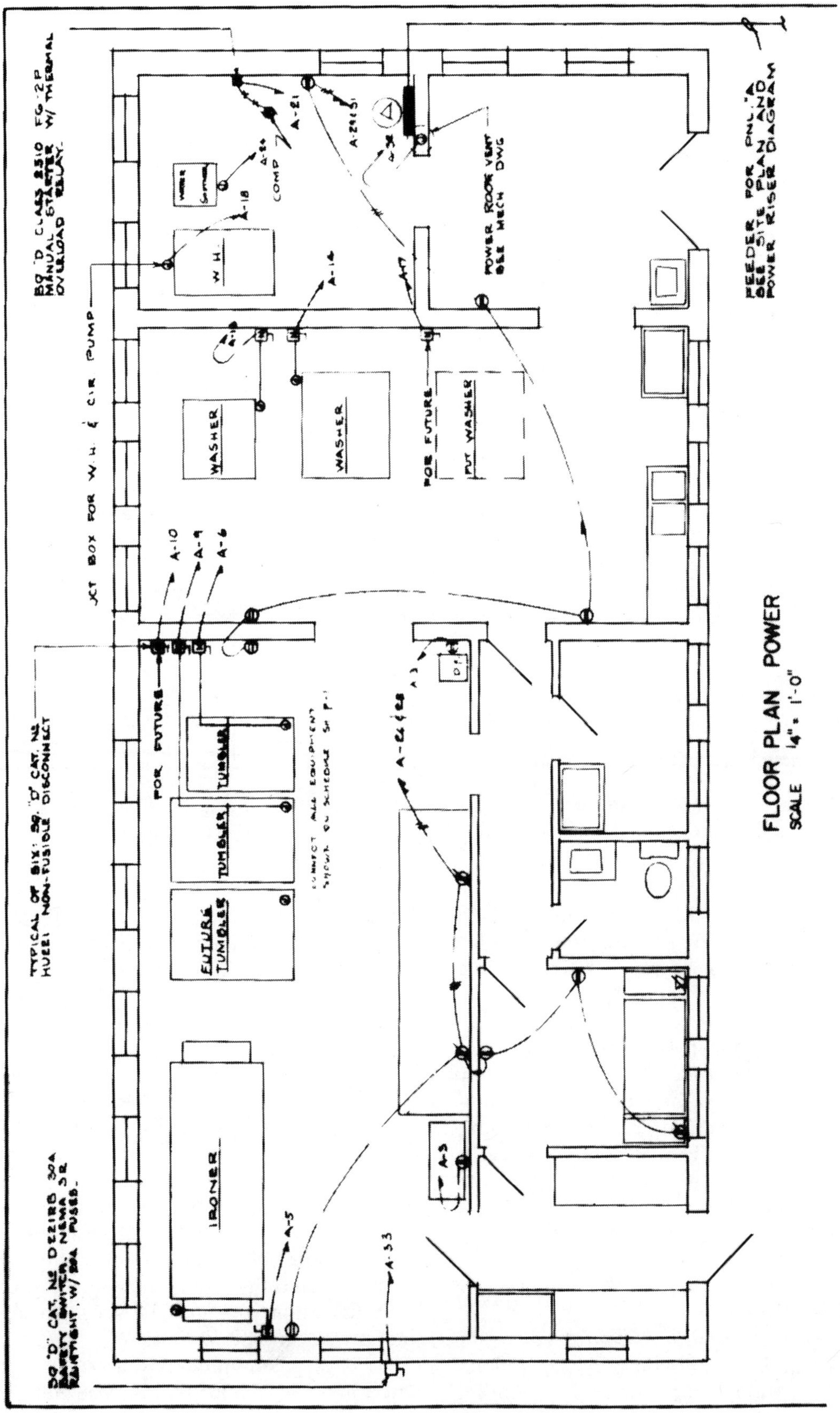

Figure 10-7. Floor plan of a small commercial building.

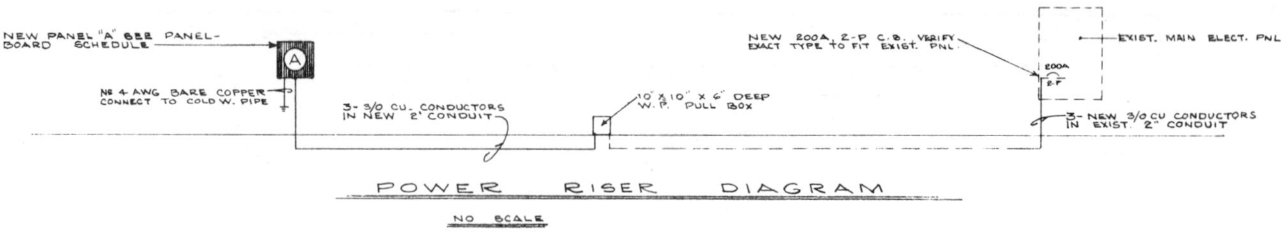

Figure 10-8. A power-riser diagram begins to clarify the components of the service entrance.

PANELBOARD SCHEDULE

PNL Nº	TYPE CABINET	PANEL MAINS AMPS	VOLTS	PHASE	BRANCHES 1P	2P	3P	PROT.	FRAME	ITEMS FED
A	SURFACE	200A.	120/240V	1φ, 3W	12	—	—	20 A	70A	LTS., RECEPTS, W. SOFTNER, ETC.
SQUARE "D" TYPE NQO M.L.O.					—	7	—	20A	70A	TUMBLERS, COMP., FAN, IRONIER, ETC.
					—	3	—	30A	70A	WASHERS
					7	—	—	—	—	PROVISIONS ONLY

Figure 10-9. When a panelboard schedule—like the one shown here—and written specifications are provided, complete information is provided to install the electrical service properly.

10-9) and the written specifications are added, complete information is provided so that the electrical service and related equipment can be installed properly.

To summarize the procedure for showing the service-entrance equipment on working drawings, first locate the main panel board, subpanels, and other service equipment on the building floor plans. Each item should be identified and drawn approximately to scale. Calculate the wire and pipe sizes for the various components and show them in a power-riser diagram. A schedule should then be provided to give an exact description of the service equipment and the size and number of overcurrent devices contained in each. See Chapter 6 for further examples of panelboard schedules.

Each electrical system is required to be grounded in a manner as prescribed by the National Electrical Code, although the details of grounding rarely appear in the power-riser diagram; rather, a paragraph usually appears in the written specifications and may read as follows:

> d. GROUNDING: The Electrical Contractor shall properly ground the electrical system as required by the National Electrical Code. The ground wire for the service entrance shall be run in conduit and made to the main water service and connected ahead of any valve or cut-off.

The paragraph above will serve in most cases, but the electrical designer or draftsman should carefully coordinate all electrical drawings with the plumbing drawings and the plumbing contractor, because PVC (plastic) water pipe may be specified, and it would not provide a ground.

It is recommended that exact details of the grounding system for each service entrance be given.

Such details may appear in the specification, such as the following for a 200-ampere service entrance.

> Furnish and install a No. 4 AWG bare copper conductor in a ¾″ rigid galvanized conduit from the neutral bus in the main distribution panel to a 1″ or larger cold water pipe in the mechanical room on the ground floor. Water pipe connections shall be with an approved ground fitting that bonds conduit and conductor to water pipe.

10.2 Electric Feeder Drawings

Electrical feeders, for our purposes, will be considered as circuits or raceways containing two or more conductors of size number 8 AWG and larger when not used for service-entrance conductors. For most projects, they usually consist of those conductors feeding:

1. Subpanels
2. Electric motors
3. Heating and cooling equipment
4. Electric welders
5. Dry-type transformers
6. Electrical appliances requiring conductors of number 8 AWG or larger

On electrical drawings, feeders are sometimes shown on the floor plans by single solid lines (Fig. 10-10), similar to the way that conventional branch circuits are shown. They are, however, most often shown in an equipment schedule (Fig. 10-11) or in power-riser diagrams, as shown in Fig. 10-12. Of these three types, the power-riser diagram method is the type most preferred by electricians who must interpret the work on the job. Therefore, a brief analysis of the power-riser diagram in Fig. 10-12 is in order.

The service entrance on this project is provided

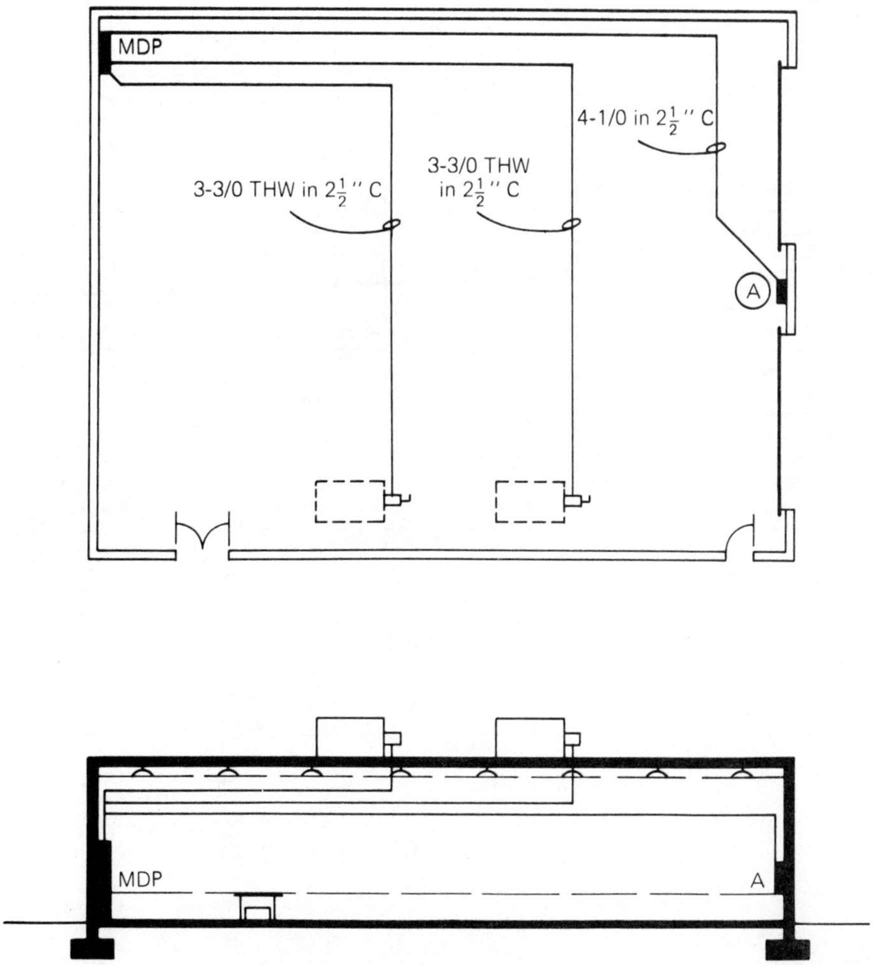

Figure 10-10. Feeders are sometimes shown on floor plans by single solid lines.

KITCHEN — EQUIPMENT SCHEDULE								
EQUIP. NO.	DESIGNATION	H.P. OR K.W.	VOLTS	CONNECTION WIRE	CONNECTION CONDUIT	CONNECTION PROT.	FURNISHED BY	REMARKS
1	FREEZER	1 H.P.	240V	NO. 12	3/4"	20A	EQUIP BY OTHERS OUTLET BY CONT.	24" A.F.F; 1 ϕ
2	MIXER	1/3 H.P	120V	NO. 12	3/4"	20A		48" A.F.F; 1 ϕ
3	ICE MACHINE	1 H.P	120V	NO. 12	3/4"	20A		72" A.F.F; 1 ϕ
4	REFRIGERATED DISPLAY CASE	1/4 H.P	120V	NO. 12	3/4"	20A		24" A.F.F; 1 ϕ
5	RECEPTICAL		120V	NO. 12	3/4"	20A		42" A.F.F; 1 ϕ
6	REFRIGERATOR		120V	NO. 12	3/4"	20A		6" A.F.F; 1 ϕ
7	ICE STORAGE CHEST		120V	NO. 12	3/4"	20A		39" A.F.F; 1 ϕ
8	COFFEE MAKER		240V	NO12	3/4"	20A		39" A.F.F; 3 ϕ
9	MILK DISPENSERS		120V	NO. 12	3/4"	20A		39" A.F.F; 1 ϕ
10	WAFFLE MAKER		240V	NO. 10	3/4"	30A		39" A.F.F; 3 ϕ
11	WAFFLE MAKER		240V	NO. 10	3/4"	30A		39" A.F.F; 3 ϕ
12	REFRIGERATOR	1/4 H.P	120V	NO. 12	3/4"	20A		24" A.F.F; 1 ϕ
13	TOASTER		240V	NO. 12	3/4"	20A		39" A.F.F; 3 ϕ
14	RECEPT		120V	NO. 12	3/4"	20A		72" A.F.F; 1 ϕ
15	MIXER		120V	NO. 12	3/4"	20A		6" A.F.F; 1 ϕ
16	UPDRAFT UNIT	33.8	240V	NO. 1	1 1/2"	125A		26" A.F.F; 3 ϕ
17	UP RIGHT REFRIGERATOR	1.5	120V	NO. 12	3/4"	20A		74" A.F.F; 1 ϕ
18	ICE STORAGE CHEST		120V	NO. 12	3/4"	20A		39" A.F.F; 1 ϕ
19	COFFEE MAKER		240V	NO. 12	3/4"	20A		39" A.F.F; 3 ϕ
20	REFRIGERATOR		120V	NO. 12	3/4"	20A		6" A.F.F; 1 ϕ
21	DUPLEX RECEPTACLE		120V	NO. 12	3/4"	20A		72" A.F.F; 1 ϕ
22	REFRIGERATED DISPLAY CASE	1/4 H.P	120V	NO. 12	3/4"	20A		24" A.F.F; 1 ϕ
23	FLOOR RECEPTACLE		120V	NO. 12	3/4"	20A		FLOOR OUTLET; 1 ϕ
24	FLOOR RECEPT.		120V	NO. 12	3/4"	20A		FLOOR OUTLET; 1 ϕ
25	REFRIG.		120V	NO. 12	3/4"	20A	↓	72" A.F.F; 1 ϕ

Figure 10-11. Feeders are also shown in an equipment schedule.

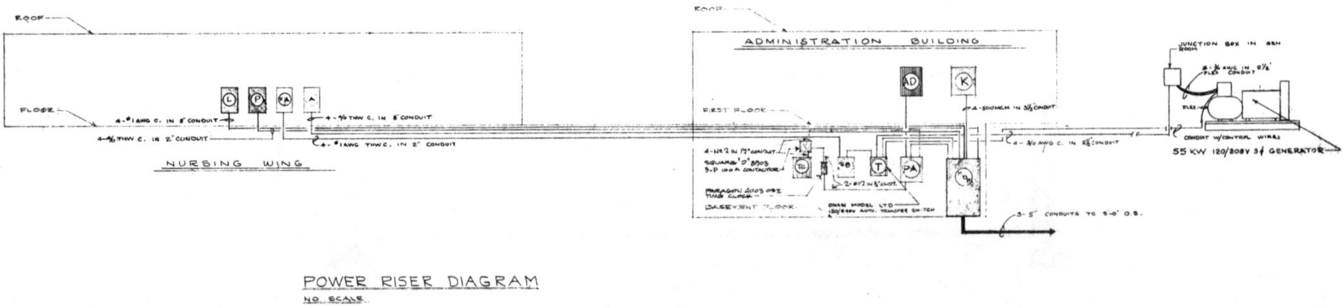

Figure 10-12. Feeders are most often shown in a power-riser diagram.

underground, and the riser diagram calls for three 5″ conduits to be installed from the main distribution panel (MDP) to three feet outside of building (O.B.). The remaining circuits in this diagram may be classified as *feeders*. Note that one feeder leaves the MDP and terminates in the transfer switch housing (T), which is also fed by four 3/0 conductors from the 55-kW generators. The feeder from the transfer switch feeds panel "EB," which in turn feeds panels "EA" and "AD" with four number 1 AWG THW conductors pulled in 2″ conduit.

Under normal operating conditions, panels "EA," "AD," and "EB" are fed from the main distribution panel. However, if the main electric service is interrupted for any reason, the automatic transfer switch will activate the gas generator and at the same time transfer the load on panels "EA," "AD," and "EB" to the emergency generator. When the normal electric service is restored, the automatic transfer switch will shut the emergency generator off and transfer the load back to the main distribution panel.

Other feeder circuits include four 500MCM conductors in 3½″ conduit from MDP to feed subpanel "K"; four 4/0 THW conductors in 3″ conduit to feed panels "P" and "A," and four number 1 AWG conductors in 2″ conduit feeding panel "L." The only remaining panel boards are panel "PA" and panel "TC." The latter panel ("TC") serves the outside lighting and is controlled by a time clock and a 100-ampere magnetic contactor. The time clock is set to energize the magnetic contactor at predetermined times; in doing so, every circuit in panel "TC" becomes energized and all outside lights fed from this panel come on. At another predetermined time, the time clock "breaks" the magnetic contactor, turning off all the circuits on the panel. A panelboard schedule showing the various loads connected to this panel—including overcurrent protection and load on each circuit—is shown in Fig. 10-13.

The modern approach to flexible power distribution in industrial electrical systems is the trend to carry *feeder* conductors in troughs or trays or to use bus bars enclosed in either a ventilated or a nonventilated duct from the source of supply to points within the system where distribution of electric power is needed.

Bus ducts provide for fast initial installation; it is also very fast and convenient to add on electrical feeders at a later date. Straight lengths are manufactured in 1½-, 3-, 5-, 6-, and 10-foot lengths. These are connected together with bolt and bus bar joints.

The broad array of feeder and plug-in duct types and standardized fittings will satisfy almost any low-voltage requirements (480 volts and lower) for local circuit protection and control. Plug-in devices—fuses, fused disconnect switches, circuit breakers, transformers, etc.—are readily available which feature rating and characteristics to match nearly any load condition.

A carefully planned network of bus ducts is necessary to provide maximum economy, flexibility, and low voltage drop characteristics. Also because of physical limitations and electrical requirements, each design of a bus duct system requires a customized solution.

A duct system can be planned easily and quickly in most cases. In order to electrically plan and execute a compatible, efficient, and economical system, one must:

1. Determine connected load and future growth requirements.
2. Determine the load demand and duty cycles of the various loads.
3. Calculate voltage drop.
4. Calculate power factors.

The owners or equipment suppliers furnish 1 and 2 above.

The voltage drop can be calculated by the following formula:

$$\text{Maximum voltage drop per 100 ft} \times \frac{\text{actual load amps}}{\text{rated amps}}$$

$$\text{percentage drop @ worst P.F.} = \text{V.D. per 100 ft}$$

A typical method of showing a bus duct on a working drawing is shown in Fig. 10-14. This system was designed for a small industry utilizing a main bus duct as the feeder with direct-mounted switches on the main duct to supply or feed various pieces of electrical equipment. The feeders to these pieces of equipment, however, are run in conventional conduit and conductors.

ASSIGNMENT

Secure a piece of 11″ × 17″ tracing paper to your drawing board and draw in border lines and a title block in the usual manner. Draw the power-riser diagram shown in the sketch in Fig. 10-15, using straightedge, templates, etc. Neatly letter all notes. In other words, provide a neat, accurate, finished drawing to a scale approximately twice the size of the drawing shown in the book.

PANEL "TC" 120/208 V., 3φ, 4-WIRE, SURFACE MOUNTED
100A. M.L.O. SQUARE "D" TYPE NQOB

CKT No	CIRCUIT BKS. POLE	FRM	TRIP	WIRE SIZE	CONNECTED LOAD IN WATTS Aφ	Bφ	Cφ	ITEMS FED OR REMARKS
1	1		20A	No 10	1050			NORTHEAST POST LTS.
2	1		20A	No 12	1200			OUTSIDE BLDG. LTS.
3	1		20A	No 10		1050		SOUTHEAST POST LTS.
4	1		20A	No 12		900		OUTSIDE BLDG. LTS.
5	1		20A	No 12			600	CANOPY LTS.
6	1		20A	No 12			1500	OUTSIDE BLDG. LTS.
7	1		20A	No 12	1200			OUTSIDE BLDG. LTS.
8	1		20A	No 12	900			OUTSIDE BLDG. LTS
9	1		20A	No 12		1500		OUTSIDE BLDG LTS.
10	1		20A	No 12		1500		OUTSIDE BLDG. LTS.
11	1		20A	No 12			1200	OUTSIDE BLDG. LTS.
12	1		20A	No 12			1320	OUTSIDE BLDG. LTS.
13	1		20A	No 12	900			OUTSIDE BLDG. LTS.
14	1		20A	—	500			SPARE
15	1		20A	—		500		
16	1		20A	—		500		
17	1		20A	—			500	↓
18	1		20A	—			500	SPARE
19								
20								
21								
22								
23								
24								
25								
26								
27								
28								
29								
30								
31								
32								
33								
TOT	CONNECTED LOAD				5750	5950	5620	

Figure 10-13. A panelboard schedule showing the various loads connected to panel TC.

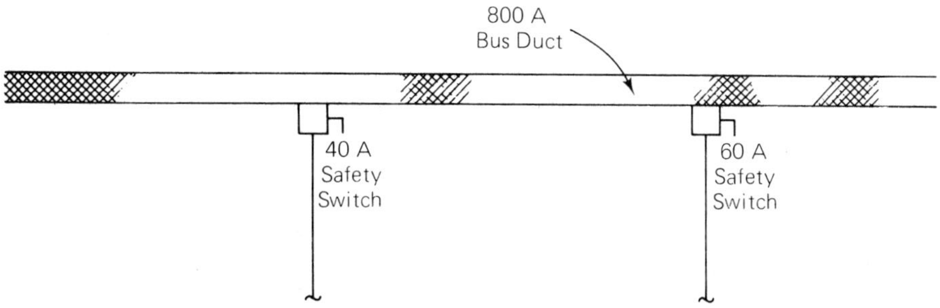

Figure 10-14. Method of showing bus duct on a working drawing.

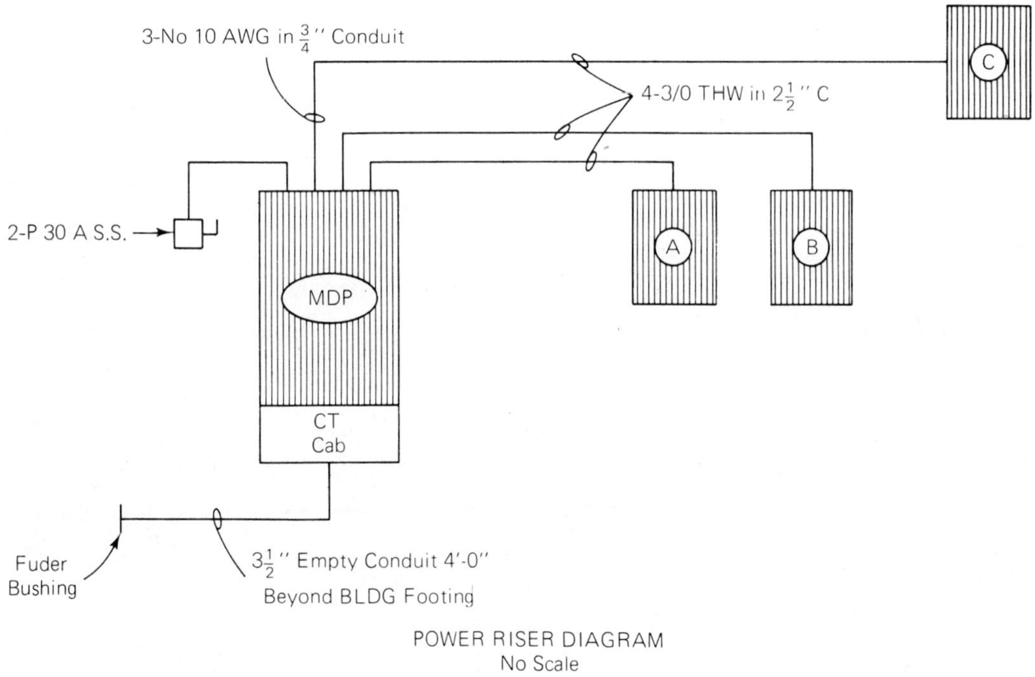

Figure 10-15. Power-riser diagram for use in assignment.

Chapter 11

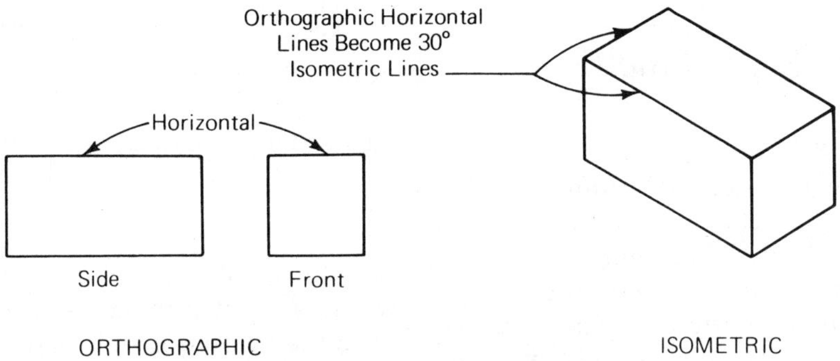

ISOMETRIC DRAWING

The objectives of this chapter are to teach the basic fundamentals of isometric drawing and to apply these basic fundamentals to electrical and electronic drafting. The reader will further learn to lay out and draw power-riser diagrams as they should appear on working drawings, as well as the procedures involved in making electrical and electronic equipment drawings.

11.1 Fundamentals of Isometric Drawing

Isometric drawing comes under the heading of pictorial drawing in that the objects drawn in isometric are pictured in one view only. Although the use of isometric drawings is limited for electrical construction drawings, they are used extensively in electronic drafting—chiefly to illustrate pieces of equipment. Isometric drawings usually are more difficult to draw than orthographic drawings, because intricate parts cannot be pictured clearly and circles and angles are difficult to draw correctly; they are also more difficult to dimension correctly than the orthographic views. Still, isometric drawings can be very helpful to describe objects and to convey information when one is dealing with highly complex systems or making drawings that must be interpreted by inexperienced workmen.

To illustrate the use of pictorial drawings on plans for electrical building construction, look at the floor plan of a factory in Fig. 11-1. The electrical system in this building consists of a cable tray system from several motor control centers to feed and control motors in other parts of the building. The trays run from the motor control centers, offset to miss beams and other runs of cable tray, and then branch off to various parts of the factory.

Although an experienced workman in the industrial electrical trade will have little trouble in reading the information in this drawing, apprentice or journeyman electricians new at industrial work may have some difficulty in visualizing this system. However, if a supplemental drawing such as the one in Fig. 11-2 were provided with the floor plan drawing in Fig. 11-1, it would provide a clearer picture of the system and leave little doubt as to how the cable tray system is to be installed. Even new workmen in the trade should be able to see how the system should be installed; this would take some of the load off experienced workmen and give them more time to accomplish other tasks.

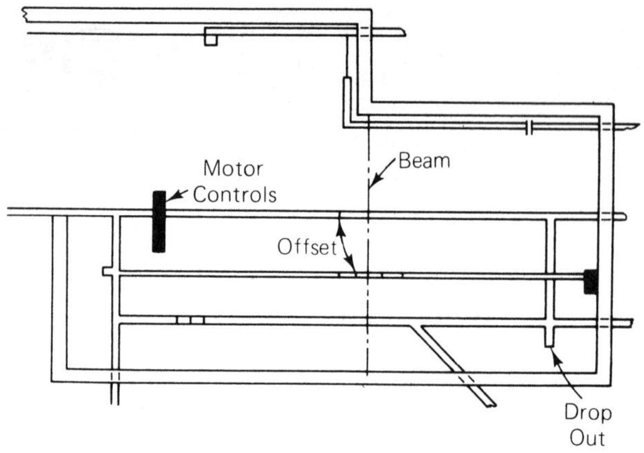

Figure 11-1. Floor plan of a factory, showing layout of a partial cable tray system.

In actual practice, consulting engineering firms would seldom furnish the isometric drawing in Fig. 11-2; they would merely show the layout, as shown in Fig. 11-1. However, the electrical contractor installing the system may have his own draftsman prepare supplemental working drawings (like the one in Fig. 11-2) to aid his workmen, thereby saving installation time. The engineering/drafting department of the factory itself may also prepare both types of drawings for the plant maintenance personnel to install the system.

By the same token, an electronic technician can take the schematic diagram in Fig. 11-3 and a component list and construct a circuit board with very little difficulty. However, someone who is new in the field may require a pictorial drawing of the circuit board, how the components are arranged on it, and an exact description of the housing and how the circuit board should be fitted into it. A drawing such as the one shown in Fig. 11-4 would provide sufficient information so that almost anyone could construct the electronic instrument—especially if step-by-step notes were also provided. Drawings of this type (Fig. 11-4) are used extensively by manufacturers of electronic do-it-yourself kits.

Electronic pictorial drawings are also used frequently in vocational school instruction materials as well as in training manuals for the U.S. armed forces and industrial firms.

11.2 Isometric Projection

A view projected onto a vertical plane in which all of the edges are foreshortened equally is called an isometric projection. The edges in the view that are

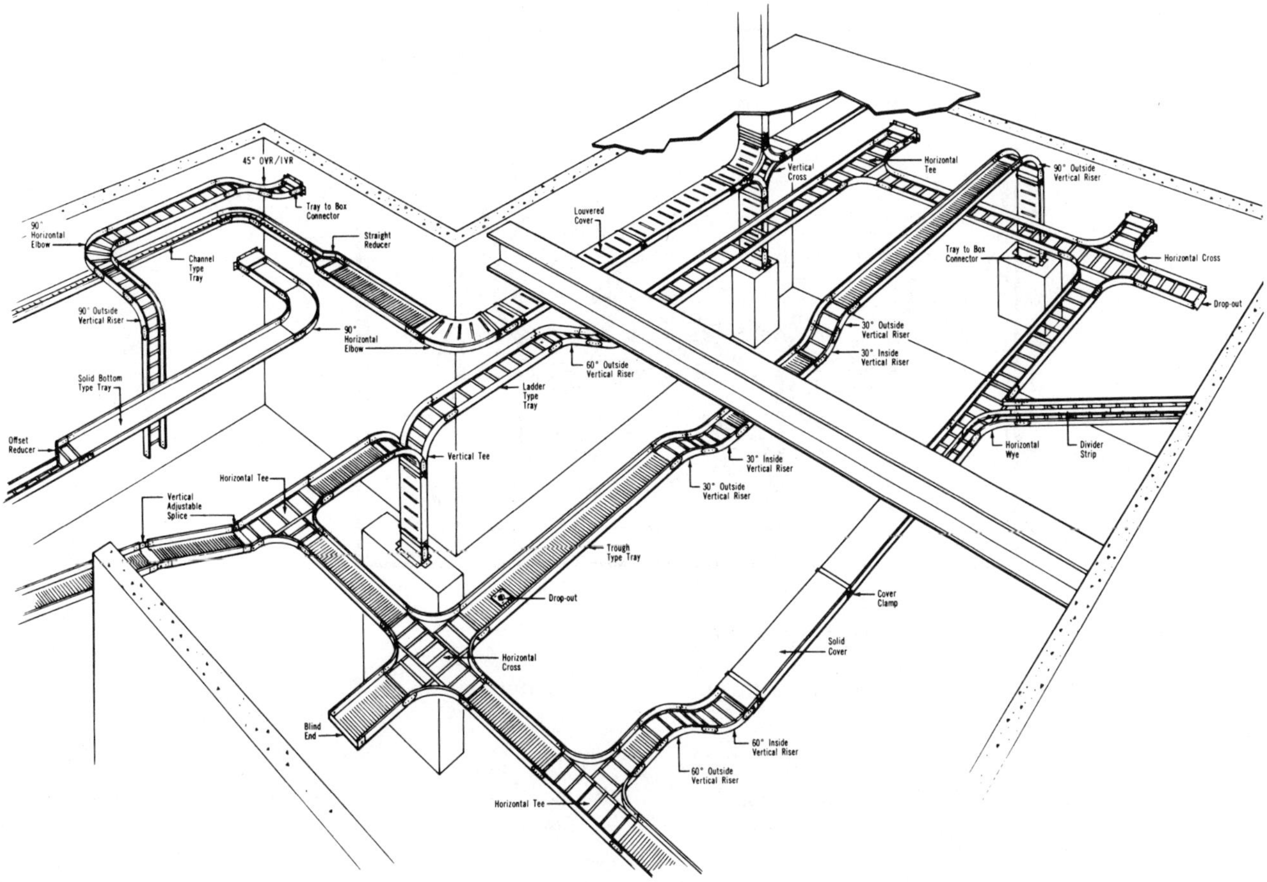

Figure 11-2. Supplemental drawing of the cable tray system shown in Fig. 11-1.

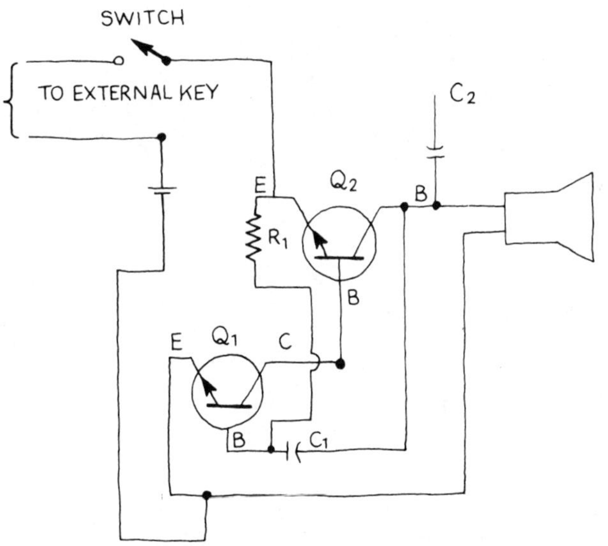

Figure 11-3. Schematic diagram of the type used by electronic technicians.

120 degrees apart are called the *isometric axes*, while the three surfaces shown in the view are called the *isometric planes*. Lines parallel to these axes are called *isometric lines*.

Because the isometric axes are always 120 degrees apart and the edges are foreshortened equally, the isometric axes and lines parallel to them can be drawn with the 30-60-90-degree triangle, and the edges can be measured with an isometric scale such as the one shown in Fig. 11-5.

Electrical drawings in isometric, however, are usually measured with a standard architect's scale rather than with the isometric scale. This results in a drawing that is somewhat larger than an isometric drawing made with the isometric scale. Figure 11-6 shows a comparison between a motor control center drawn as an isometric projection and drawn as an isometric drawing; that is, (a) is drawn with the isometric scale, and the larger view, (b), is drawn with the architect's scale.

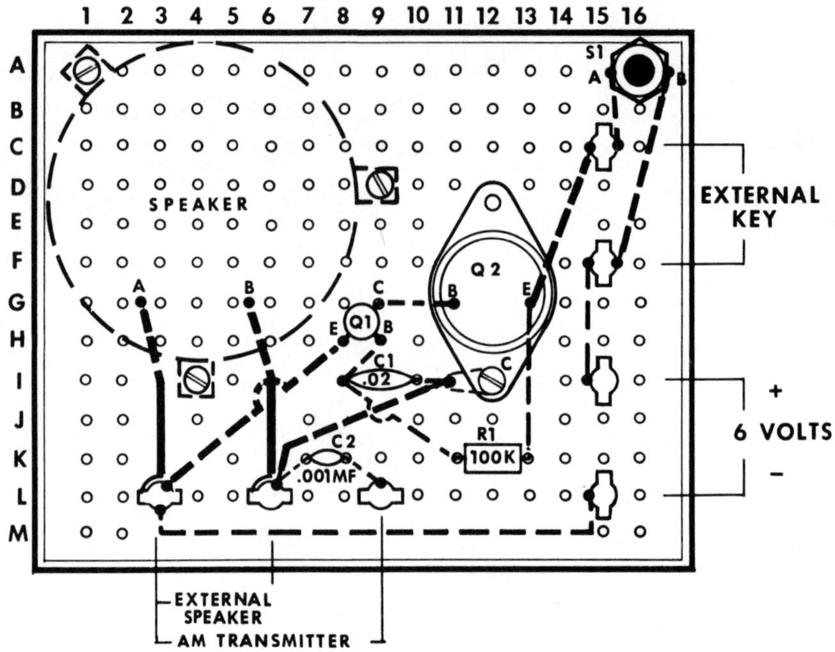

Figure 11-4. Pictorial drawing of the same electronic circuit shown in Fig. 11-3.

The advantage of using the architect's scale on electrical drawings for building construction is that the isometric views can be drawn and scaled the same as the building lines.

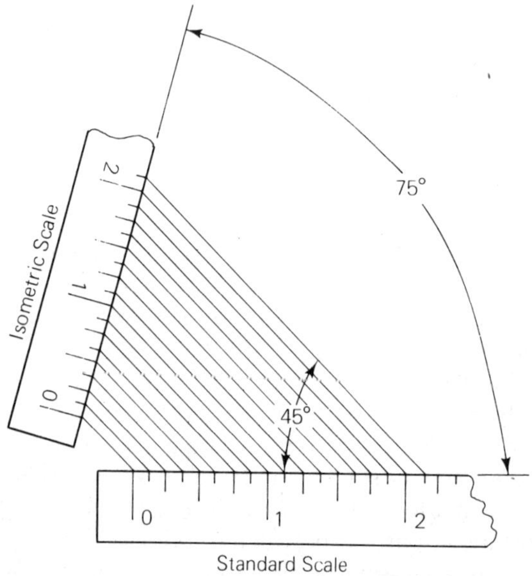

Figure 11-5. Comparison between standard and isometric scales.

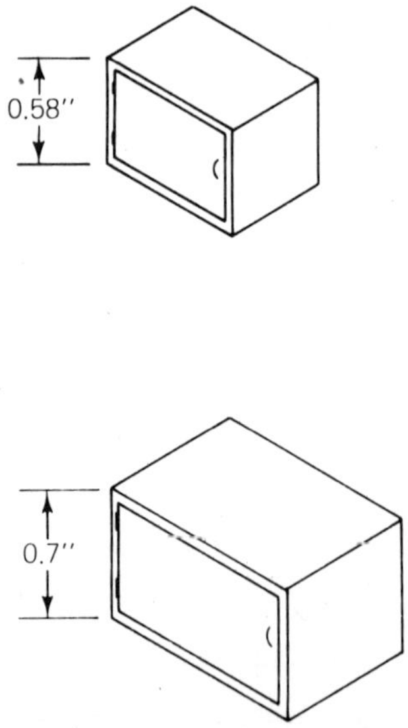

Figure 11-6. Comparison between an isometric projection and an isometric drawing.

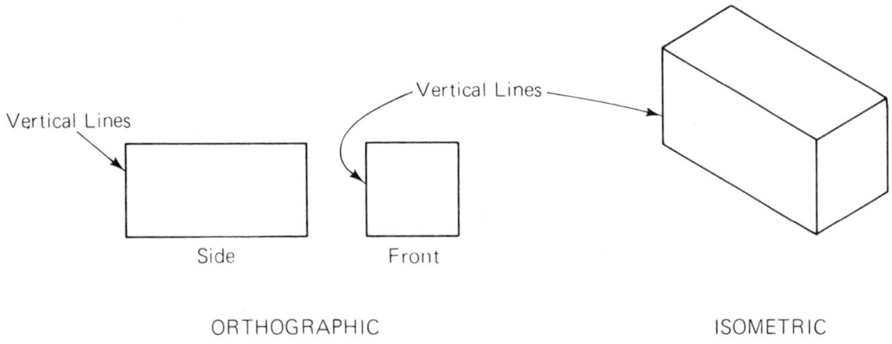

Figure 11-7. Vertical lines remain the same in both orthographic and isometric views.

11.3 Isometric Drawing

Isometric drawing is a method of showing a three-dimensional view on a flat surface such as a drawing sheet. This method of drawing is preferred for certain electrical and electronic details and diagrams, because it is possible to draw isometric lines to scale in the same manner as floor plans are drawn on orthographic views. The basic rules for isometric drawing are as follows:

1. Vertical lines remain the same in both orthographic and isometric views, as shown in Fig. 11-7.
2. Horizontal lines in an orthographic view are projected at an angle of 30 degrees in an isometric view, as shown in Fig. 11-8.

Figure 11-9 shows three orthographic views of a typical transformer vault prior to the placement of the electrical equipment, while Fig. 11-10 shows this same vault in an isometric view.

Notice that the vertical lines in Fig. 11-9 remain vertical in the isometric view also, while the horizontal lines in the orthographic view project at 30-degree angles in the isometric view, and are no longer horizontal lines. The lengths of the lines—both vertical and horizontal—remain the same in both views, since the same scale is used for drawing both views.

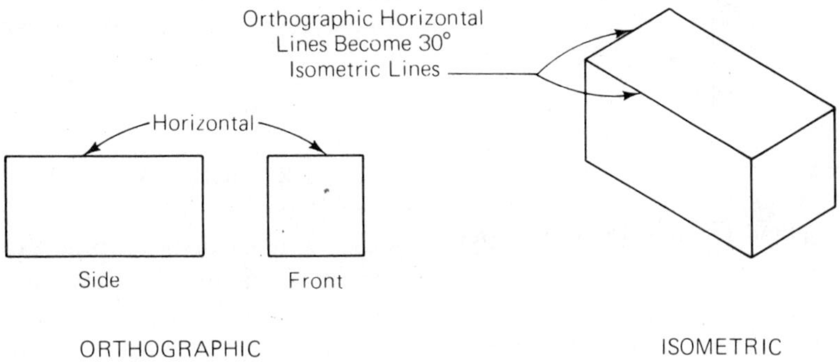

Figure 11-8. Horizontal lines in an orthographic view are projected at an angle of 30 degrees in an isometric view.

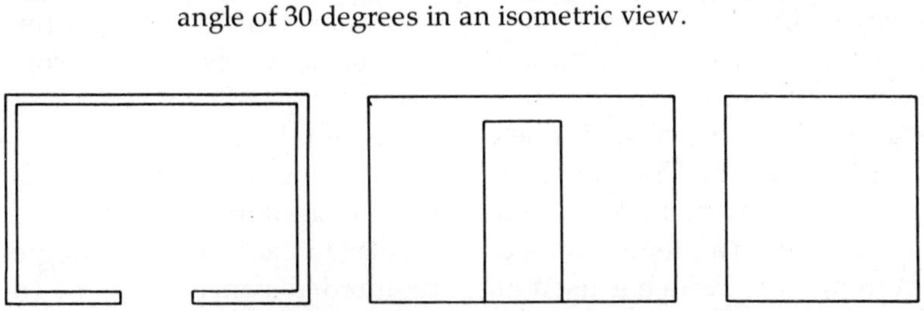

Figure 11-9. Three orthographic views of a typical transformer vault prior to the placement of the electrical equipment.

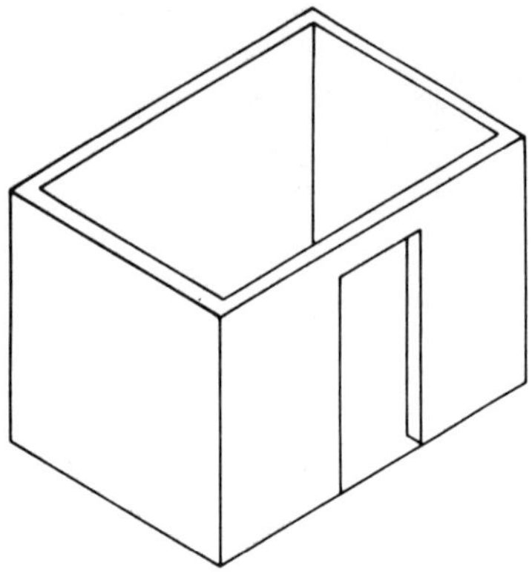

Figure 11-10. Isometric view of the transformer vault shown in Fig. 11-9.

11.4 Isometric Angles and Curves

Angles and curves are the most difficult parts of isometric drawing, but if the following methods are carefully followed, the reader will be able to draw both angles and curves with neatness and accuracy.

A diagonal line drawn through a perfect square as shown in Fig. 11-11 forms an angle of 45 degrees. The sides of the square are drawn at right angles of 90 degrees. If this angle is then divided in half as with the diagonal line, the two resulting angles at each corner of the square are 45 degrees each.

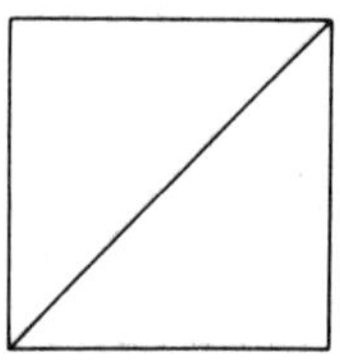

Figure 11-11. A simple method for drawing a 45-degree angle.

A similar method is used to lay out 45-degree angles in isometric drawings. For example, in order to draw a 45-degree angle in a wiring diagram, use the following method (see Fig. 11-12). This figure shows a feeder conduit running from a main distribution panel board in the basement of a building to a subpanel on the first-floor level. Draw the conduit lines with the heavy solid line (as described previously), using the 30-60-90-degree triangle, to point A. Extend the line (using a light broken line) to a predetermined distance beyond point A to point B; ½ inch is used in this case. Now measure down vertically the same distance (½ inch) from point B to point C. This establishes the two sides of a square. Connecting points A

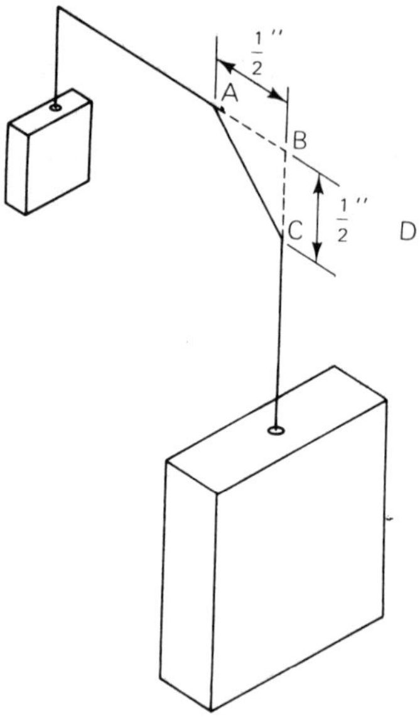

Figure 11-12. Procedure for determining position of a line set at a 45-degree angle in an isometric drawing.

and C establishes the diagonal line, which is a 45-degree angle. Then connect points C and D with a straightedge.

It is well to note that the distance from point A to point C (line AC) is not drawn to true scale, as measurements can only be made on isometric; measurements cannot be made on nonisometric lines such as line AC. All lines that are not 30, 60, or 90 degrees from the horizontal plane are nonisometric lines.

11.5 Isometric Circles

A circle in an orthographic drawing becomes an ellipse in an isometric drawing. Semicircles, likewise, become parts of an ellipse. Although an ellipse can be drawn with a compass or circle template by using four different centers, an ellipse or isometric circle template is much faster, and the final results will also be neater. However, all draftsmen should know how to make an ellipse using a compass only. You may be in a position—as for a job examination—where it will be required.

Figure 11-13 shows the four basic steps in drawing an isometric circle. First, draw two lines perpendicular to each other (90 degrees apart), mark off the required distances (size of the circle) and draw an isometric view of one side of a cube, as shown in Fig. 11-13(a); these lines should be 30 degrees from the first lines. Then draw four lines within this isometric

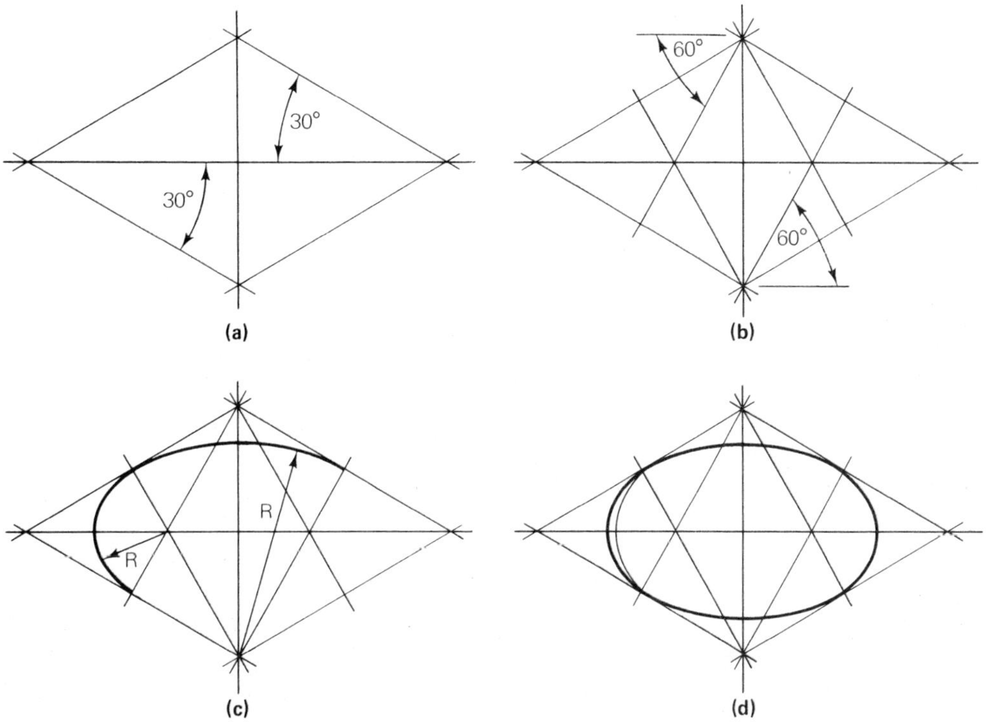

Figure 11-13. Procedure for drawing isometric circle with a compass.

square as shown in Fig. 11-13 (b). Where these lines intersect the sides of the square, so will the edge of the isometric circle. Now use a compass or circle template to form the isometric circle, as shown in Fig. 11-13 (c) and (d).

Isometric circles are used on electrical and electronic drawings for a number of applications, but we will use a resistor as an example to show the use of isometric circles.

In order to draw the resistor shown in Fig. 11-14, begin by drawing a light horizontal line to represent the centerline of the resistor. Place the appropriate size ellipse template on the centerline and draw a full ellipse; this represents the left end of the resistor. Continue by drawing the horizintal lines AB and CD—again by scaling the appropriate distances. Again, place the ellipse template on the centerline and draw arc BD. If the code markings are desired, use this same arc to draw these in also. Two horizontal lines (representing the resistor leads) complete the drawing.

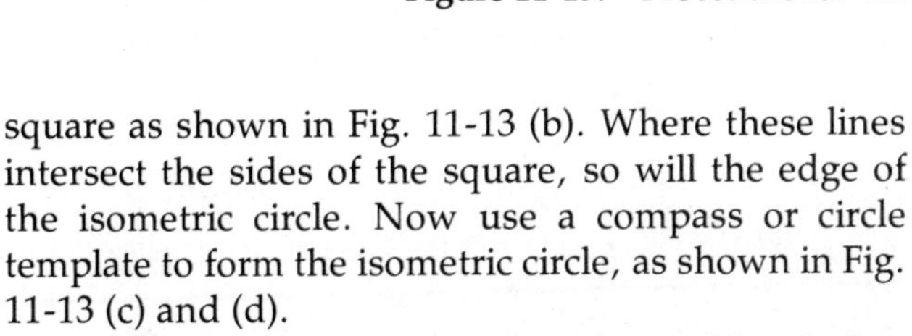

Figure 11-14. Isometric view of resistor.

11.6 Dimensioning Isometric Drawings

An isometric drawing is dimensioned with extension lines in much the same manner as an orthographic drawing, but isometric dimensioning is somewhat more difficult, because less room is available on the single view. However, in all but a few isometric drawings used in electronic and electrical drafting, the dimension lines are usually omitted, since the isometric drawings usually are accompanied by conventional orthographic drawings. The orthographic drawings have all the details of construction—including dimension—and the isometric drawings are used as supplemental drawings to give the "reader" a better picture of the object or project.

In the rare instances where isometric drawings are dimensioned, Chapter 4 gives the proper methods of doing so. Alternatively, the lines can be merely drawn to scale and the user can then use the architect's scale (or whatever scale is used) to measure the lines on the drawings.

SUMMARY OF 11.1 THROUGH 11.6

It is necessary to remember the direction of the three principal isometric planes when one is drawing isometric views. Figure 11-15 shows these directions.

All lines parallel to one of the three axes shown in Fig. 11-15 are isometric lines, and the lines not parallel to one of these axes are nonisometric lines. All isometric lines may be drawn with the 30-60-90-degree triangle.

Nonisometric lines are drawn by first locating their points on isometric lines and then connecting the end points with lines made with a straightedge.

The best way to accurately draw circles and arcs on isometric drawings is by using an ellipse or isometric template. The use of such templates is fully described in Chapter 6.

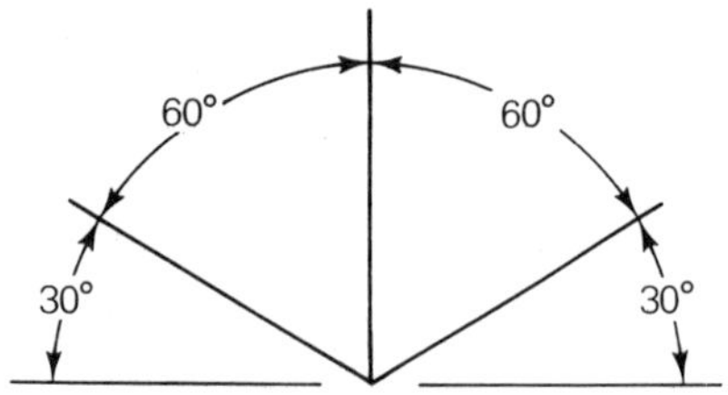

Figure 11-15. Directions of the three principal isometric planes.

ASSIGNMENT

Make an isometric drawing of a main distribution panel board, using the dimensions given in the orthographic views in Fig. 11-16. Draw this view to a scale of 1/8″ = 1′0″. The following is the procedure to follow:

1. Secure a sheet of 11″ × 17″ tracing paper to your drawing board and lay out the title block and draw border lines as described earlier.
2. Locate your reference lines approximately in the center of the sheet.
3. Use 2H drawing lead for all main-object line work; use 4H lead for all dimension lines, and H lead for all lettering. Make certain that you use guidelines to letter all of the notes.
4. Dimension the drawing according to the rules given in Chapter 4 for dimensioning isometric drawings.

QUESTIONS

1. What are isometric axes?
2. What are isometric lines?
3. What are nonisometric lines?

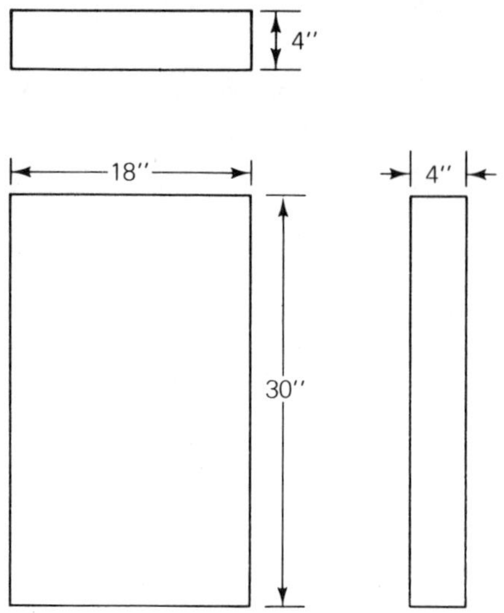

Figure 11-16. Orthographic views of panel board for use in assignment.

4. What is the best method for drawing circles on isometric drawings? Why?
5. Name one instance where an isometric drawing would be helpful in either an electrical or electronic drawing.

11.7 Power-riser Diagrams

Power-riser diagrams are used on almost every electrical drawing for building construction. Such diagrams are great time-saving devices for giving a picture of what components are used in a system and how they are connected in relation to one another. In most instances, these diagrams are drawn by using orthographic views, but sometimes—as, for example, when obstacles in the pipe runs are encountered—they are drawn in isometric views.

Figure 11-17 shows a typical power-riser diagram used on a branch bank. The diagram shows that the electrical contractor is to provide an empty 2½-in. conduit from the wall-mounted meter base (electric meter), running underground to three feet beyond the building footing. The local power company will then install the service wires underground—using a type of cable suitable for direct burial—and then run them through the empty conduit provided in order to connect them to the electric meter. The "W/BUSHING" in the note indicates that the electrical contractor is to provide a suitable bushing on the end of the electrical conduit to protect the wires as they enter from underground. Without the bushing, the metal edges of the conduit could cut the conductors.

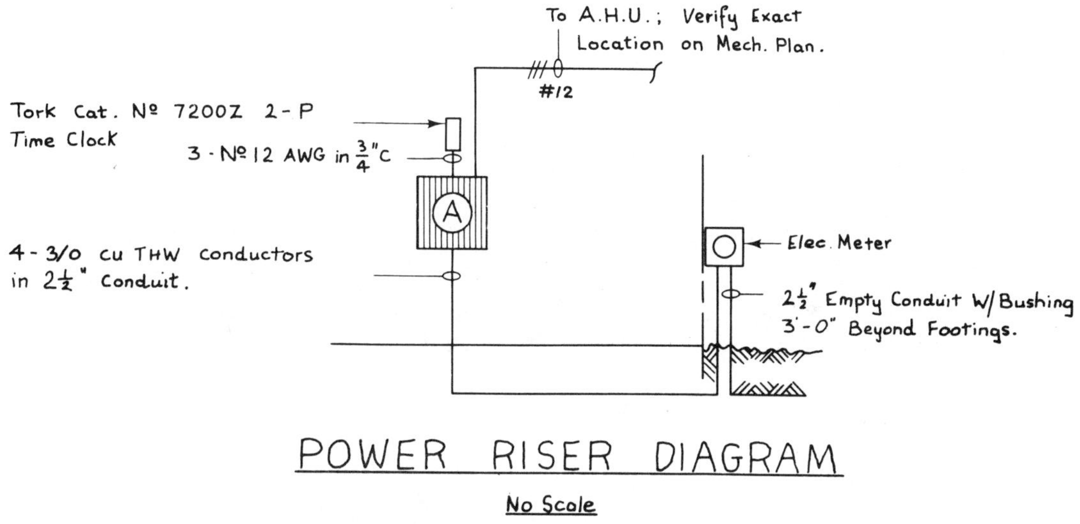

Figure 11-17. Typical power-riser diagram used to describe the electric service for a branch bank.

The data given in this diagram also show that the electrical contractor is responsible for installing a 2½-in. conduit from the electric meter to panel board A; this time with four 3/0 cu (copper) conductors. Connections, of course, will be made at both ends of the run—at the electric meter and at the panel board.

Exact details of the panel board will be given elsewhee; that is, the location will be shown on the floor plans, and the panel's components and the circuit they feed will be given in a panelboard schedule.

The remaining items shown in the diagram in question consist of a conduit containing three number 12 AWG wires to feed an air handling unit (A.H.U.), the location of which is shown on the mechanical (heating, ventilating, and air conditioning) drawings. The Tork Cat. No. 7200Z time clock is fed with three number 12 AWG wires in ¾" conduit. This time clock will be used to control outside lights; it will turn them on at dusk and off at dawn.

This power-riser diagram under consideration is relatively simple and does not require an isometric drawing unless there happen to be certain obstacles that an isometric drawing would bring out. However, in order to give the reader an example of how such a diagram would appear in isometric, this same diagram appears in isometric form in Fig. 11-18. Note that both Fig. 11-17 and Fig. 11-18 are labeled "Power-Riser Diagram" and are drawn to "no scale." They are diagrammatic only.

Another power-riser diagram is shown in Fig. 11-19. This is of a somewhat larger system than the one shown in Figs. 11-17 and 11-18. Note that the service entrance in this diagram is overhead, as opposed to the underground service on the previous example. The diagram calls for three 3½" conduits, each containing four 500MCM (five hundred thousand circular mils) conductors that terminate in a C/T cabinet (current transformer). Figure 11-20 shows how how this service entrance looks after it has been installed.

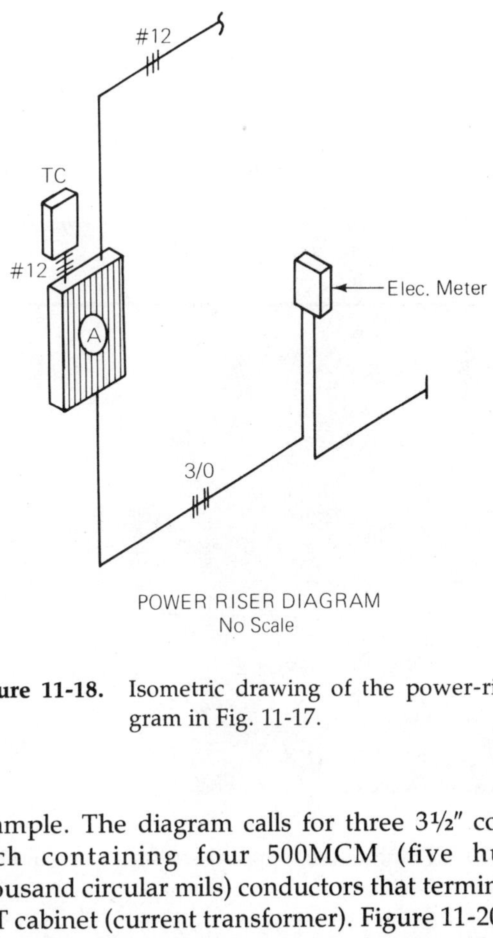

Figure 11-18. Isometric drawing of the power-riser diagram in Fig. 11-17.

A 1" empty conduit is called for between the C/T cabinet and the electric meter. This is for the power company's metering wires.

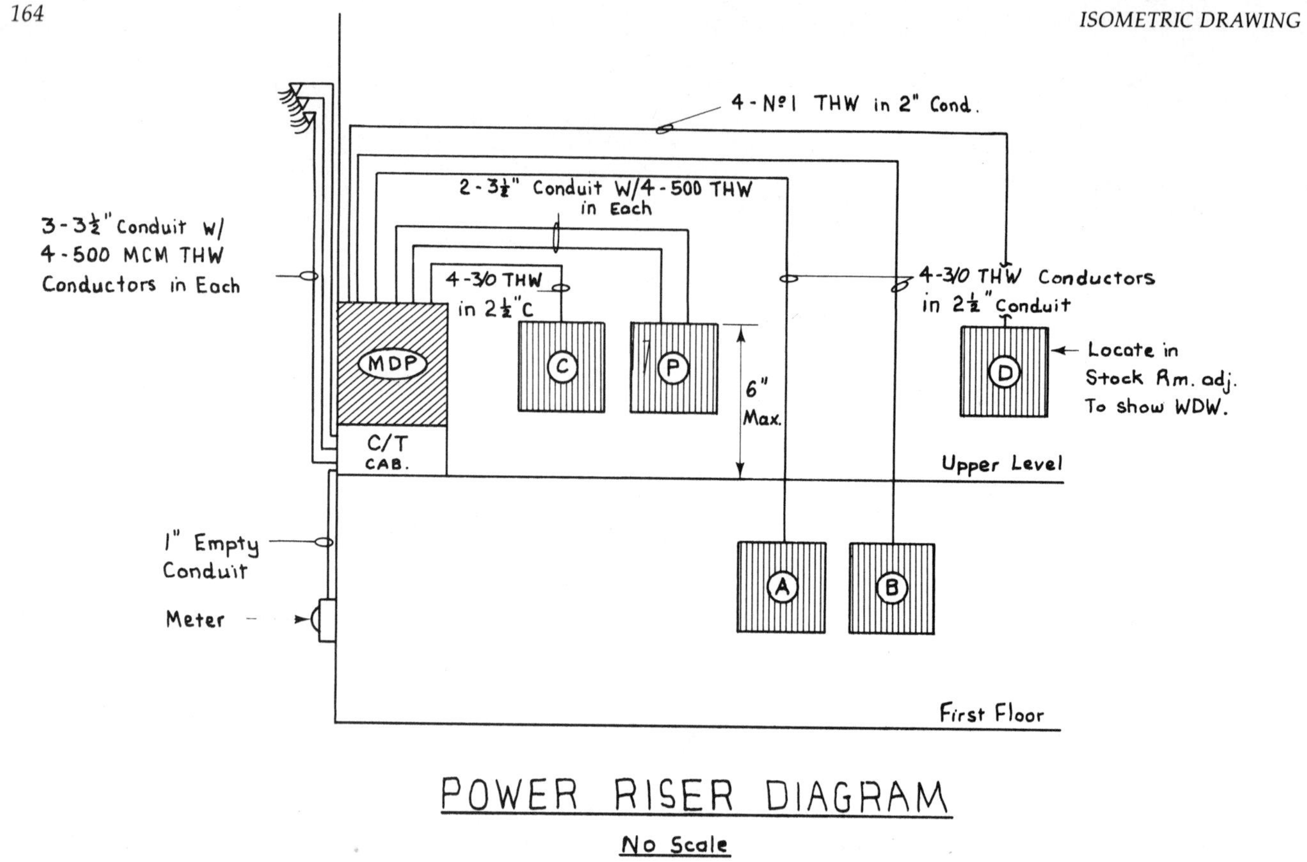

Figure 11-19. A larger power-riser diagram used on a department store project.

Figure 11-20. Photo of the electrical service shown in Fig. 11-19.

Although the exact connection details are not shown in Fig. 11-19, the service wires from the outside pass through the C/T cabinet and are connected to the main distribution panel (MDP). This panel contains overcurrent protection for the six feeder circuits feeding subpanels A, B, C, E, and P. Note that conduit and wire size (as well as number of wires in each) are indicated. One dimension line is given to remind the contractor that the top of the panel boards should be less than 6 feet above the finished floor to comply with National Electrical Code regulations. Also notice that panels D, C, and P are located on the upper level, while panels A and B are located on the lower level, or first floor. The note affecting panel D tells the installers that this panel is mounted in the stock room adjacent to the show window; it contains overcurrent protection for the show window lights, details of which appear elsewhere on the drawings or in the written specifications.

Again, this power-riser diagram is titled "Power-Riser Diagram" and is drawn to no specific scale. However, the various components should be in reasonable proportion to each other.

Riser diagrams are used for other systems besides service and power equipment. The telephone riser diagram in Fig. 11-21, for example, shows a telephone cabinet installed on the second floor of a building. A 2" empty conduit runs from the cabinet and terminates into the bar joists of the sales area; an inverted T-bar ceiling with lay-in panels conceals the bar joists in this area. When the telephone company installs the various phones in this area, they will pull

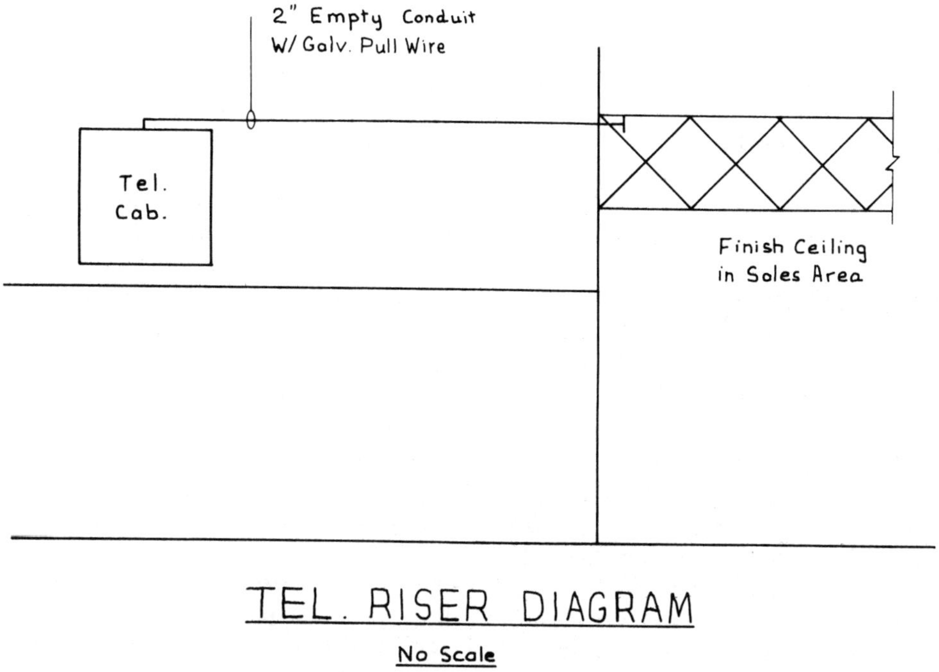

Figure 11-21. A typical telephone riser diagram.

the wires through the bar joists, and then all of the leads will be connected to the telephone cabinet via the 2" conduit. The galvanized wire is used by the telephone workmen to pull the wires through the conduit.

Another example of a riser diagram's being used on an electrical working drawing is shown in Fig. 11-22—a fire alarm system riser diagram used in a nursing home. The hexagons with "SD" lettered inside indicate smoke detectors; circles with the letter "G" indicate a gong or bell; squares with the letter "F" represent manual fire alarm stations, and the squares within squares represent door release switches to be installed by others. These various components are shown connected by circuit lines, with the slash marks indicating the number of conductors in each. Note that all systems terminate in a fire alarm panel, which is fed by a 120-volt circuit connected to the emergency generator system.

Other systems that may require the use of a riser diagram on the electrical drawings would include intercom systems, burglar alarm systems, nurses' call systems, and other types of communication and alarm. In all of these, symbols are used to indicate the various components, and lines are used to connect them. Notes are lettered on these drawings to provide sufficient data so that the systems may be installed properly without further instruction.

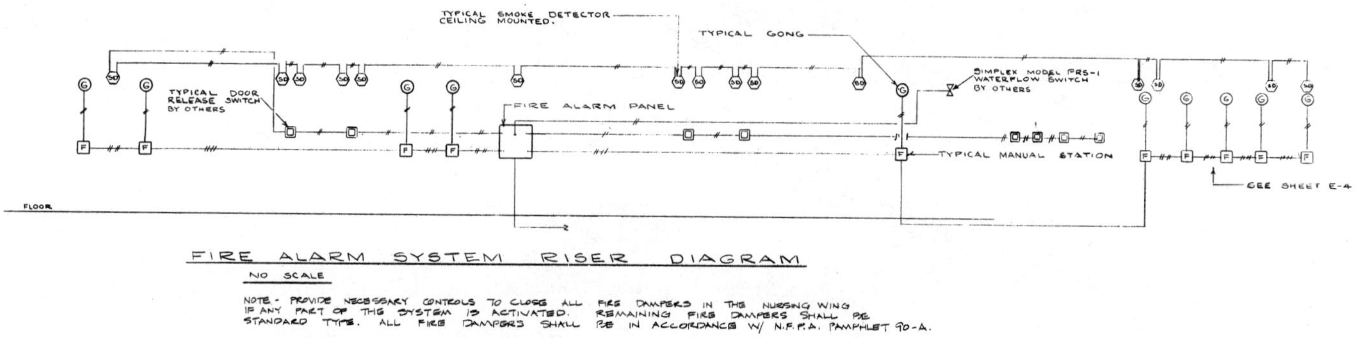

Figure 11-22. Fire alarm system riser diagram.

In most cases, orthographic views will be sufficient, but if difficult obstacles are present within the path of the circuits, it may be best to provide an isometric drawing to aid the contractor, the estimator, and the workmen on the job to better visualize the situation.

11.8 Electrical and Electronic Equipment Drawings

Isometric and other pictorial drawings are used mostly by equipment manufacturers, and electrical or electronic draftsmen employed by these firms can expect to encounter much of this type of drawing during the course of their employment. The following examples will give you a brief sampling of what types of drawings you will be expected to draw and the quality that will be required. Since straightedges and templates are used for practically all such drawings (very little freehand drawing is required), you should be able to turn out drawings of the same quality at this time.

There are preprinted forms on the market to aid in the drawing of pictorial views; better yet, drafting paper is available with printed light blue lines for practically all types of pictorial drawing (isometric, oblique, perspective, etc.). These lines are so laid out that they form the proper angle for the type of drawing, and scales are also provided so that proportioning is easy—merely let one measuring unit on this paper equal 1 inch, 1 foot, 10 feet, etc. When a blueprint is made from the tracing, the light blue guidelines will not show up; all you see is the drawing made with lead.

The perspective drawing in Fig. 11-23 shows the proper method of grounding a cable tray; that is, a grounding strap is bolted to the tray and secured to the grounding electrode, which in this case is the building steel. A splice plate at connection points insures continuity along the entire run of the cable tray. When such a tray is properly grounded, if a short circuit occurs between the live cable and the tray, the overcurrent protection (fuse, circuit breaker, etc.) will trip immediately. If the metal tray were not properly grounded, a live wire touching the tray could cause the entire tray to become "live" with voltage equal to that of the conductor touching it—causing harm to life and property.

The drawing in Fig. 11-23 is really not difficult at all, provided that perspective drawing paper is used. Most of the lines are drawn with a straightedge—following the guidelines on the perspective drawing paper—and the circles (cable end, splice plate, screw heads, etc.) are made with an ellipse template. The strap connecting the cable tray to the building steel is outlined with a French curve; only the holes in the strap, the screw slots, the shaded area of the cable in-

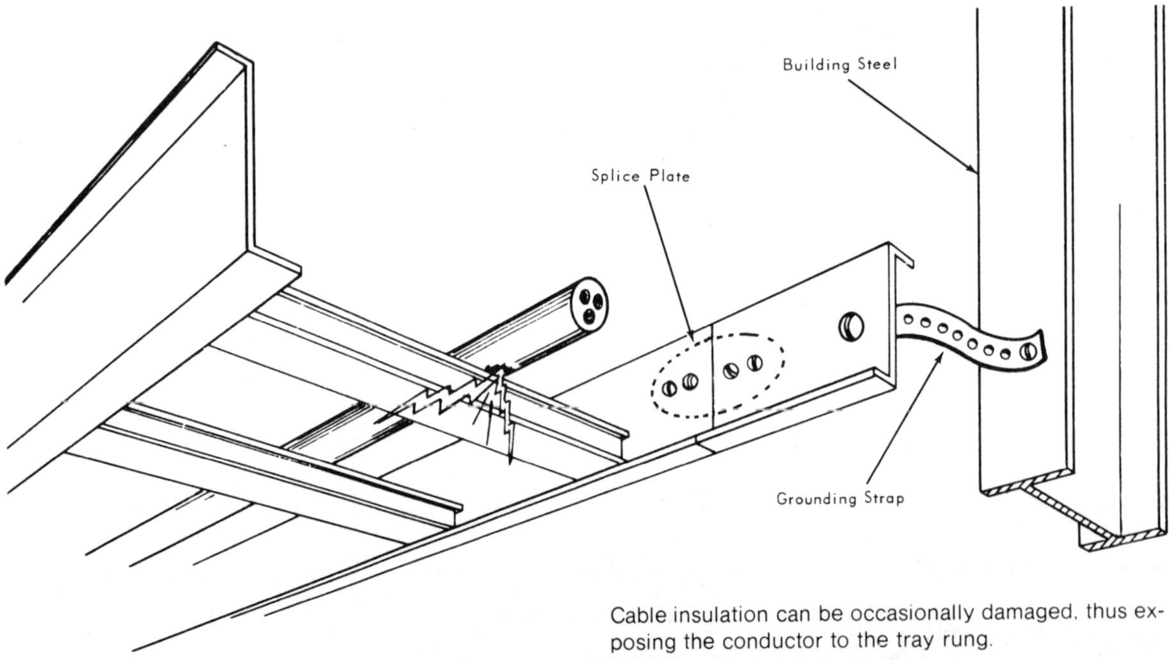

Figure 11-23. Perspective drawing showing method of grounding a cable tray.

dicating damaged insulation, and the "lightning" were drawn freehand. It would take several notes as well as many orthographic views to convey this same information to a newcomer in the electrical field, and even then, it is doubtful if the information would be conveyed as clearly as the perspective drawing indicates.

Another pictorial drawing of a cable tray is shown in Fig. 11-24. This time, the drawing shows three methods of supporting cable trays. Again, all of the details of this drawing can be made with a straightedge and templates, but in this case, the nuts, rod threads, and the conduit clamp were drawn freehand. The drawing indicates clearly how the tray is supported and also how a conduit may be secured to the tray.

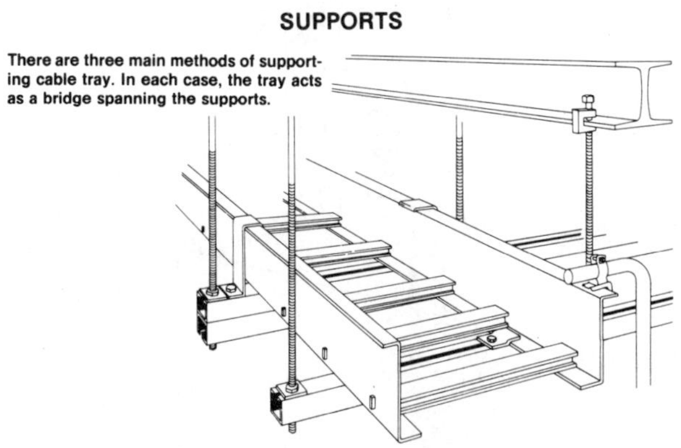

Figure 11-24. Pictorial drawing of a cable tray, showing methods of supporting the tray.

You should now begin to see just how valuable pictorial drawings are to the electrical trade in instructing workmen in new techniques.

Pictorial drawings are also extremely helpful to those working in the electronics industry. For example, the pictorial drawing in Fig. 11-25 clearly shows the installation of this portion of an electronic instrument. Can you imagine the amount of notes, schedules, and even orthographic drawings that would be required to convey this same information? Even those who have had no previous experience in assembling electronic components should be able to assemble this portion of the instrument from the drawing alone. In most cases, however, step-by-step instructions are also provided as an extra precaution.

Electrical draftsmen working for lighting manufacturers are often required to make pictorial drawings of lighting fixtures in order to give their customers an idea of what a special fixture will look like prior to its manufacture. Such a drawing appears in Fig. 11-26. This drawing looks rather complicated at first glance, but if pictorial drawing paper is used, the drawing will be greatly simplified. To illustrate, the greatest portion of the lighting fixture is made with the straightedge to form the blocks. These blocks are then connected with an ellipse template. The canopy (where the chains are connected to the hanger) is drawn with an ellipse template, and the chain links can be drawn with a straightedge. Only the wood grain and a few other minor details are drawn freehand.

The drawing in Fig. 11-27 shows a pictorial drawing of another lighting fixture. Pictorial drawing paper is used; the cubes are drawn with a straightedge, the circles with an ellipse template, and the decorative patterns on the sides of the fixture are made with a combination of templates and a straightedge. An experienced draftsman—after the drawing has been laid out—should easily finish such a drawing in a hour or even less.

ASSIGNMENT

Prepare a sheet of 11″ × 17″ tracing paper with border lines and title block, after securing it to your drafting board.

The drawing in Fig. 11-28 shows an orthographic view of a power-riser diagram. Although the depth of the electrical panels is not shown on this drawing, assume them to be 4″ deep. Then draw this same riser diagram in an isometric drawing, enlarging the view from this book to twice its size on the drawing sheet. Remember that this is a diagram only, and exact dimensions are not critical, but try to keep the various components proportional to each other in size.

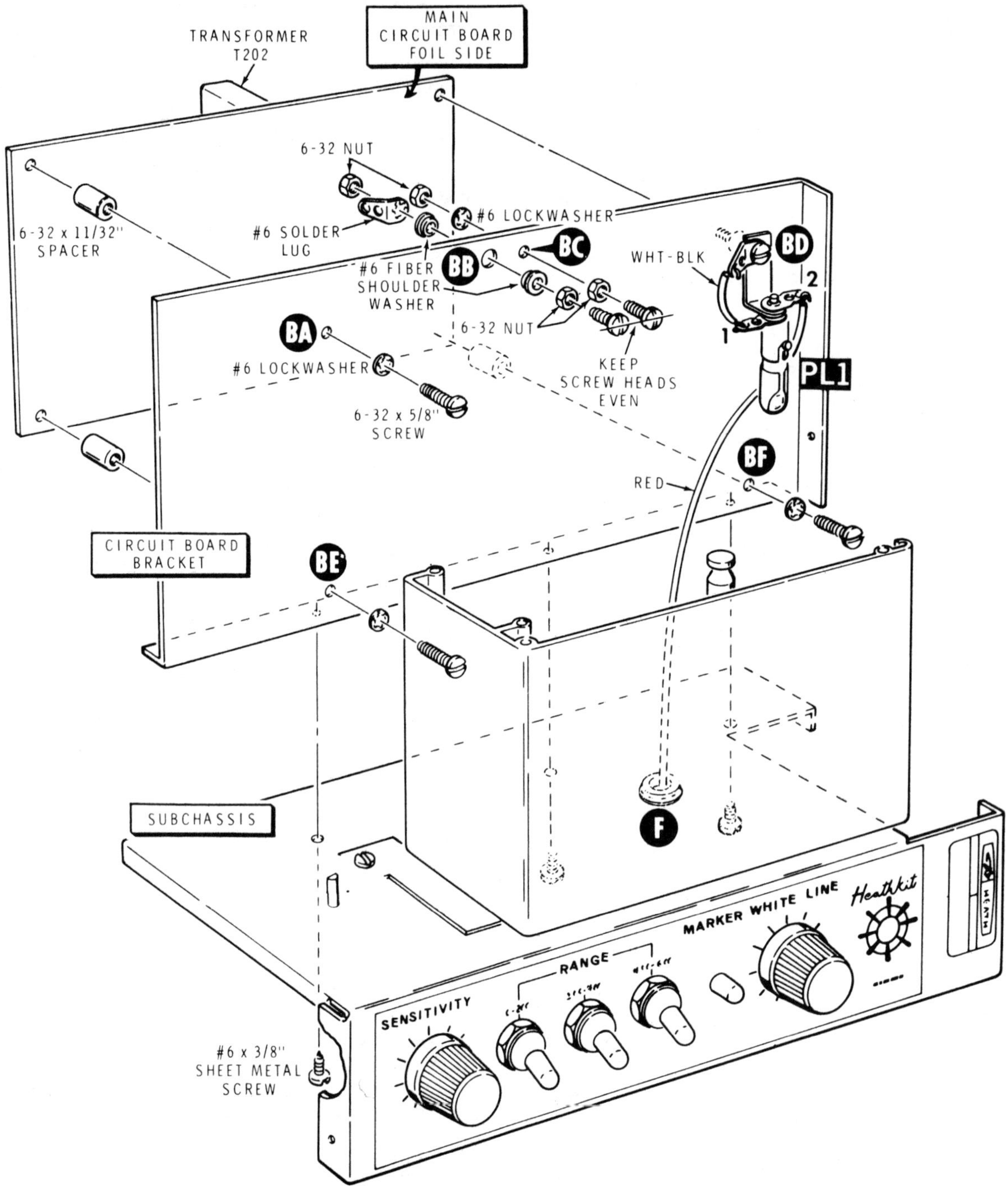

Figure 11-25. Pictorial drawing of an electronic instrument.

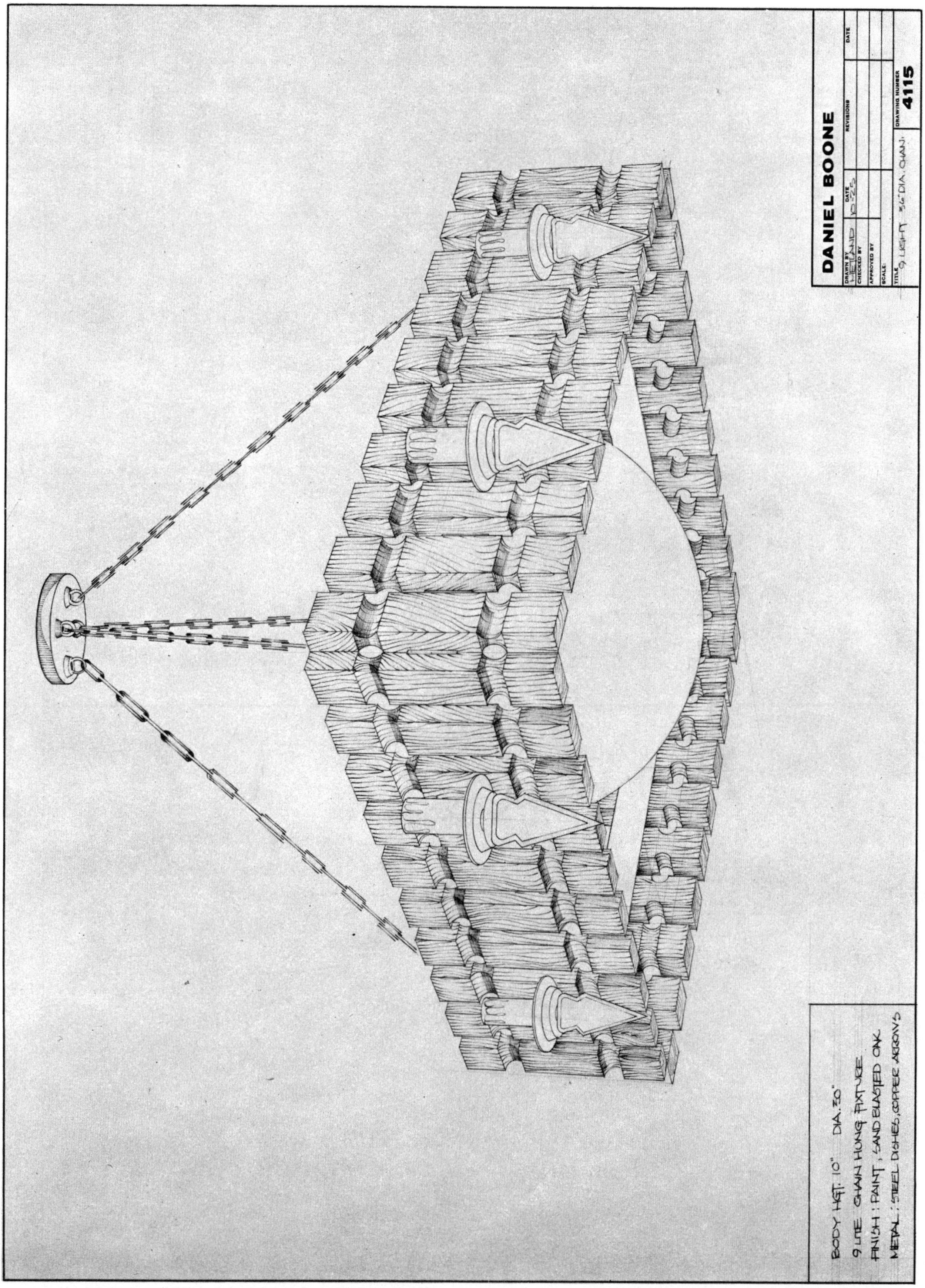

Figure 11-26. Pictorial drawing of a lighting fixture as it would be produced by a lighting manufacturer.

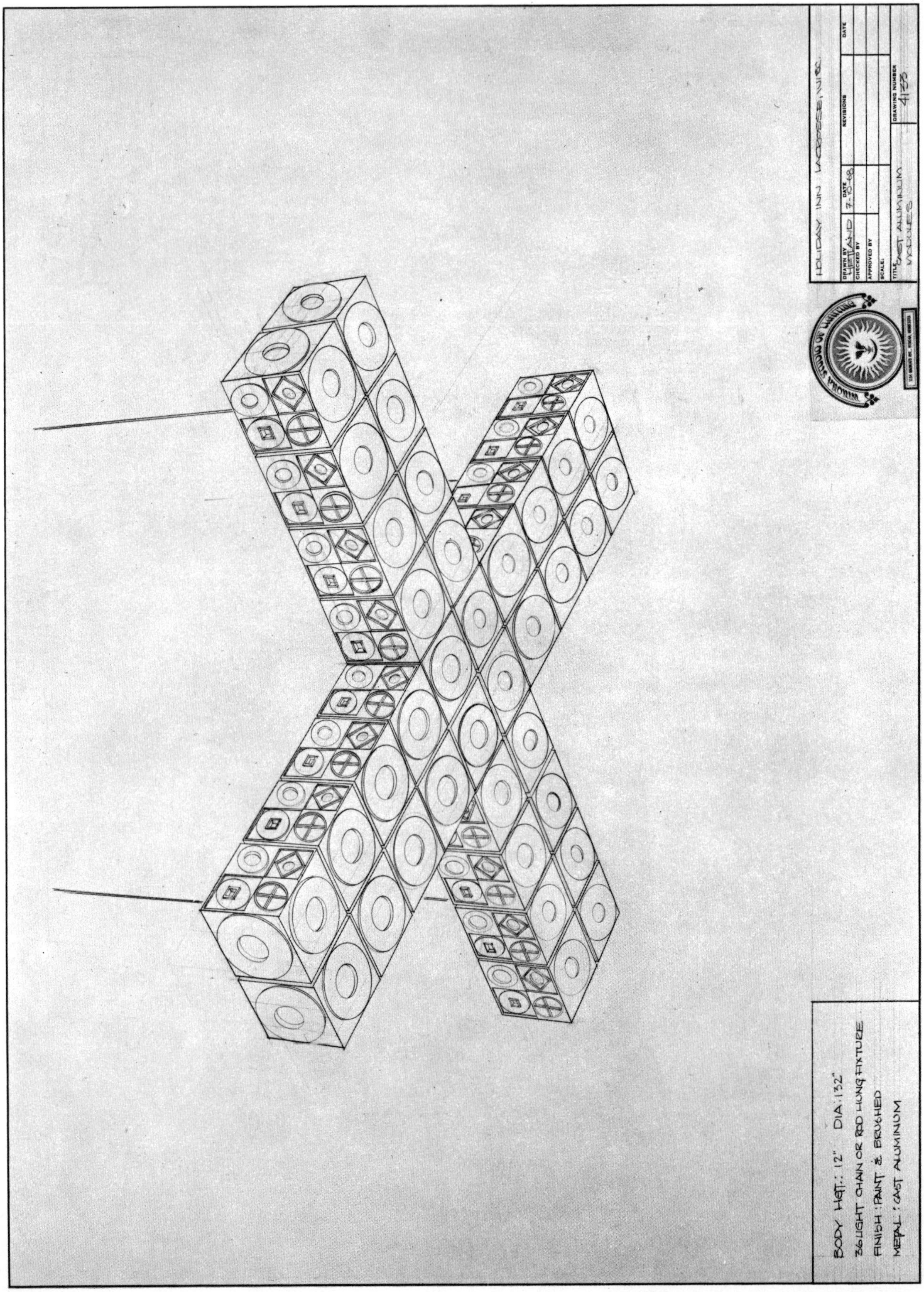

Figure 11-27. Another pictorial drawing of a lighting fixture.

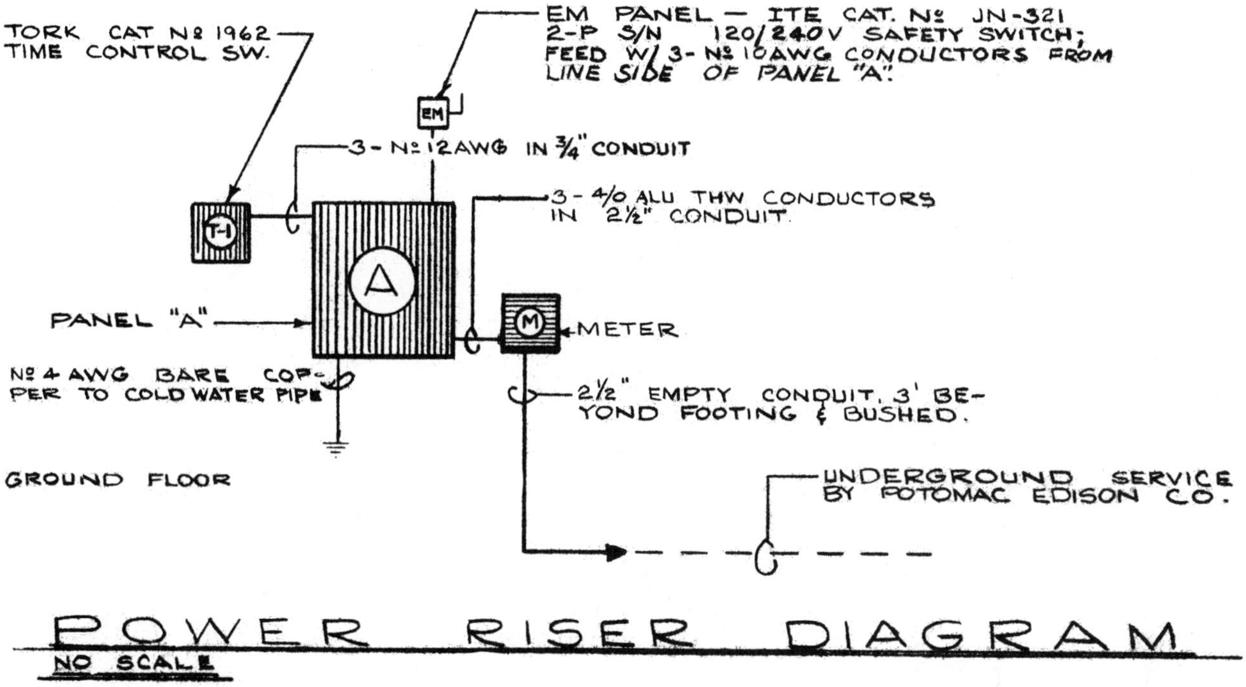

Figure 11-28. Drawing for assignment.

Chapter 12

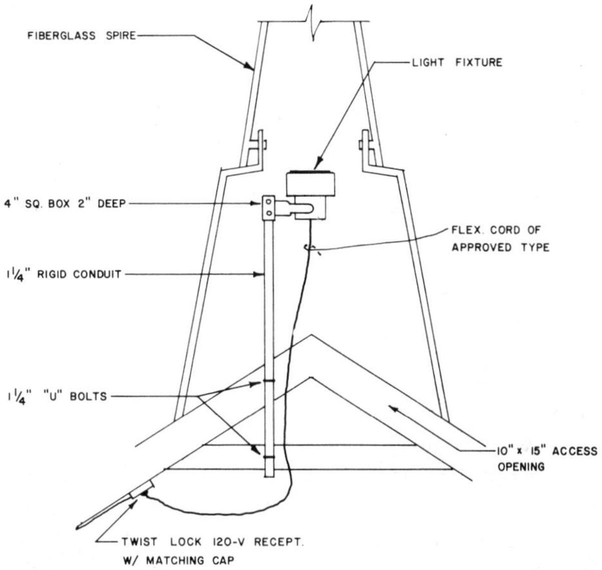

SECTIONAL DRAWINGS AND DETAILS

The main objective of this chapter is to give the reader instructions on building sections so that he will obtain an overall mental picture of the buildings as they are shown on architectural drawings. The importance of using and understanding sectional and detail drawings will also be pointed out.

12.1 Sectional Views

A section of any object—as applied to drafting—is what could be seen if the object were sliced or sawed into two parts at the point where the section was taken. For example, if you wanted to see how a capacitor is constructed, you could place the capacitor in a vise or clamp and saw it in half with a hacksaw. When the two parts are separated, you can easily see how the capacitor is constructed, or you would at least have a view of its internal construction—even if you did not understand it. This typifies the need for sectional views in electrical and electronic drawings.

In dealing with sections, the draftsman must use a considerable amount of visualization. Some sections are very easy to visualize, but some are extremely difficult. There are no given rules for determining what a section will look like. For example, an electrical conduit, cut vertically as shown in Fig. 12-1, will be in the shape of a rectangle; cut horizontally, as shown in Fig. 12-2, it will be a circle; cut on the slant, as shown in Fig. 12-3, it will be an ellipse. Without such views, however, all invisible details within an object could not be shown—except for the use of dotted hidden lines, which are seldom very clear and which cause the drawing to be difficult to read.

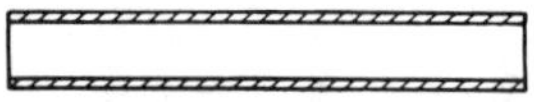

Figure 12-1. An electrical conduit, cut vertically, will be in the shape of a rectangle.

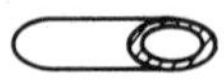

Figure 12-2. (*Left*) Cut horizontally, the conduit will be a circle.

Figure 12-3. (*Right*) Cut on the slant, the conduit will be an ellipse.

The theory of the construction of a sectional view is illustrated in Figs. 12-4 through 12-6. Figure 12-4 shows an electrical conduit viewed from the end (a) and side (b). In Fig. 12-5(a), a cutting plane is

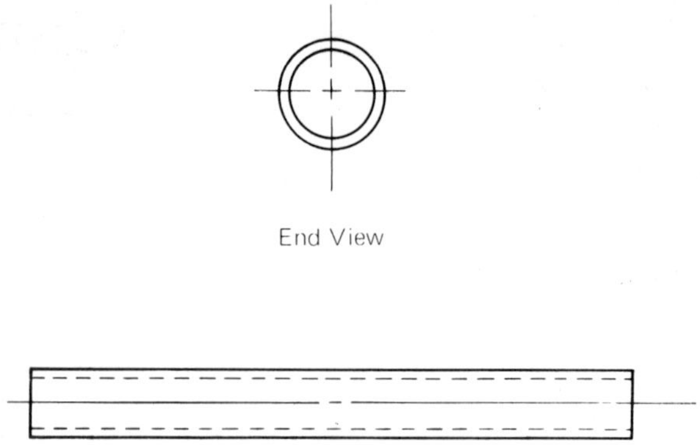

Figure 12-4. An electrical conduit viewed (a) from the end and (b) from the side.

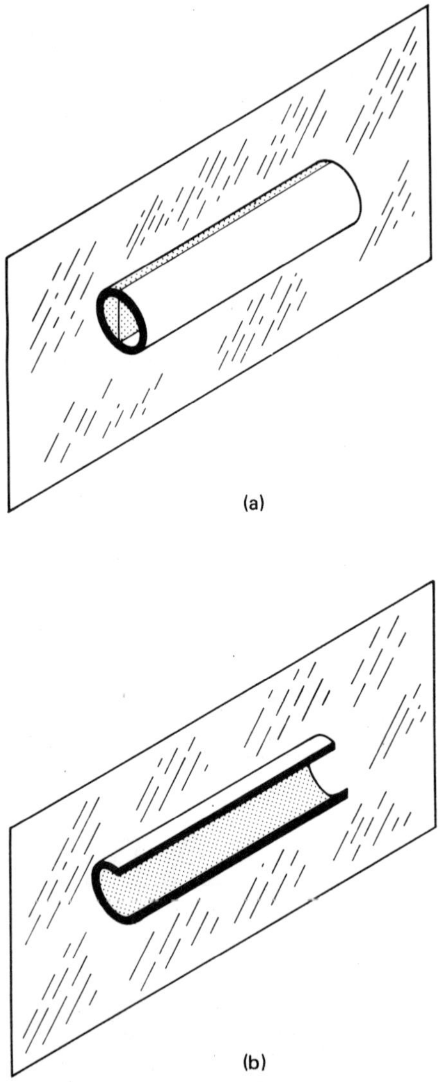

Figure 12-5. (a) A cutting plane line passing through the conduit. (b) The portion of the conduit section between the viewer and the cutting plane has been removed to reveal the interior details of the conduit.

shown passing through the conduit; this is where the section is taken and is indicated by the cutting-plane line. In Fig. 12-5(b), the portion of the conduit section between the viewer and the cutting plane has been removed to reveal the interior details of the conduit. Then in Fig. 12-6, the cutting plane is removed, and the section would be drawn in an orthographic view.

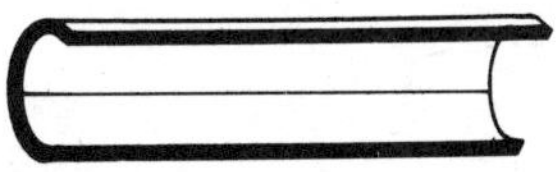

Figure 12-6. Cutting plane removed from the conduit.

Notice in the sectional view in Fig. 12-4 that sectional lining or cross-hatching is drawn across those surfaces that are in contact with the cutting plane. This section lining is used in sectional views to indicate the various materials of construction. For the common metal—such as the galvanized steel of the conduit—lining is made with fine lines, usually drawn at angles of 45 degrees. A few of the common section-lining symbols are given in Fig. 12-7.

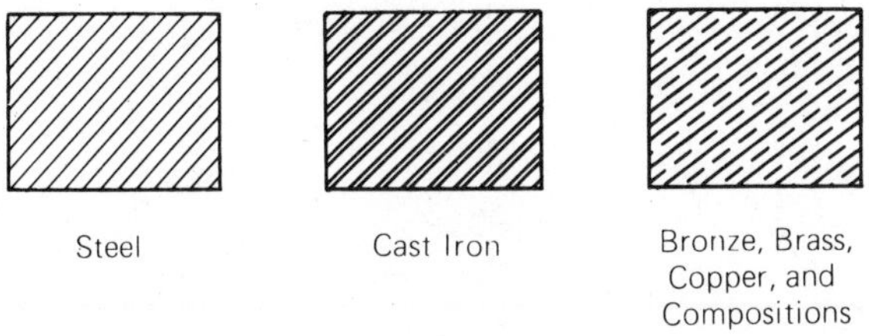

Figure 12-7. Common section-lining symbols.

Care should be taken to draw section lines or groups of section lines (whatever the case) with uniform spacing. In the smaller views, the lines should be approximately 1/32 inch apart; on larger drawings, they should be about 3/16 inch apart. The beginning draftsman should measure these distances until he has acquired enough skill to space the lines uniformly by eye.

Sectional views of conduit, electrical cable, manholes, equipment chassis, and other electrical and electronic equipment are classified as:

1. Full section
2. Half section
3. Offset section
4. Revolved section
5. Removed section
6. Aligned section
7. Broken-out section
8. Assembly section

Full section. A full section is a sectional view in which the cutting plane is assumed to pass entirely through the object. The sectional view in Fig. 12-8 is called a full section.

Half section. A half section is a sectional view in which the cutting plane passes halfway through the object. One half of the view is shown in section, while the other half is shown from the exterior. Figure 12-9 shows a cutting plane passing halfway through a piece of underfloor round duct which is constructed of tile. Figure 12-9 shows a portion in front of the cutting plane removed, and this portion is indicated by the cross-hatched lines.

Offset section. An offset section is a sectional view in which the cutting plane is bent, or offset, as shown in Fig. 12-10.

Revolved section. A revolved section is a cross section that has been revolved through 90 degrees. It is used to show the true shape of the cross section of bars and other elongated parts. Figure 12-11 shows an example of such a section.

Removed section. A removed, or detail, section is a cross section that has been removed from its original position to a convenient space near the principal view, such as the section shown in Fig. 12-12.

Aligned section. An aligned section is a sectional view in which a sloping part is rotated parallel to the cutting plane to show its true shape, such as the section shown in Fig. 12-13.

Broken-out section. A broken-out section, as shown in Fig. 12-14, is used when less than a half section is sufficient to show some interior detail. An irregular break line separates the section from the exterior view.

Assembly section. Assembly drawings often are drawn in section to show how the interior parts are fitted together. However, parts that lie in the path of the cutting plane, such as bolts, screws, etc., are not drawn in section. See Fig. 12-15.

The views in Figs. 12-14 and 12-16 through 12-19 are typical sectional views used on electrical and electronic drawings. They should be carefully studied and all details of each should be noted.

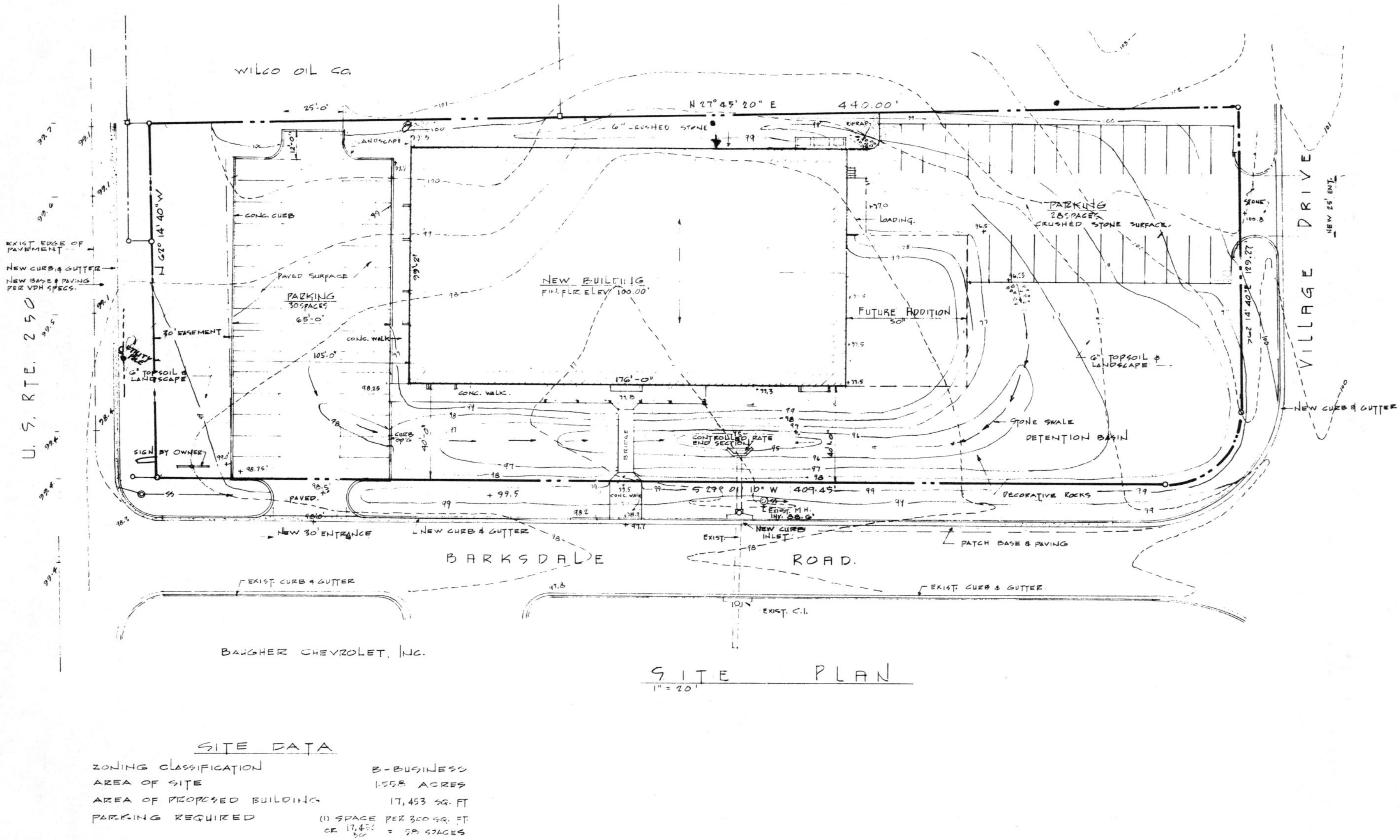

Figure 12-8. Full sectional view.

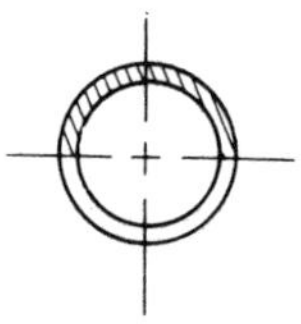

Figure 12-9. A half section.

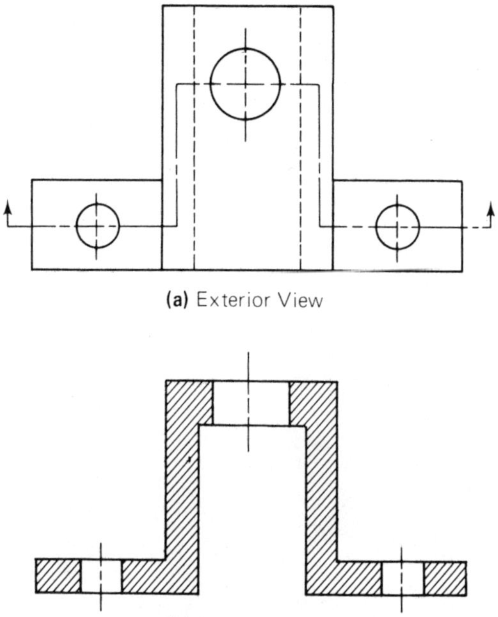

(a) Exterior View

(b) Sectional View

Figure 12-10. An offset section.

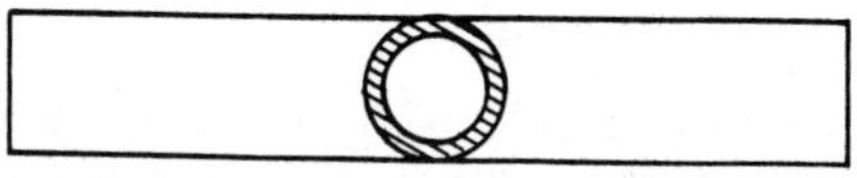

Figure 12-11. A revolved section.

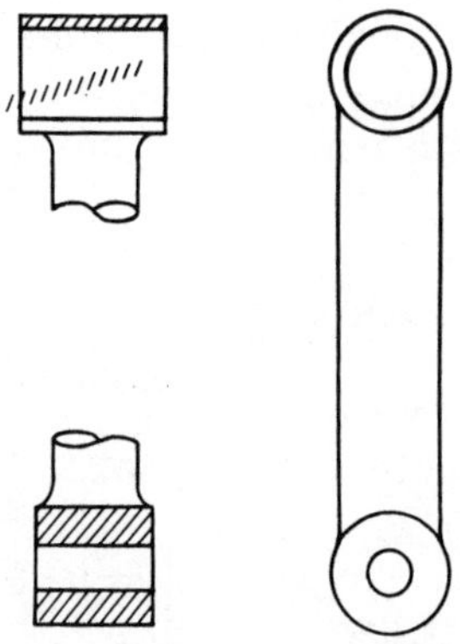

Figure 12-12. A removed or detailed section.

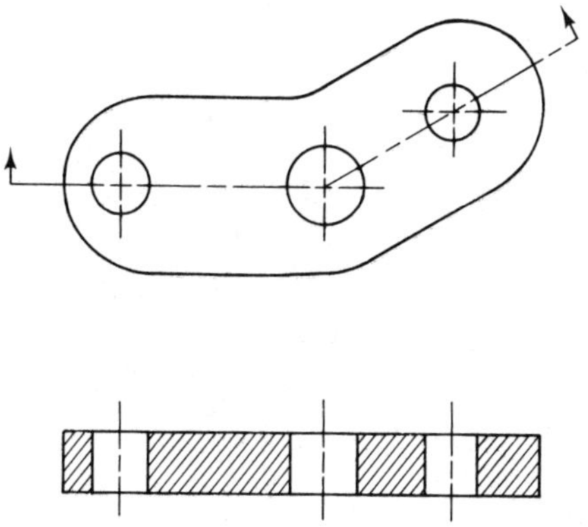

Figure 12-13. An aligned section.

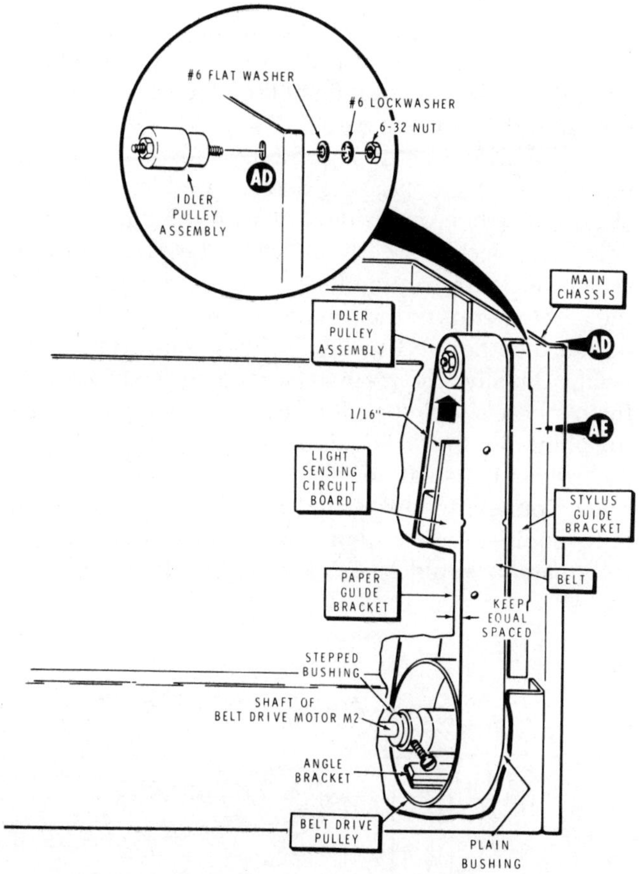

Figure 12-14. A broken-out section.

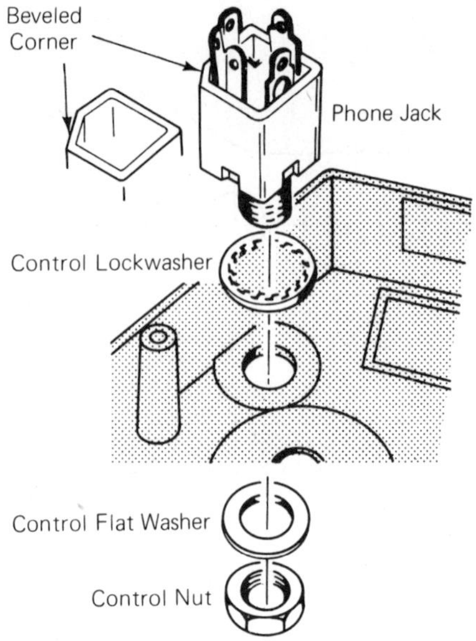

Figure 12-15. An assembly drawing.

Figure 12-16 shows (a) an elevation drawing of an outdoor lighting fixture mounted on an aluminum pole and (b) a plan view and sectional view of the base. Note that the section cutting plane in the plan view is indicated by the use of a cutting plane line. This cut is then shown in the sectional view, which gives details on the location of the mounting bolts installed in the concrete base.

The drawing in Fig. 12-17 shows a section through a cable splice, the details of which are very important when one is dealing with the higher voltages. This section clearly indicates the placement of the cable clamps on the high-voltage armored cable, the tapered ends of the conductors, and how they are butted together and secured with a crimp-type connector. Finally, the method of tapping is shown. (A further note should be added to this drawing to stress the point that the taped joints should equal or surpass the original insulation of the conductors.) Written notes conveying this same information would not only require much writing, but it is still doubtful if the information would be as clear as the drawing indicates.

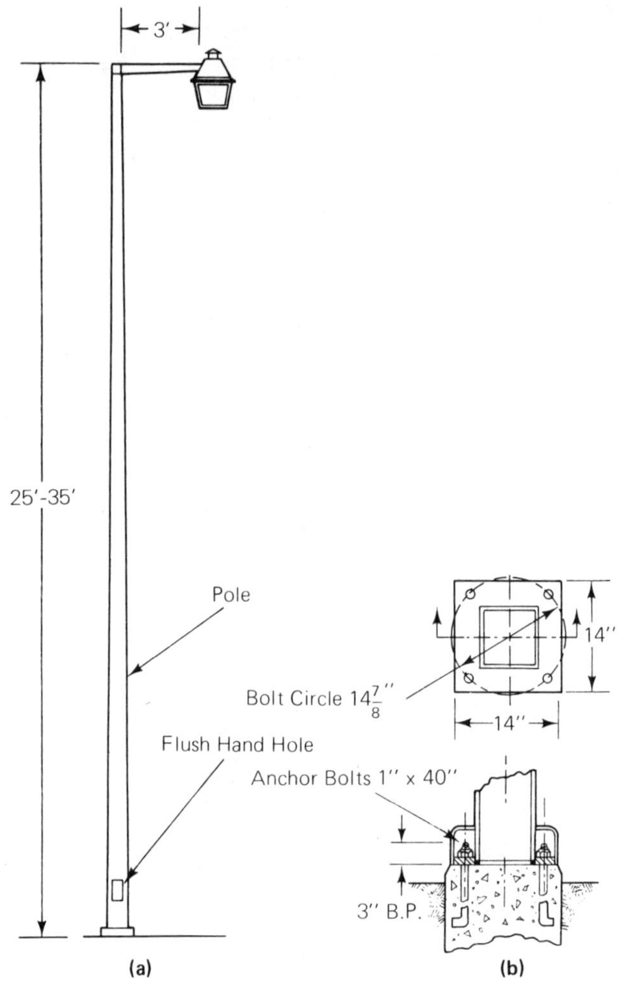

Figure 12-16. (a) An elevation drawing of an outdoor lighting fixture with (b) a plan and sectional view of the base.

An example of a broken-out section is shown in Fig. 12-14. A portion of the electronic housing (main chassis) is "broken out" to reveal the details of the belt drive pulley, the idle pulley assembly, the paper guide bracket, the belt, and other related items. Although this is shown in a pictorial view, similar sections are often shown in orthographic views.

Sectional views are extremely valuable in showing how a certain portion of a project should be in-

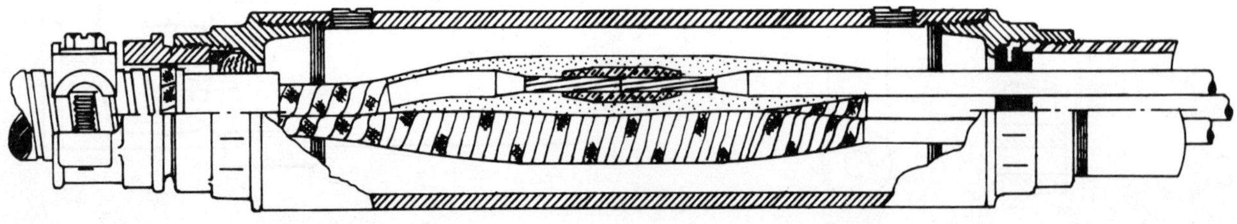

Figure 12-17. A section through a cable splice.

stalled. Consider, for example, the mounting of a transducer for an electronic depth sounder used on many boats. It is important that the transducer be mounted on the bottom of a boat so that the water flow across the face of it will be smooth. This requirement is not difficult where the transducer is mounted on a flat surface, but if the surface is not flat, fairing blocks will be required to achieve the requirement. The section through a boat hull in Fig. 12-18 shows the correct method of mounting a transducer on an angled surface—making the installation clearer than it would be without the drawing.

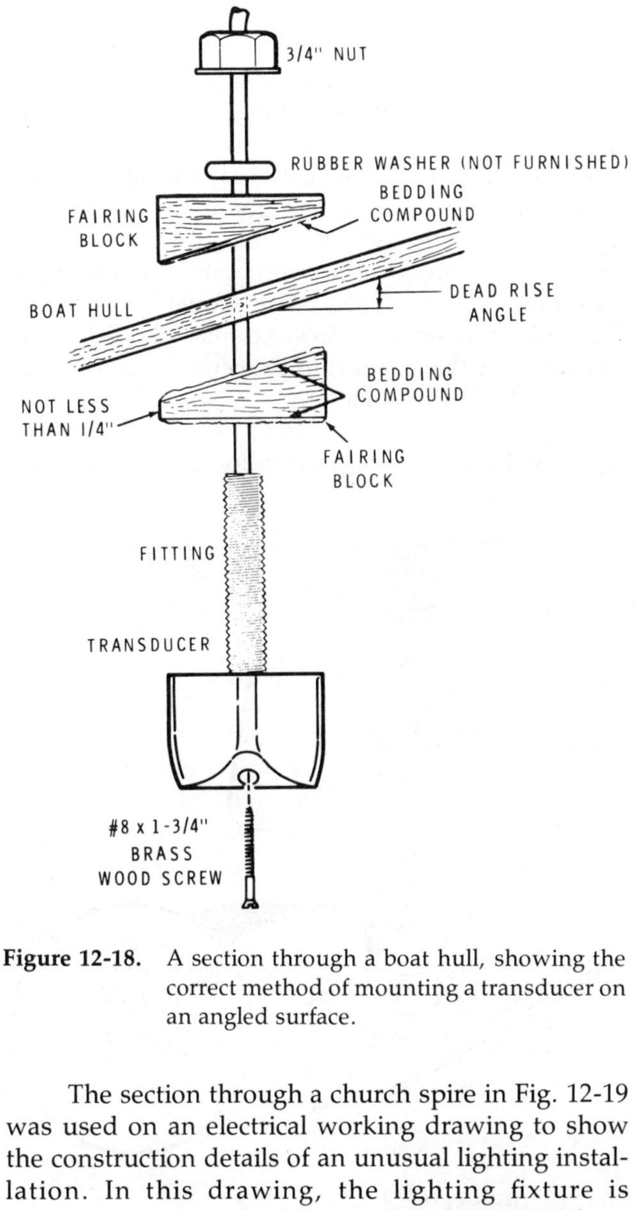

Figure 12-18. A section through a boat hull, showing the correct method of mounting a transducer on an angled surface.

The section through a church spire in Fig. 12-19 was used on an electrical working drawing to show the construction details of an unusual lighting installation. In this drawing, the lighting fixture is mounted on an outlet box which is supported by a length of 1¼-inch conduit. This outlet box and conduit are for support only, since the electrical supply is provided by a flexible cord plugged into a twist-lock

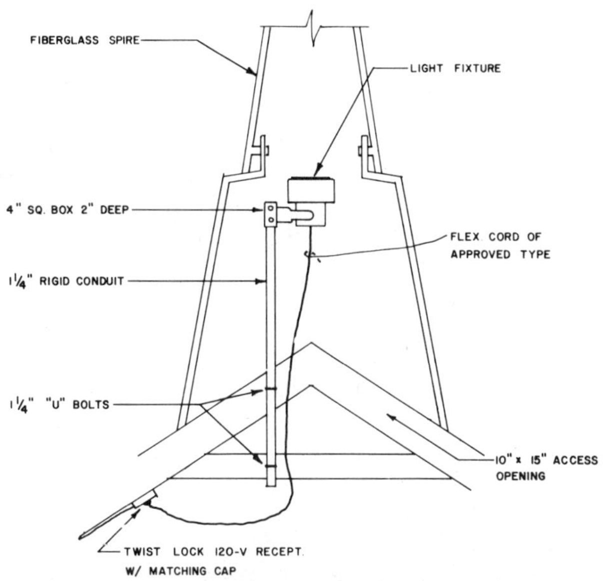

Figure 12-19. Section through a church spire used to show the construction details of a lighting installation.

120-volt receptacle. With this arrangement, the narrow-beam spotlight will shine up into the spire and give it the effect of glowing during the hours of darkness.

12.2 Electrical and Electronic Details

Electrical and electronic drawings frequently require supplemental detail drawings of a particular item or portion of the system, giving a complete and a more exact description of the item's use and all details needed to show the workmen or technicians exactly what is required for the installation. To illustrate the use of such details on working drawings, look at the floor plan of a vacation cabin in Fig. 12-20. It is obvious—from the drawing—that several outlet boxes are to be installed in the ceiling for electric lighting. However, the ceiling is constructed of tongue-and-groove solid wood deck. In order to conceal the outlet boxes, conduit, and other related electrical components, certain procedures are necessary at various stages of the building construction and an enlarged detail drawing—like the one in Fig. 12-21—would be extremely helpful to the workmen installing the system. With this detail and a few supplemental notes, the installation of the outlet boxes and branch circuits feeding the outlet boxes would consist of the following.

Immediately after the wood decking has been

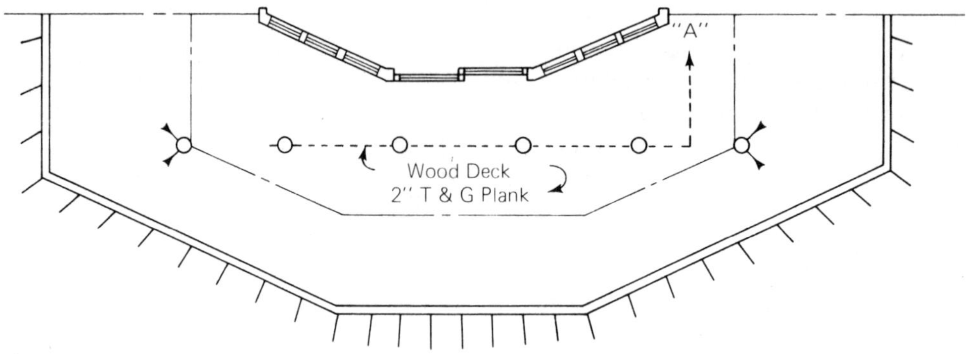

Figure 12-20. Floor plan of a vacation cabin.

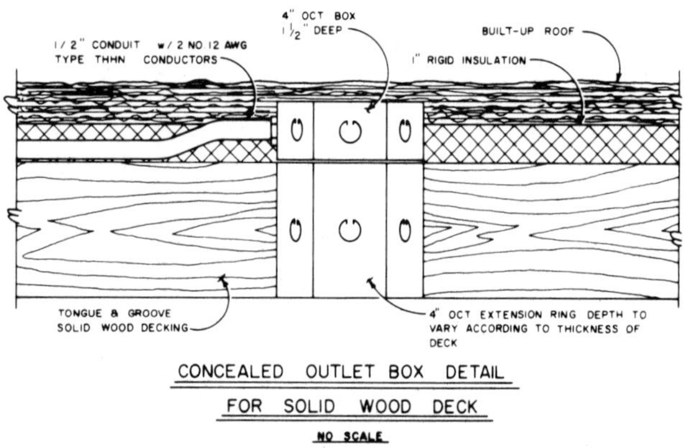

Figure 12-21. An outlet box mounting detail used to aid the workmen installing the system.

installed, all required outlets should be located and marked. A 4⅛" hole saw is then used to drill through the wood deck at each outlet location.

A 4" octagonal outlet box with an extension ring is then inserted in the opening; it should be flush with the underside of the deck. Branch circuit conduit feeding each box and strapped against the wood decking will support the box. However, if only one conduit enters the box (the last outlet on the circuit), a piece of "dummy" conduit about 12" long should be connected to the outlet box to provide support for the box on two sides.

THHN conductors should be pulled in the conduit due to the high temperatures expected on the roof from direct sunlight.

When the raceway system is completed, the roofing contractor will install the rigid insulation—leaving channels for the conduit—and then proceed with the built-up roof.

The detail shown in Fig. 12-22 appeared in a set of assembly drawings for an electronic instrument. It gives details of the installation of the circuit's cathode. Note that the insulator is cut where indicated and the funnel-shaped piece is discarded. The insert shows that a ¼" flat washer is pushed down into the insulator as shown. The cathode (part SCR) is mounted along with the insulator on the circuit board as shown; that is, with a ¼" lockwasher and ¼-20 nut. A note stating *not* to overtighten the nut to prevent breaking the circuit board would be all the information necessary for an electronics technician (or a beginner for that matter) to properly install this portion of the electronic kit.

A set of electrical drawings—from time to time—will also require large-scale drawings of certain

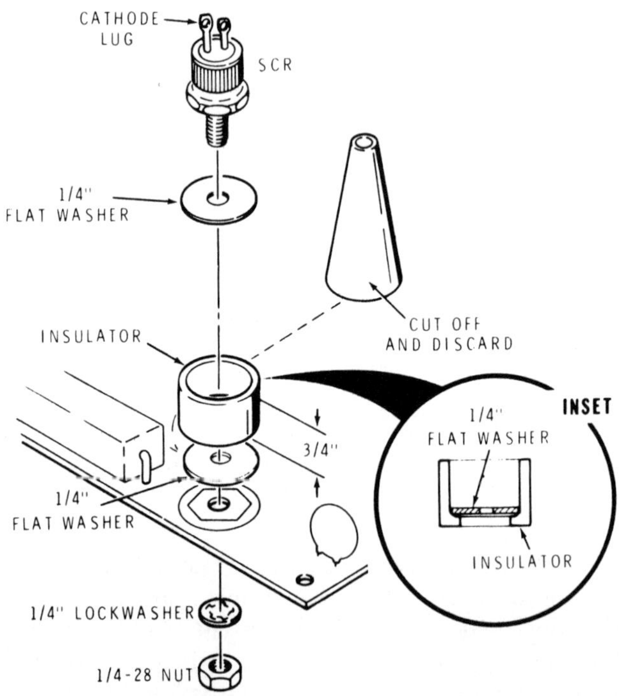

Figure 12-22. Detail which appeared in a set of assembly drawings for an electronic instrument.

areas that are not indicated with sufficient clarity. For example, the floor plan of the electrical room in Fig. 12-23 was originally drawn to a scale of 1/16″ = 1′0″. At this scale, the equipment and related wiring is quite difficult to read, but a larger-scale drawing (Fig. 12-24) is much clearer.

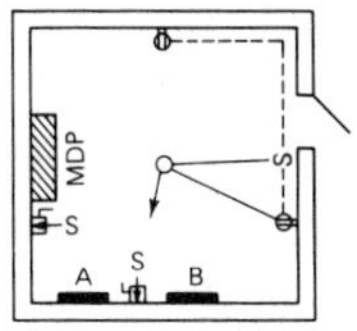

Figure 12-23. Floor plan of an electrical equipment room.

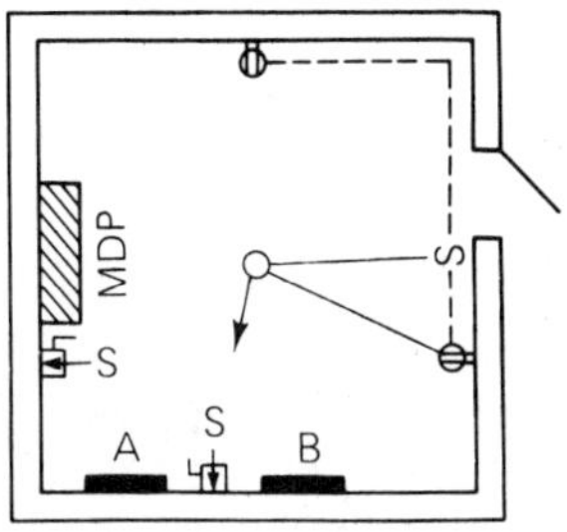

Figure 12-24. A larger-scale drawing of the electrical room in Fig. 12-23.

The electrical details in Figs. 12-25 through 12-29 are typical of those encountered on electrical and electronic drawings. Some were taken directly from actual working drawings, and some were made available by electrical equipment manufacturers. All should be studied, and every detail should be carefully noted.

Both sectional and detail drawings are made with standard drafting instruments such as straightedge, triangles, templates, etc. In most cases the draftsman will be furnished with a rough sketch—like the one in Fig. 12-30—to be used as a guide. However, sometimes the draftsman himself will be required to make a judgment as to when these sections and details are necessary on a drawing, and then what views would be appropriate. Good judgment can be gained only by experience in the electronic and electrical field, but any draftsman should have the capability of at least drawing any sectional view or detail when it is laid out by others.

SUMMARY

In working with any types of electrical and electronic drawings, a considerable amount of visualization is required. Sectional drawing can lessen the amount of

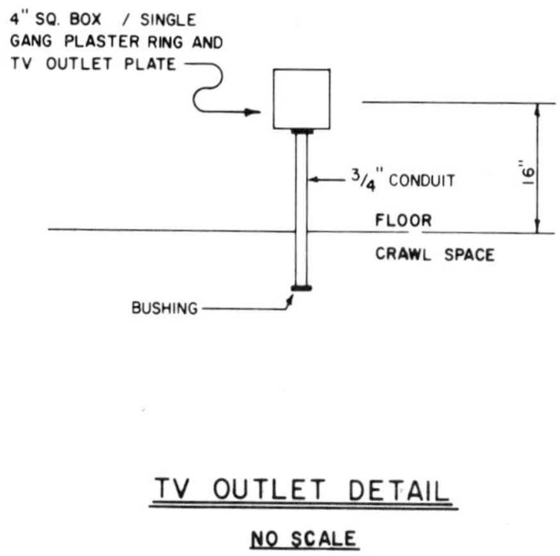

Figure 12-25. TV outlet detail.

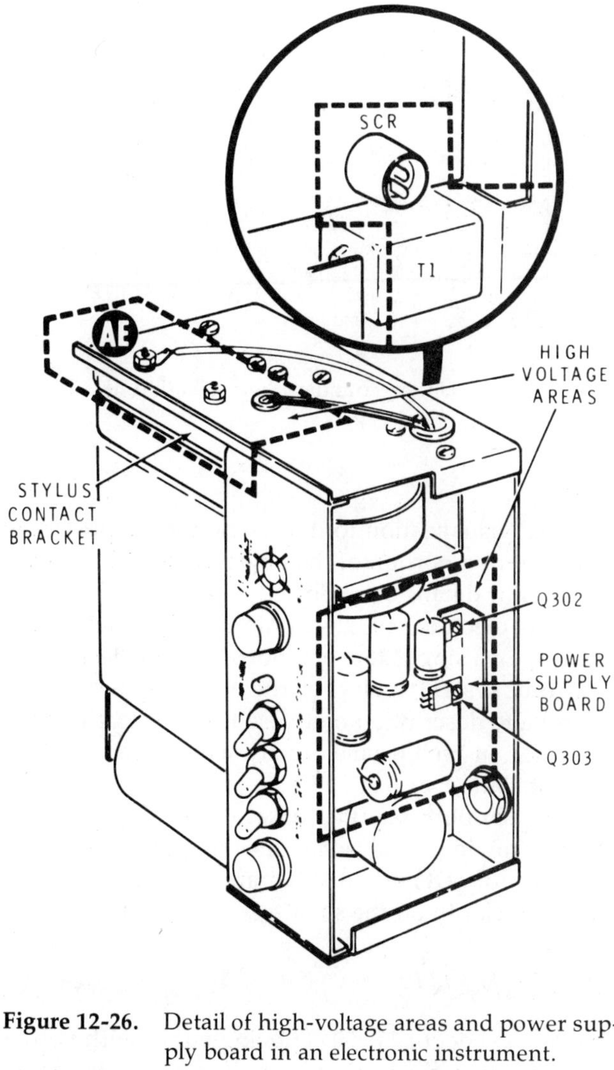

Figure 12-26. Detail of high-voltage areas and power supply board in an electronic instrument.

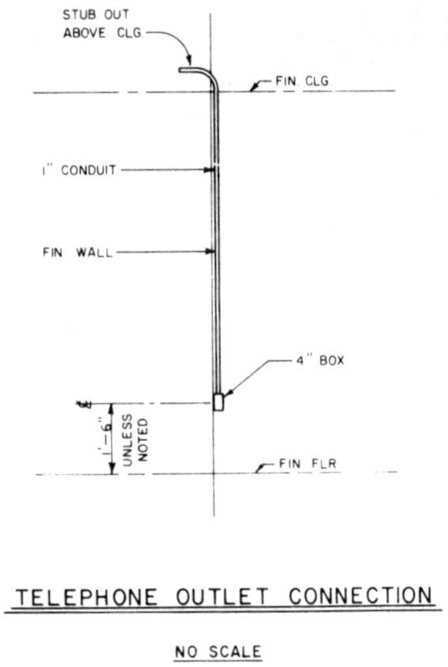

Figure 12-27. Telephone outlet connection detail.

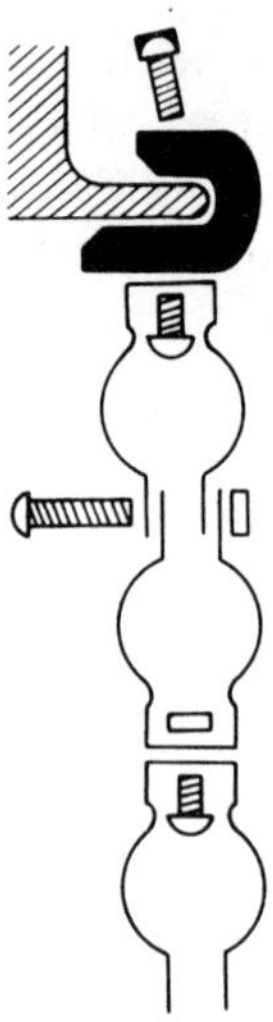

Figure 12-29. Beam clamp detail.

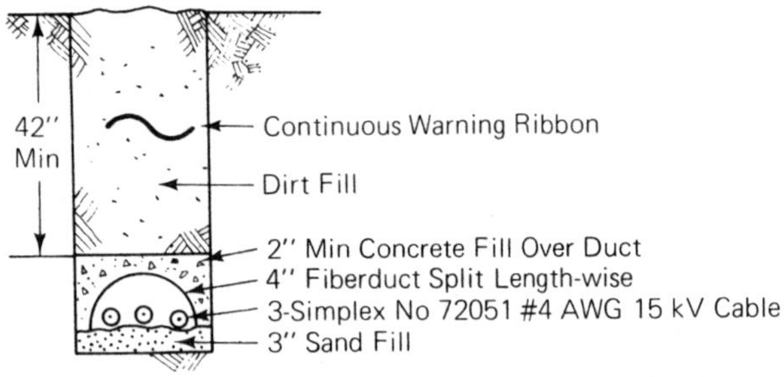

Figure 12-28. Underground cable installation detail.

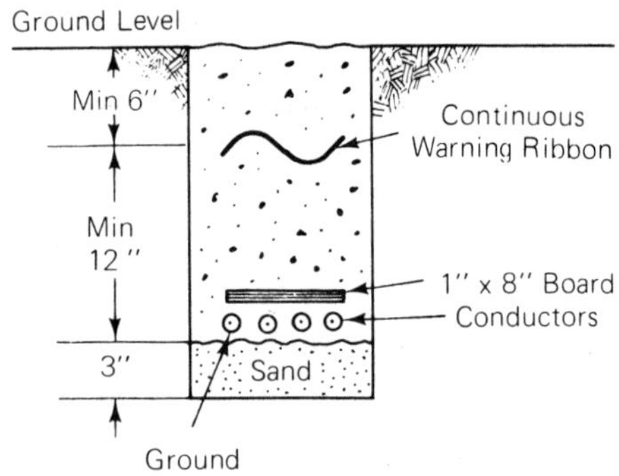

Figure 12-30. A rough sketch typical of those furnished to draftsmen by designers or engineers.

required visualization for the workmen by providing a view with a portion of the object removed to reveal the interior details of an object.

An electrical or electronic detail drawing is a drawing of a single item or a portion of an electrical or electronic system; it gives all the necessary details and a complete description of its use in order to show the workmen or technicians exactly what is required for the installation.

To simplify reading, a set of electrical or electronic drawings will sometimes require large-scale drawings of certain items that are not indicated with sufficient clarity on the small-scale drawings.

ASSIGNMENT

Prepare a sheet of 11"× 17" tracing paper with border lines and title block. Divide the length of the paper into two sections with a light line and then neatly draw the sectional drawing shown in Fig. 12-31 to the dimensions shown on the left-hand side of your paper; draw the detail in Fig. 12-32 on the right-hand side of your paper. Use guidelines for all lettering for notes; however, it is not necessary to show dimensions on these drawings—merely draw them to scale.

QUESTIONS

1. A section of an object is what could be seen if the object were _____ or sawed into _____ parts at the point where the section was taken; then the portion between the viewer and the cutting plane line is _____ to reveal the interior details of the object.
2. In dealing with sections, the draftsman must use a considerable amount of _____.

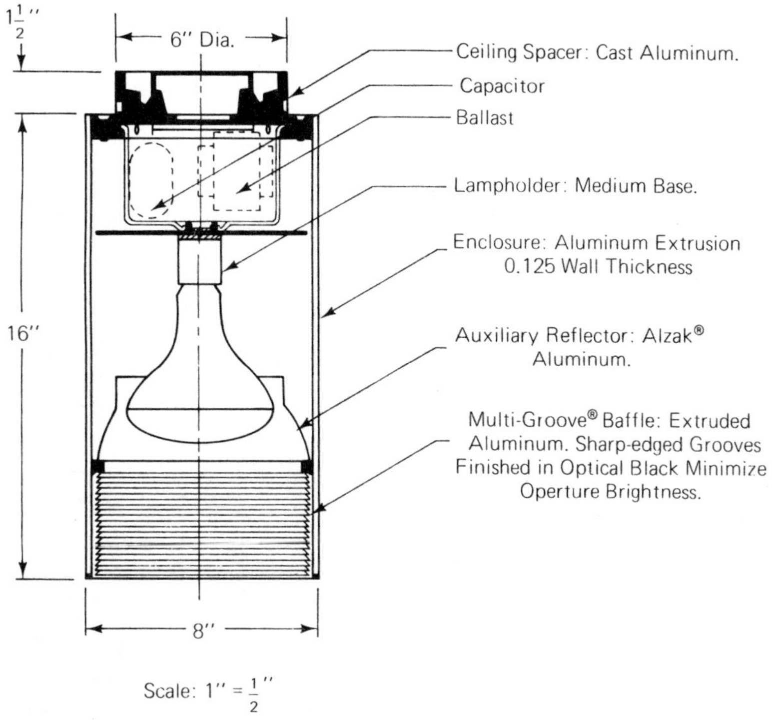

Figure 12-31. Sectional drawing for assignment.

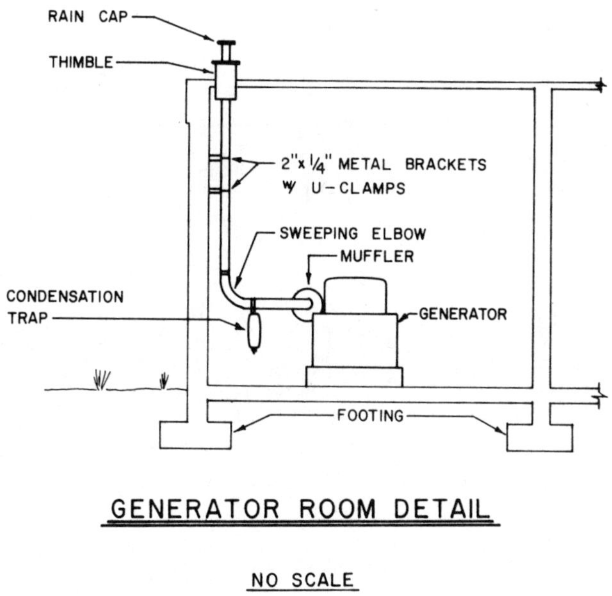

Figure 12-32. Detail drawing for assignment.

3. A half section is a sectional view in which the cutting plane passes ____ through the object.
4. A revolved section is a cross section that has been revolved through ____ degrees.
5. A supplemental drawing used with conventional electrical and electronic drawings that gives a complete and more exact description of the item's use is called a ____ drawing.
6. A set of electrical or electronic working drawings will sometimes require ____ drawings of certain items that are not indicated with sufficient clarity on the regular drawings.

Chapter 13

SITE PLANS

The material in this chapter will describe building site plans and the methods of drawing electrical requirements on them. This chapter will further describe the use of the engineer's scale—the most common form of scaling site plans.

13.1 Building Site Plans and Their Use

A *site plan* is a plan view (as if viewed from an airplane) that shows the property boundaries and the building(s) drawn to scale and in its (their) proper location on the lot. Such plans will also include (where applicable) sidewalks, drives, streets, and similar details. Utilities such as water lines, sanitary sewer lines, telephone lines, and electrical power lines must also appear on site plans—sometimes on the original site plan furnished by the architect and other times on a separate site plan prepared by the engineering firm.

In actual practice, the initial site plan will be prepared by a certified land surveyor from information obtained from a deed description of the property. This *property survey*, however, will show only the property lines and their lengths as if the property were perfectly flat. If additional information is necessary, a complete field or topographic survey will have to be made. This topographic survey, in addition to showing the property lines, will also show the physical characteristics of the land by using contour lines, notes, and symbols. The physical characteristics may include:

1. All property lines.
2. Pertinent landmarks.
3. The direction of the land slope.
4. Whether the land is flat, hilly, wooded, swampy, high or low, and other features of its physical nature.

All of the above information is necessary so that the architect can properly place the building on the property. The electrical designer, however, needs additional information in order to prepare his electrical design for the building in question. Such additional information includes the available public utilities and their location on or near the property lines. Once in possession of this information, the designer can lay out the new electrical, telephone, etc., services (in sketch form) so that the electrical draftsman can prepare neat and accurate working drawings, which will later be used by the workmen installing the systems.

Figure 13-1 shows a typical site plan of a project as it might be received from the architect for the engineer's or designer's use in laying out his portion of the electrical system. Note that the existing utility lines are shown on this drawing running along the east boundary lines of the property; this is where the designer must obtain power for the new electrical service that will feed the building.

Another architect's site plan is shown in Fig. 13-2. This plan is complete in that it shows the property lines, the existing contour lines, the new contour lines (after grading), the location of the buildings on the property, new and existing roadways, all utility lines, etc. Descriptive notes appearing on the plan also list the names of adjacent property owners, the land surveyor and the date of the survey, the direction of north, and other pertinent data, such as bench marks, monuments, and the like. A legend or symbol list is also provided so that anyone familar with site plans can readily read the information provided. This is a high-quality site plan (called "Plot Plan" on the drawing), and the reader should study this plan until he is thoroughly familiar with all details described on it.

13.2 Using the Engineer's Scale

Earlier in this book, we found that most buildings are drawn to scale with a scale known as the architect's scale; its use was fully explained in Chapter 4. Site plans are drawn to scale also, but in most instances, a scale known as the "engineer's scale" is used rather than the architect's scale. The engineer's scale is similar to the architect's scale, except that an inch is divided into 10, 20, 30, 40, 50, or 60 equal parts (Fig. 13-3). The engineer's scale marked 10 is read 1″ = 10′ or some other integral power of 10; for example, 1″ = 100 feet; 1″ = 1000′; etc.

Usually, for small buildings on small lots, a scale of 1″ = 10′ or 1″ = 20′ is used. This means that 1 inch (actual measurement) on the drawing is equal to 10 or 20 feet—whichever the case may be—on the actual lot. Since the engineer's scale is the chief means of making scaled site plans, its use should be thoroughly understood.

The engineer's scale is used by placing it on the drawing with the working edge away from the user, as shown in Fig. 13-4. The scale is then aligned in the direction of the required measurement—such as along a property line. Then, looking down over the scale, the user reads the dimension, in the case of an

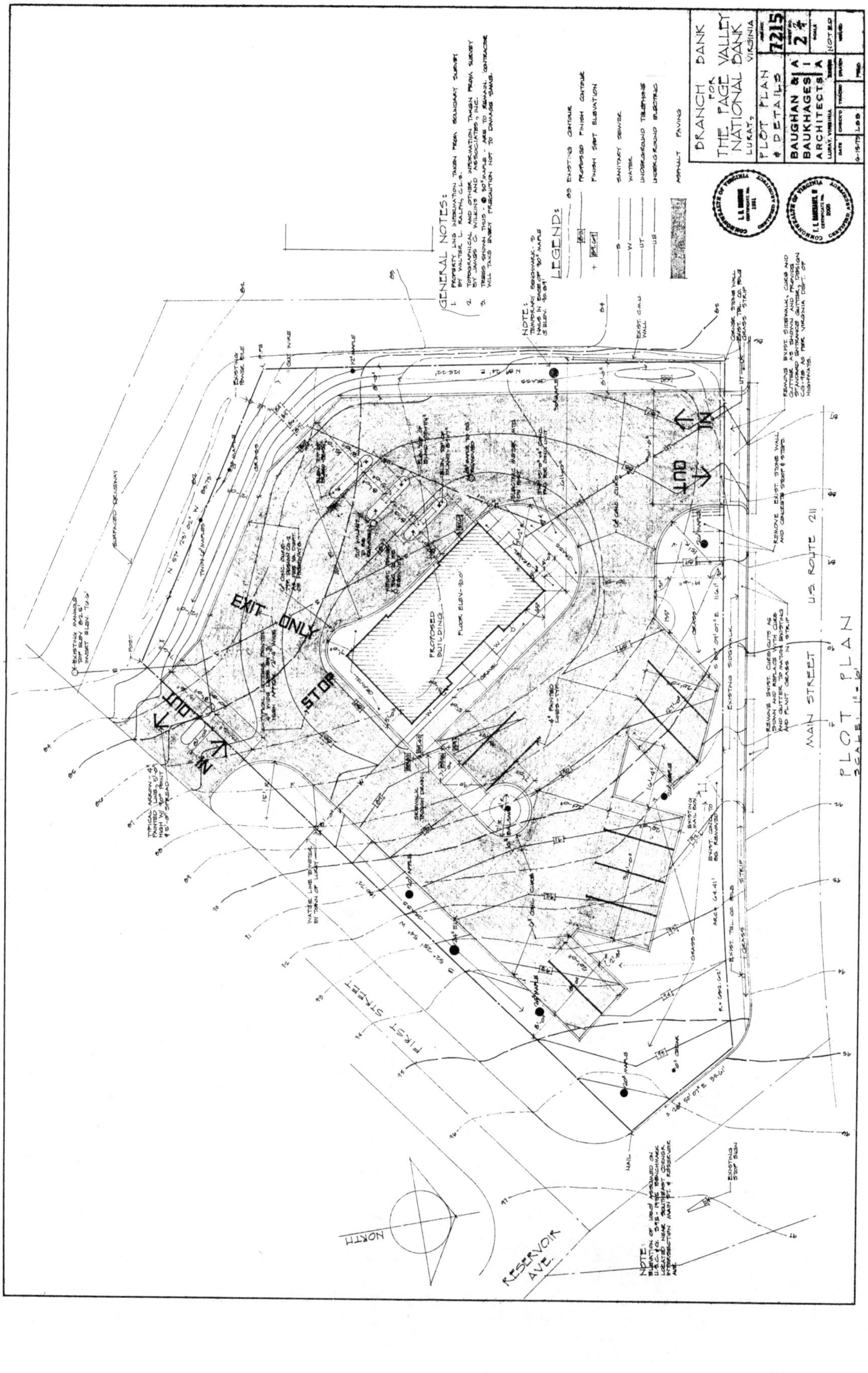

Figure 13-1. A typical architect's site plan.

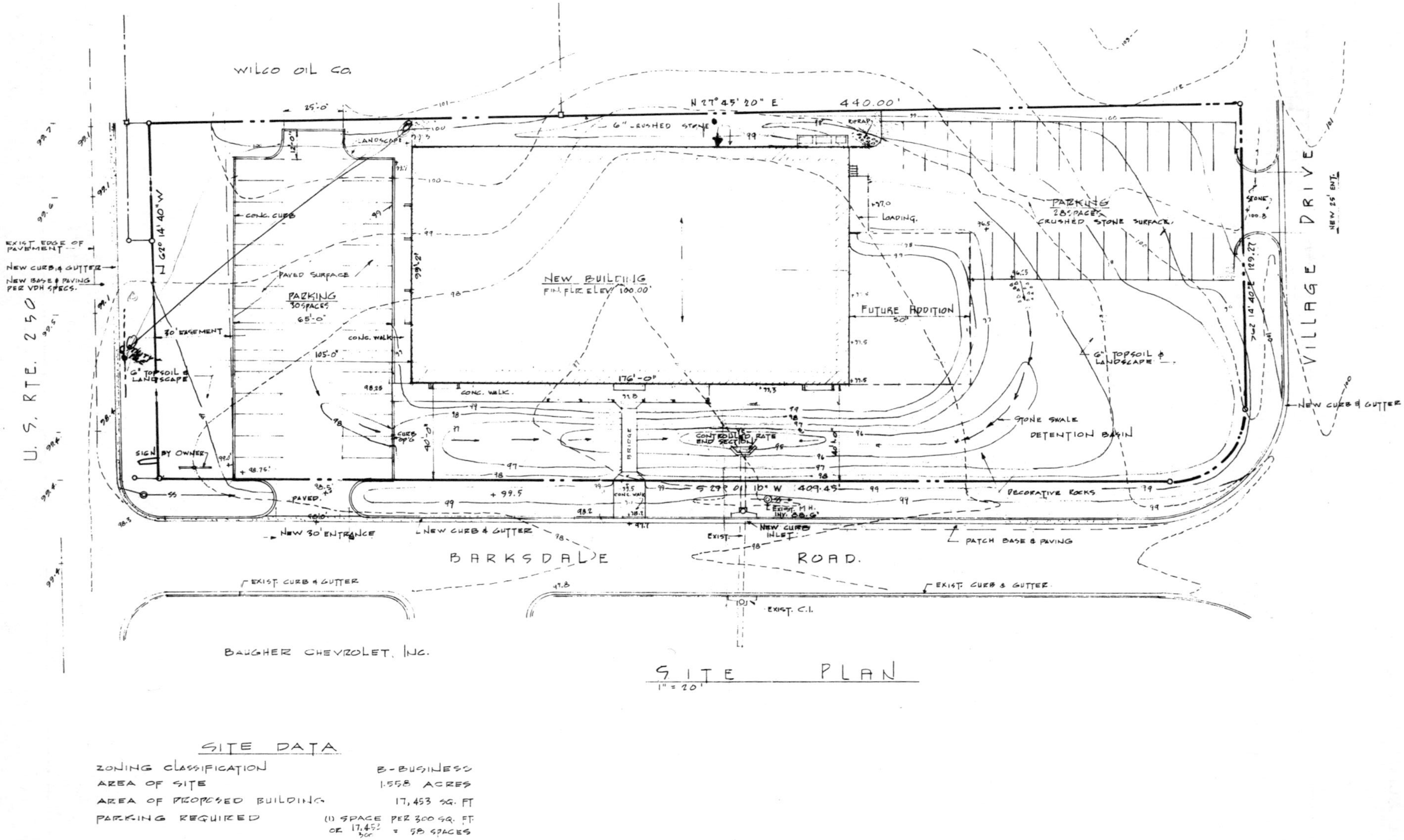

Figure 13-2. Another architect's site plan showing contour lines, building, etc.

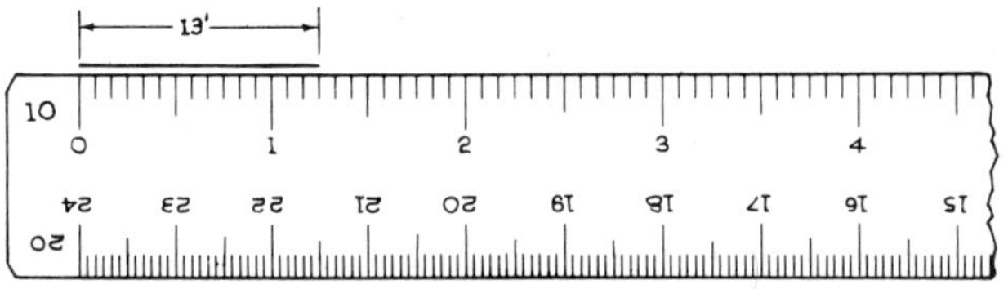

Figure 13-3. Method of using engineer's scale to measure line dimensions.

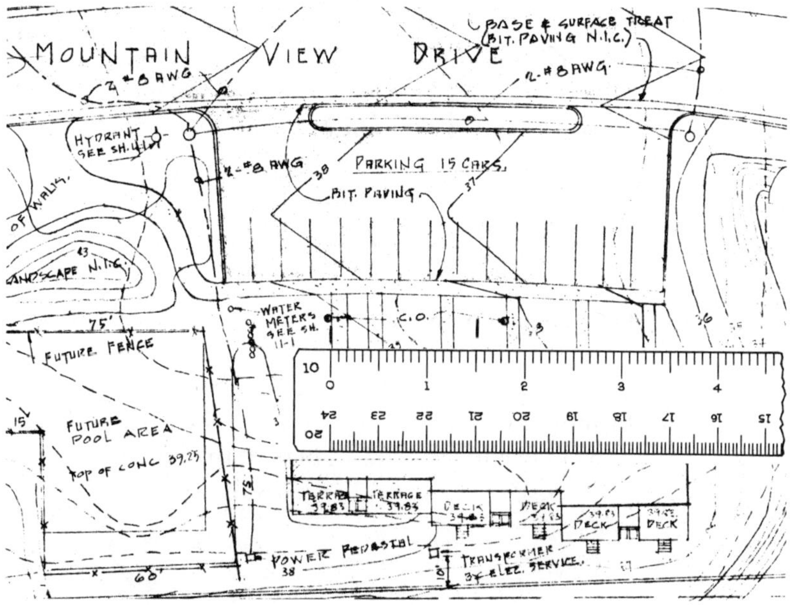

Figure 13-4. The engineer's scale is used by placing it on the drawing with the working edge away from the user.

existing drawing, or marks off the required dimension, in the case of a line that is to be drawn. The purpose of this scale, therefore, is to transfer the relative dimensions of an object to the drawing, or vice versa.

Although the drawing itself may appear reduced in scale, depending on the size of the object and the size of the sheet to be used, the actual true-length dimensions must be shown on the drawings at all times. When reading or drawing plans to scale, remember to think and speak of each dimension in its full size and not in the reduced size that it happens to be on the drawing. For example, if using a scale of 1″= 10′ and the actual measurement on the drawing happens to be 3½″, you do not say that the particular line is three and one-half inches; rather, read the measurement as thirty-five feet (3½″ × 10 = 35).

The practice problems in Fig. 13-5 will acquaint you with the use of the most commonly used graduations on the engineer's scale. For each problem, use the scale indicated just below the line. Determine the length of each line and write it just above the line. When all the lengths have been determined, compare your answers with the ones given at the end of this chapter.

If it is desirable to draw a given line to a given scale, first mark off the distance with the appropriate scale; this is indicated by two light dots on the drawing. Then use a straightedge to draw the line between the dots.

Refer again to the site plan in Fig. 13-2 and note that the scale for this drawing is 1″ = 40′. This means that every inch (actual measurement on the drawing) is equal to 40 feet on the building site. Every drawing should have the scale to which it is drawn plainly marked upon it either as shown in Fig. 13-2 or as a part of the title block.

QUESTIONS

Answer the following questions by filling in the blanks.

1. Building site plans will normally show electrical power lines as well as ____ and ____ (as related to the electrical work).
2. Site plans are usually prepared by a certified land surveyor from information obtained from a ____ description of the property.

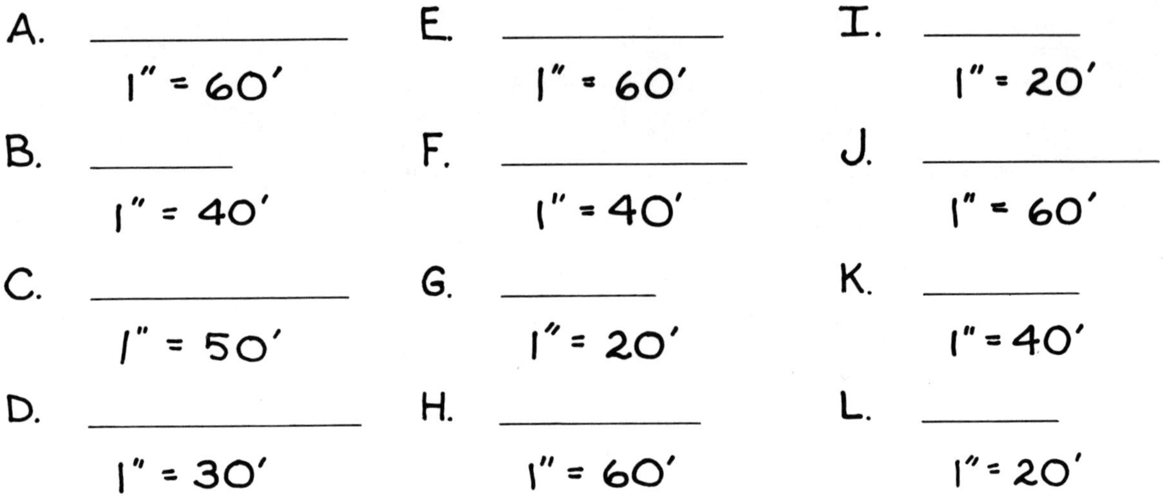

Figure 13-5. Line lengths for practice in using the engineer's scale.

3. If contour lines are necessary to show the land slope, a complete _____ or _____ will have to be made.

4. A _____ is normally provided on site plan drawings so that anyone can read the symbols.

Chapter 14

LIGHTING FIXTURE SCHEDULE

FIXT. TYPE	MANUFACTURER'S DESCRIPTION	LAMPS No.	LAMPS TYPE	VOLTS	MOUNTING	REMARKS
R-1	STARLIGHT CAT. NO. 456-7950	1	100W	120	WALL	6'-6" Above Finish Floor
R-2	Moonlight CAT. NO. 341267	1	150W	120	Recessed	
R-3	EASILITE CAT NO. & X-$%&*	1	150W	120	Recessed	
R-4	P & S CAT. NO. 110	1	100W	120	Surface	
R-5,6	Starlight Cat. No. 3476	1	150W	120	Wall	6'-6" A.F.F.
R-7	MORELITE Cat. No. RTY-60	1	100W	120	Surface	
R-8	MORELITE Cat. No. 60-X-8	1	60W	120	Wall	
R-9	Same as R-7					
R-10	Customlite Cat. No. 809	1	100W	120	Recessed	
R-11	Same as R-7					
R-12	Same as R-7					
	ETC.					

RELATED CONSTRUCTION PRACTICES

Electrical draftsmen are sometimes called upon to perform other tasks for their employer than actual drafting; that is, taking measurements and making surveys of existing buildings and their related wiring systems, checking over conditions of a drawing or project to ensure that it complies with the National Electrical Code, or perhaps finding certain data in written specifications. The draftsman may even be called upon to make a preliminary cost estimate of the project on which the draftsman is working.

This chapter, therefore, is designed to acquaint the reader with the basic procedures of electrical estimating; it also provides an introduction to the National Electrical Code, and a sampling of written electrical specifications. The knowledge of these items is considered necessary for a well-rounded and complete course in electrical drafting.

14.1 Basic Electrical Estimating

Sound electrical estimating consists of a complete take-off (or quantity survey) of all materials and equipment required for a complete installation, and then the calculation of the total labor required to install equipment and materials. To the cost of materials, equipment, and labor are added all direct expenses, variable job factors, taxes, overhead, and profit to determine a "selling price" for the project. If these procedures are intelligently performed and then combined with good job management, the estimate should compare very favorably with the actual construction cost.

The steps necessary to prepare a cost estimate for a given electrical installation will normally run as follows:

Take-off. The count of all lighting fixtures, outlets, and similar items and the measurement of all branch circuit wiring, feeders, service raceways, and similar items.

Listing the material. All items accounted for in the take-off should be listed in an orderly sequence on a standard pricing sheet, as will be described later.

Applying labor units. Determining the proper labor unit from proven labor-unit tables and applying them to the various materials or labor operation under the labor-unit column on the pricing sheet.

Finalizing. The summation of material dollars and labor hours, the assignment of job factors and dollar values to labor hours, and the determination of overhead and profit.

A take-off and pricing sheet such as the one in Fig. 14-1 should be used to assist the estimator in this phase of the estimate. In the first column (reading from left to right), a description of an item—such as a lighting fixture or an outlet box—is entered by name and perhaps by symbol also. For example, a duplex receptacle would have its normal symbol (⊜) in the column, and then "duplex recept " written after it.

The quantity of that particular item, represented by a description and symbol, is listed in the next column under "Quantity." When counting the quantity of items from a plan, most estimators prefer to use some type of mechanical counting device in the left hand (for a right-handed person) and a colored pencil in the right hand. The estimator then begins counting each individual item, checking them off the drawings with the colored pencil as they are counted, and recording them on the counting device. When all of a particular type of device are counted, the total is then entered in the "Quantity" column next to a description of the device.

After all the electrical items are accounted for, the electrical circuits are identified on the pricing sheet in the "Description" column as to wire size, number of conductors in run, etc., as shown in Fig. 14-2.

When the circuits appear on drawings that are drawn to scale, a rotometer (Fig. 14-3) is the best method of determining the length of the runs. However, on plan views, the runs obviously can only be scaled in one plane; the vertical risers are not shown. To allow for the vertical risers, such as turn-ups to wall outlet boxes, a cross section of a wall should be drawn to scale on a vacant spot on the drawing sheet, showing the height of receptacles, wall switches, wall-mounted light fixtures, etc. If this cross section is drawn to the same scale as the floor plan, all vertical risers can be taken off with the rotometer at the same time as the horizontal runs.

To illustrate, the estimator first measures all two-wire, number 12 AWG circuits with the rotometer. He then begins at the last outlet on the circuit and runs the wheel of the rotometer along the path of the circuit drawing. When another outlet is encountered, he jumps to the cross-section drawing, runs either up or down for the riser, and then back to the floor plan for the horizontal run to the next outlet. This procedure is continued until all two-wire circuits are accounted for.

JOHN M. WATSON INC.

RECAPITULATION SHEET

PROJECT ____________________ SECTION __________ SEC. NO. __________

LOCATION ____________________ SHEET NO. __________

ARCHITECT ____________________ DATE __________

SUMMARY BY ____________ PRICES BY ____________ CHECKED BY __________

DESCRIPTION	QUANTITY	UNIT	UNIT PRICE	MATERIAL COST	UNIT PRICE	LABOR COST

Figure 14-1. The pricing sheet shown here is used in the take-off phase of the estimate.

M-30.1

LABOR FACTORS
0 NONE 5 + 50%
1 + 10% 6 - 10%
2 + 15% 7 - 15%
3 + 20% 8 - 20%
4 + 30% 9 MINOR PRICE

TAKE-OFF AND PRICING SHEET

DATAMATIC SYSTEM

SHEET NO. 2 OF 2
DIVISION NO. 0 1
CONTRACTOR NO. 4 2 0
ESTIMATE NO. 0 4 1 7
PRICE LEVEL 3
DATE 1-1-71

JOB NAME CRESENT MFG. CO.

SYMBOL	DESCRIPTION	TAKE-OFF	DATA NUMBER	QTY.	FAC.	MATERIAL UNIT	MATERIAL EXTENSION	LABOR UNIT	LABOR EXTENSION
	GRC EMB		87	187		174	32 54	.056	10 47
	GRC EMB		88	96		195	18 72	.058	5 57
	GRC EMB		89	50		255	12 75	.065	3 25
	EMT EXP		57	200		118	23 60	.063	12 60
	EMT EXP		58	100		140	14 —	.065	6 50
	EMT CONC		72	224		108	24 19	.045	10 08
	EMT CONC		73	80		130	10 40	.047	3 76
	EMT CONC		74	16		.179	2 86	.051	82
	WIRE ONLY	'G' FIXTURE	549	320		043	13 76	.017	5 44
$	EXPOBID		408	1		1 93	1 93	.71	71
	FLOOR		326	1		7 95	7 95	1.80	1 80
	3½ GRC		180	10		18 90	18 90	.241	2 41
	4 NO 500 MCM		539	10		59 40	59 40	.190	1 90
		FROM SHEET 1					2962 08		301 12
							3203 08		366 43

Figure 14-2. Pricing form after all the electrical circuits have been identified in the description column.

Figure 14-3. A rotometer is a great aid to the estimator for determining circuit lengths.

In order that none of the circuits will be overlooked or counted more than once, the circuit lines on the drawings should be marked with a colored pencil as they are being measured. Once the measurements are completed, the total figure is entered in the proper space on the form. Then he goes to the next type of circuit, say, a three-wire number 12 AWG, measuring all of these runs in a like manner.

On straight runs of conduit such as feeders, some estimators prefer to use a scaled tape measure, such as the one shown in Fig. 14-4. Also, rotometers are available in direct-reading scales such as ⅛″ = 1′0″, ¼″ = 1′0″, etc., but most are calibrated in inches. In this case, it is necessary to convert the readings to coincide with the drawing scale before the final reading can be entered on the take-off form.

Figure 14-4. On straight runs of conduit, some estimators prefer to use a scaled tape measure.

For example, assume that the final reading on the rotometer (calibrated in inches) for two-wire, number 12 circuits shows 57′2″, and the drawing is drawn to a scale of ¼″ = 1′0″. First multiply 57 (the number of feet) by 12 to convert the feet to inches. Then add the 2 inches on the meter for a total of 686 inches. Since the scale of the drawing is ¼″, multiply the 686 inches by 4 for a total of 2744 feet in the run; this is the actual amount of conduit and two-wire cable required for the installation.

If the scale of the drawings were ⅛″ = 1′0″, the total number of inches (obtained from the rotometer reading) would be multiplied by 8; a scale of 1/16″ = 1′0″ would be multiplied by 16, etc.

When the take-off is complete, and the estimator is satisfied that all items are accounted for, the summation of those figures shown in the "Quantity" column is checked for accuracy.

The next step is to assign labor units to each item in the "Description" and "Quantity" columns. A reliable guide is mandatory in this step. A manual furnished by Estimatic Corporation, Denver, Colorado 80223 is noted as a good source for labor units, as are several others on the market. When assigning the labor units or material costs, make certain that the form used matches the method that you are using to obtain your final figures—there are several different methods.

After the labor units have been assigned, the unit prices for all materials on the take-off sheet are obtained from a reliable source. Again, the method used in the estimate dictates the method of pricing. For example, some estimators use unit prices for a given item—including all the necessary components in one price unit. Others require that each item be priced separately.

Once all unit prices have been entered in the proper place on the form, the material and labor units are extended to obtain the total for each item. Figure 14-5 shows a typical take-off form with all of the items completed with material and labor units entered in appropriate places.

With the pricing complete, the estimate is ready to be finalized. This procedure includes: (1) totaling all material dollars and labor man-hours, (2) assigning job factors (if any) to labor hours, (3) assigning dollar values to labor hours, (4) totaling all job expense items, and (5) determining overhead and profit.

To provide a better understanding of electrical estimating in general, we will take an actual job and perform a cost estimate and selling price from start to finish, using the floor plan in Fig. 14-6 as a guide. This plan is a small residence.

In general, the wiring system of this residence consists of duplex receptacles, wall switches, lighting fixtures, branch circuits, feeders for water heater and electric range, service entrance with a circuit-breaker panel board, etc. The types of lighting fixtures, along with the quantity of lamps for each, may be found in the Lighting Fixture Schedule in Fig. 14-7.

These plans, schedules, etc., should be carefully reviewed until all details of the project are fully understood. We are now ready to begin the estimate, which will consist of the counting of all outlets (by

LABOR FACTORS		
0 NONE	5 + 50%	
1 + 10%	5 - %10%	
2 + 15%	7 - 15%	
3 + 20%	8 - 20%	
4 + 30%	9 MINOR PRICE	

TAKE-OFF AND PRICING SHEET

DATAMATIC SYSTEM

SHEET NO. 1 OF 1
DIVISION NO. 0 1
CONTRACTOR NO. 4 2 0
ESTIMATE NO. 0 4 1 7
PRICE LEVEL 3
DATE 1-1-71

JOB NAME CRESENT MFG. CO.

SYMBOL	DESCRIPTION	TAKE - OFF	DATA NUMBER	QTY.	FAC.	MATERIAL UNIT	MATERIAL EXTENSION	LABOR UNIT	LABOR EXTENSION
	FIX-A	4	224	4		40.00	160.–	1.13	4.52
	-B	12	209	12		18.37	220.44	.96	11.52
	-C	7	211	7		41.50	290.50	1.45	10.15
	-D	1	260	1		10.64	10.64	1.40	1.40
	-E	3	215	3		18.99	56.97	.89	2.67
	-F	2	256	2		49	98	.23	46
	-G	40	237	40	9	2.00	80.–	1.05	42.–
	-H	2	243	2		8.79	17.58	1.33	2.66
	EXPOSED	8	283	8		74	5.92	.72	5.76
	FLUSH	22+4	285	26		1.50	26.–	.69	17.94
$	FLUSH	8	416	8		2.27	18.16	.81	6.48
	FLUSH	2	418	2		2.83	5.66	.93	1.86
	FLUSH	18	322	18		1.87	33.66	.80	14.40
	FLUSH	6	296	6		93	5.58	.64	3.84
	FLOOR	1	292	1		14.27	14.27	1.85	1.85
	30X18X4	1	29	1		38.49	38.49	4.31	4.31
	24 CKT W/FDR BKR	1	300	1		199.13	199.13	26.21	26.21
	400A	1	342	1		405.53	405.53	41.31	41.31
	CIR. BKR	2	11	2		153.98	307.96	4.21	8.42
	4x4 WIREWAY	2+2+2+2=8	563	2		6.63	13.26	1.12	2.24
	25 HP	4	354	4		130.03	520.12	6.72	26.88
	15 HP AC	1	353	1		78.64	78.64	3.92	3.92
	NF	4	402	4		22.84	91.36	2.42	9.68
	NF	1	401	1		9.86	9.86	2.17	2.17
/	2" GRC EMB	MOTORS LPA 50'+30'+50'	167	130		550	71.50	.079	10.27
	3-No 2/0 THW	90'	514	90		1274	114.66	.075	6.75
	4-No 2 THW	50'	529	50		847	42.35	.058	2.90
/	1¼" EXP	TEL SER 10'	175	10		380	3.80	.084	84
/	¾" CONCEALED	TEL 65'	163	65		169	10.99	.044	2.86
/	1" CONCEALED	TEL 87'	164	87		241	20.97	.053	4.61
/	1" EXP	HVAC 30'	174	30		277	8.31	.067	2.01
	3 No 6 THW	HVAC 40	508	40		297	11.88	.036	1.44
	20A EXP	5	327	5		2.39	11.95	.79	3.95
	TO UNIT HTRS	6	349	6		9.16	54.96	2.14	12.84
							2962.08		301.12

Figure 14-5. A typical take-off form with all of the items completed with material and labor units entered in appropriate places.

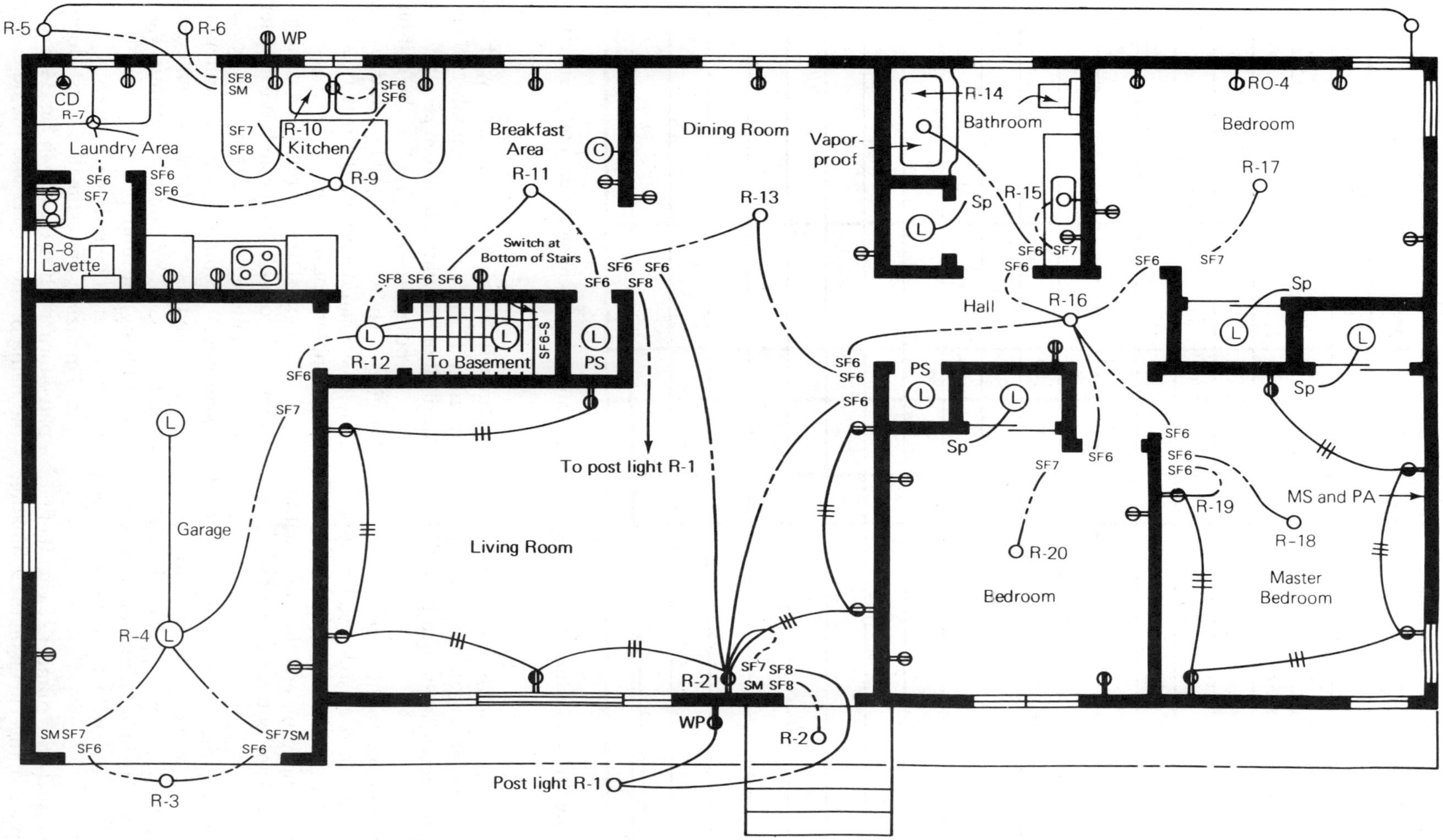

Figure 14-6. Floor plan used to demonstrate the process of electrical estimating.

LIGHTING FIXTURE SCHEDULE

FIXT. TYPE	MANUFACTURER'S DESCRIPTION	LAMPS		VOLTS	MOUNTING	REMARKS
		NO.	TYPE			
R-1	STARLIGHT CAT. NO. 45687950	1	100W	120	WALL	6'-6" Above Finish Floor
R-2	Moonlight CAT. NO. 341267	1	150W	120	Recessed	
R-3	EASILITE CAT NO. & X-$%&*	1	150W	120	Recessed	
R-4	P & S CAT. NO. 110	1	100W	120	Surface	
R-5,6	Starlight Cat. No. 3476	1	150W	120	Wall	6'-6" A.F.F.
R-7	MORELITE Cat. No. RTY-60	1	100W	120	Surface	
R-8	MORELITE Cat. No. 60-X-8	1	60W	120	Wall	
R-9	Same as R-7					
R-10	Customlite Cat. No. 809	1	100W	120	Recessed	
R-11	Same as R-7					
R-12	Same as R-7					
	ETC.					

Figure 14-7. Lighting fixture schedule giving all pertinent information about lighting fixtures on the project.

type) and entering the total on the take-off form, determining the total amount of each size wire required, and the components necessary to complete the service entrance.

In the completed estimating form for this project, quantities are obtained as follows: The type of each lighting fixture is first entered under the "Description" column in the form, and then the quantity of each type is counted and entered under the "Quantity" column.

Now go to the duplex receptacles, wall switches, junction boxes, etc., and enter the quantities in the "Quantity" column opposite the appropriate description line; then go to the various circuits, which are first measured with a rotometer as described previously.

We shall take all 2-wire number 12 circuits first, entering a description of this circuit on the estimating or take-off form and then entering the total feet in the "Quantity" column. Continue with the 3-wire, number 12 circuits, 2-wire, number 10 circuits, etc., until all of the wiring has been accounted for.

A power-riser diagram of the complete electric service is shown in Fig. 14-8, and using this as a guide, we can determine the materials necessary for this complete installation.

The next step is to refer to a reliable source for labor units for each item listed. When determined, all units are entered in the appropriate space under the "Man-hour" column. Then the unit price for each material item is also priced and entered under the "Material Cost" column. Continue by extending (multiplying) all unit prices (or man-hours) by the total number of items in each category.

Finalizing the estimate is best done on a prepared form such as the one in Fig. 14-9. On this form, the name of the project and the date are entered in their proper spaces; the total labor and materials costs

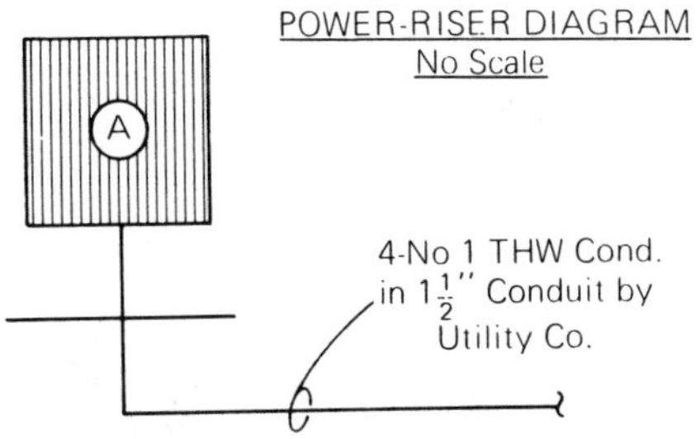

Figure 14-8. A power-riser diagram of the electric service used on the given project.

are entered next. Continue completing the form by entering sales tax on materials, overhead, and desired profit. The sum of all these items will then give a selling price or bid for the project.

It should be pointed out that this chapter is not meant to be a crash course in electrical estimating, as there are numerous other factors to cover. Rather, it is meant to give the reader a basic knowledge of using estimating tools and forms and also a basic sampling of a complete estimate.

BID SUMMARY SHEET

JOB NAME ______________________ ______________________ DATE __________

(1) TOTAL MATERIAL (FROM PRICING SHEET) $ __________

LABOR (FROM PRICING SHEET) __________ HRS.

JOB FACTORS __________ HRS.

(2) TOTAL LABOR $ __________ HR. __________ HRS. $ __________

JOB EXPENSE

SALES TAX ______ % OF LINE 1 $ __________

PAYROLL TAX ______ % OF LINE 2 $ __________

SUBCONTRACTS $ __________

RENTAL OF TOOLS, EQUIP., ETC. $ __________

OTHER JOB EXPENSE $ __________

(3) TOTAL JOB EXPENSE $ __________

(4) PRIME COST (SUM OF LINE 1, 2 & 3) $ __________

(5) OVERHEAD ______ % OF LINE 4 $ __________

(6) NET COST (SUM OF LINES 4 & 5) $ __________

(7) PROFIT ______ % OF LINE 6 $ __________

(8) TOTAL COST (SUM OF LINES 6 & 7) $ __________

(9) BOND $ __________

(10) PERMIT $ __________

TOTAL SELLING PRICE (SUM OF LINES 8, 9, & 10) $ __________

PREPARED BY: ______________________ CHECKED BY: ______________________

Figure 14-9. A form used for finalizing the estimate.

14.2 National Electrical Code

Owing to the potential fire and explosion hazards caused by the improper handling and installation of electrical wiring, certain rules in the selection of materials, quality of workmanship, and precautions for safety must be followed when one is working with electrical systems. In order to standardize and simplify these rules and provide some reliable guide for electrical construction, the National Electrical Code (NEC) was developed. This code, originally prepared in 1897, is frequently revised to meet changing conditions, improved equipment and materials, and new fire hazards. It is the result of the best efforts of electrical engineers, manufacturers of electrical equipment, insurance underwriters, fire fighters, and other concerned experts throughout the country.

The NEC is now published by the NFPTA (National Fire Protection Association), 60 Batterymarch St., Boston, Mass. 02110. It contains specific rules and regulations intended to help in "the practical safeguarding of persons and of buildings and their contents from the electrical hazards arising from the use of electricity for lighting, power, heat . . ." Although the NEC itself states, "This Code is to be regarded neither as a design specification nor an instruction manual for untrained persons," it does provide a sound basis for the study of electrical design and installation procedures—under the proper guidance.

The probable reason for the NEC's self-analysis contained in the preceding paragraph is that the NEC further states

> The provisions of this Code constitute a *Minimum* standard. Compliance therewith, and proper maintenance, will result in an installation reasonably free from hazard but not necessarily efficient or convenient . . . good electrical service and satisfactory results will often require larger sizes of wire (in the case of a large voltage drop), more branch circuits, and better types of equipment than the minimum which is here specified.

The NEC, however, has become the "Bible" of the electrical construction industry, and anyone involved in electrical work—in any capacity—should obtain an up-to-date copy of it, keep it handy at all times, and refer to it often.

14.3 Using the National Electrical Code

In order that electrical draftsmen may understand the language and terms used in the NEC as well as in the industry in general, the definitions given in Chapter 1, Article 100 of the Code should be fully understood. For simplicity, this article lists only definitions essential to the proper use of the Code; other definitions, however, are defined in other individual articles of the NEC where they apply.

Mandatory rules of the NEC are characterized by the use of the word "shall," and advisory rules are characterized by the use of the word "should." When statements are made using the latter word, they are stated as recommendations of that which is advised but not necessarily required. Some local ordinances, however, may change the meaning of "should" to "shall," so be certain to investigate when working with projects in a new area.

Other general requirements for electrical installations are given in Article 110—General. This is a short article, and it can be easily read in full in less than five minutes. This article—along with Article 100—should be read over several times until the information contained in both is fully understood and firmly implanted in the reader's mind. With a good understanding of this basic knowledge, the remaining portions of the NEC will be easier to understand.

Chapter 2, Wiring Design and Protection, of the NEC will be the one chapter that most draftsmen will find a need to refer to most often. It covers such data as using and identifying grounded conductors, branch circuits, feeders, calculations, services, overcurrent protection, and grounding—all necessary for any type of electrical system regardless of the building type in which the system is installed.

Since electrical draftsmen and designers must refer to Chapter 2 of the NEC so often, it does not take long for most of them to almost memorize this portion of the Code. However, the beginner will have to constantly look up various items. In doing so, remember that Chapter 2 deals with "how-to" items; that is, how to provide proper spacing for conductor supports, how to provide temporary wiring, how to size the proper grounding conductor or electrode, etc. If a problem develops pertaining to the design of installation of a conventional electrical system, you can usually find the answer in Chapter 2. Thumb through the pages and glance at the article headings until you find a heading under which your particular question or problem should appear and then read the entire article. If you cannot find the answer there, refer to the index to find other articles in the NEC that cover your particular problem.

The chapters continue on to "Wiring Methods and Materials," "Equipment for General Use," "Special Conditions," "Communication Systems," and "Tables and Examples." An example of the NEC's use is as follows:

Assume that you are drawing the wiring layout for a mobile home park that will contain several factory-wired mobile homes; the size of each unit will be 70 feet long by 10 feet wide. Your problem is to determine the size of the power cord from the home's electric panel to the service disconnect at each meter base and then to determine how to ground the entire system in an approved manner.

Begin by turning to the table of contents in the front of the Code book and glance down the headings. The first subject you will see that pertains to the situation is "Services" under Article 230; write this article number and page number on a pad for later reference. Continue until you come to "Grounding"; write this article and page number on the pad also.

Perhaps you will skip over several of the other articles until you come to Article 550—"Mobile Homes and Mobile Home Parks"—one you could certainly not pass up. The remaining articles seem insignificant for the situation, so your final list contains three articles that should be of help in designing and preparing drawings for the installation of the power cord and electric service for the mobile home.

Your next step would be to take the article that seems the most important and read this through. Obviously, Article 550 would seem to be the one to read first. In doing so, you will find that the power cord should be rated for 50 amperes, that the service equipment must be located adjacent to the mobile home and not mounted in or on the mobile home itself, and similar items of importance.

After completing this article, go to Article 230 and glance through it to see if any requirements pertain to your situation—you will find several; then go to Article 250 and do the same. You may also need to know the AWG wire size that will safely carry 50 amperes of current. If so, look in the Index under "Conductors" and read down the list of categories until you locate the article number for conductor capacities; the tables in Article 310 will tell you that you need number 6 AWG copper wire with TW insulation to carry 50 amperes of current. You will also find that a number 8 AWG copper conductor must be used for grounding a service of this size; refer to the Index, which will refer you to Article 250-94(a), "Grounding Electrode Conductor or Grounding Systems."

The best advice that can be offered to one who desires to know how to use the NEC is to become familiar with it. Read portions of the NEC frequently; look up anything that you are uncertain about; study articles pertaining to areas with which you are not familiar. All of these actions will help you to become more familiar with the regulations of the NEC in the fastest possible time.

14.4 Local Codes and Ordinances

A number of towns and cities have their own local electrical code or ordinance. In general, these are based on, or similar to, the NEC, but on certain classes of work they may have a few specific rules that are usually more rigid than the NEC.

In addition to the NEC and local ordinances of certain cities, local power companies may have some special rules regarding location of service-entrance wires, meter connections, and similar details that must be satisfied before connection can be made to a building. Therefore, before a design or installation of an electrical system is begun, it is wise for the designer to become familiar with any and all local codes and ordinances.

14.5 Electrical Specifications

The electrical specifications for a building or project are the written descriptions of work and duties required of the owner, architect, and engineer. Together with the working drawings, these specifications form the basis of the contract requirements for the construction.

Those who prepare the construction drawings and specifications must always be on the alert for conflicts between the working drawings and the written specifications. Such conflicts occur particularly when:

1. Architects or engineers use standard or prototype specifications and attempt to apply them to specific working drawings.
2. Previously prepared standard drawings are to be changed or amended by reference in the specifications only; the drawings themselves are not changed.

In such situations, the conflicts must be cleared up, preferably before the work is undertaken, in order to avoid added cost either to the owner, the architect, or the contractor performing the work.

In general, the electrical spcifications should show the grade of materials to be used on the project and the manner in which the electrical system shall be installed. A typical electrical specification follows. The exercises at the end of this chapter are designed to help the reader learn to use written specifications when working as an electrical draftsman for consulting firms, electrical contractors, and similar organizations.

DIVISION 16-ELECTRICAL

Section 16A-General Provisions

1. The "Instructions to Bidders," "General Conditions," and "General Requirements" of the architectural specifications govern work under this Section.
2. It is understood and agreed that the Electrical Contractor has, by careful examination of the Plans and Specifications, and the site where appropriate, satisfied himself as to the nature and location of the work, and all conditions which must be met in order to carry out the work under this Section of the Contract.

3. Scope of the Work

a. The scope of the work consists of the furnishing and installing of complete electrical systems—exterior and interior—including miscellaneous systems. The Electrical Contractor shall provide all supervision, labor, materials, equipment, machinery, and any and all other items necessary to complete the systems. The Electrical Contractor shall note that all items of equipment are specified in the singular; however, the Contractor shall provide and install the number of items of equipment as indicated on the drawings and as required for complete systems.

b. It is the intention of the Specifications and Drawings to call for finished work, tested, and ready for operation.

c. Any apparatus, applince, material, or work not shown on drawings but mentioned in the specifications or vice versa, or any incidental accessories necessary to make the work complete and perfect in all respects and ready for operation, even if not particularly specified, shall be furnished, delivered, and installed by the Contractor without additional expense to the Owner.

d. Minor details not usually shown or specified, but necessary for proper installation and operation, shall be included in the Contractor's estimate, the same as if herein specified or shown.

e. With submission of bid, the Electrical Contractor shall give written notice to the Architect of any materials or apparatus believed inadequate or unsuitable, in violation of laws, ordinances, rules; and any necessary items or work omitted. In the absence of such written notice, it is mutually agreed that the Contractor has included the cost of all required items in his proposal, and that he will be responsible for the approved satisfactory functioning of the entire system without extra compensation.

4. Electrical Drawings

a. The Electrical drawings are diagrammatic and indicate the general arrangement of fixtures, equipment and work included in the contract. Consult the Architectural drawings and details for exact location of fixtures and equipment; where same are not definitely located, obtain this information from the Architect.

b. Contractor shall follow drawings in laying out work and check drawings of other trades to verify spaces in which work will be installed. Maintain maximum headroom and space conditions at all points. Where headroom or space conditions appear inadequate, the Architect shall be notified before proceeding with installation.

c. If directed by the Architect, the Contractor shall, without extra charge, make reasonable modifications in the layout as needed to prevent conflict with work of other trades or for proper execution of the work.

5. Codes, Permits, and Fees

a. Contractor shall give all necessary notices, including electric and telephone utilities, obtain all permits and pay all government taxes, fees and other costs, including utility connections or extensions in connection with his work; file all necessary plans, prepare all documents and obtain all necessary approvals of all governmental departments having jurisdiction; obtain all required certificates of inspection for his work and deliver same to the Architect before request for acceptance and final payment for the work.

b. Contractor shall include in the work, without extra cost to the Owner, any labor, materials, services, apparatus, drawings (in addition to contract drawings and documents) in order to comply with all applicable laws, ordinances, rules and regulations, whether or not shown on drawings and/or specified.

c. Work and materials shall conform to the latest rules of the National Board of Fire Underwriters' Code, Regulations of the State Fire Marshal, and with applicable local codes and with all prevailing rules and regulations pertaining to adequate protection and/or guarding of all moving parts, or otherwise hazardous conditions. Nothing in these specifications shall be construed to permit work not conforming to the most stringent of applicable codes.

d. The National Electric Code, the Local Electric Code, and the electrical requirements as established by the State and Local Fire Marshal, and rules and regulations of the power company serving the project, are hereby made part of the drawings or specifications to make the work comply with these requirements. The Electrical Contractor shall so notify the Architect.

6. Shop Drawings

a. The Electrical Contractor shall submit five (5) copies of the shop drawings to the Architect for approval within thirty (30) days after the award of the general contract. If such a schedule cannot be met, the Electrical Contractor may request in writing for an extension of time to the Architect. If the Electrical Contractor does not submit shop drawings in the prescribed time, the Architect has the right to select the equipment.

b. Shop drawings shall be submitted on all major

pieces of electrical equipment, including service entrance equipment, lighting fixtures, panel boards, switches, wiring devices and plates and equipment for miscellaneous systems. Each item of equipment proposed shall be a standard catalog product of an established manufacturer. The shop drawing shall give complete information on the proposed equipment. Each item of the shop drawings shall be properly labeled, indicating *the intended service of the material*, the job name and Electrical Contractor's name.

c. The shop drawings shall be neatly bound in five (5) sets and submitted to the Architect with a letter of transmittal. The letter of transmittal shall list each item submitted along with the manufacturer's name.

d. Approval rendered on shop drawings shall not be considered as a guarantee of measurements or building conditions. Where drawings are approved, said approval does not mean that drawings have been checked in detail; said approval does not in any way relieve the Contractor from his responsibility, or necessity of furnishing material or performing work as required by the contract drawings and specifications.

7. Cooperation with Other Trades

a. The Electrical Contractor shall give full cooperation to other trades and shall furnish (in writing, with copies to Architect) any information necessary to permit the work of all trades to be installed satisfactorily and with least possible interference or delay.

b. Where the work of the Electrical Contractor will be installed in close proximity to work of other trades, or where there is evidence that work of the Electrical Contractor will interfere with the work of other trades, he shall assist in working out space conditions to make a satisfactory adjustment. If so directed by the Architect, the Electrical Contractor shall prepare composite working drawings and sections at a suitable scale clearly showing how his work is to be installed in relation to the work of other trades. If the Electrical Contractor installs his work before coordinating with other trades or so as to cause any interference with work of other trades, he shall make necessary changes in his work to correct the condition without extra charge.

c. The complexity of equipment and the variation between equipment manufacturers requires complete coordination of all trades. The Contractor who offers, for consideration, substitutes of equal products of reliable manufacturers, has to be responsible for all changes that affect his installation and the installation equipment of other trades. All systems and their associated controls must be completely installed, connected, and operating to the satisfaction of the Architect prior to final acceptance and contract payment.

8. Temporary Electrical Service

a. The Electrical Contractor shall be responsible for all arrangements and costs for providing at the site temporary electrical metering, main switches and distribution panels as required for construction purposes. The distribution panels shall be located at a central point designated by the Architect. The General Contractor shall indicate prior to installation whether three-phase or single-phase service is required.

9. Electrical Connections

a. The Electrical Contractor shall provide and install power wiring to all motors and electrical equipment complete and ready for operation including disconnect switches and fuses. Starters, relays and accessories shall be furnished by others unless otherwise noted, but shall be installed by the Electrical Contractor. This Contractor shall be responsible for checking the shop drawings of the equipment manufacturer to obtain the exact location of the electrical rough-in and connections for equipment installed.

b. The Mechanical Contractor will furnish and install all temperature control wiring and all interlock wiring unless otherwise noted.

c. It shall be the responsibility of the Electrical Contractor to check all motors for proper rotation.

10. As-built Drawings

a. The Electrical Contractor shall maintain accurate records of all deviations in work as actually installed from work indicated on the drawings. On completion of the project, two (2) complete sets of marked-up prints shall be delivered to the Architect.

11. Inspection and Certificates

a. On the completion of the entire installation, the approval of the Architect and Owner shall be secured, covering the installation throughout. The Contractor shall obtain and pay for Certificate of Approval from the public authorities having jurisdiction. A final inspection certificate shall be submitted to the Architect prior to final payment. Any and all cost incurred for fees shall be paid for by the Contractor.

12. Tests

a. The right is reserved to inspect and test any portion of the equipment and/or materials during the progress of its erection. This Contractor shall test all wiring and connections for continuity and grounds, before connecting any fixtures or equipment.

b. The Contractor shall test the entire system in the presence of the Architect or his engineer when the work is finally completed to ensure that all portions are free from short circuits or grounds. All equipment necessary to conduct these tests shall be furnished at the Contractor's expense.

13. Equivalents

a. When material or equipment is mentioned by

name, it shall form the basis of the Contract. When approved by the Architect in writing, other material and equipment may be used in place of those specified, but written application for such substitutions shall be made to the Architect as described in the Bidding Documents. The difference in cost of substitute material or equipment shall be given when such request is made. Approval of substitute is, of course, contingent on same meeting specified requirements and being of such design and dimensions as to comply with space requirements.

14. Guarantee

a. The Electrical Contractor shall guarantee, by his acceptance of the Contract, that all work installed will be free from defects in workmanship and materials. If during the period of one year, or as otherwise specified, from date of Certificate of Completion and acceptance of work, any such defects in workmanship, materials or performance appear, the Contractor shall, without cost to the Owner, remedy such defects within a reasonable time to be specified in notice from Architect. In default, the Owner may have such work done and charge cost to Contractor.

Section 16B-Basic Materials and Workmanship

1. Portions of the sections of the Documents designated by the letters "A," "B," and "C" and "DIVISION ONE—GENERAL REQUIREMENTS" apply to this Division.
2. Consult Index to be certain that set of Documents and Specifications is complete. Report omissions or discrepancies to the Architect.

3. Conduit Material and Workmanship

a. GENERAL: The Electrical Contractor shall install a complete raceway system as shown in the drawings and stated in other sections of the specifications. All material used in the raceway system shall be new and the proper material for the job. Conduit, couplings and connectors shall be a product of a reputable manufacturer equal to conduit as manufactured by Triangle Conduit and Cable or National Electric.

b. CONDUIT INSTALLATION

(1) Conduit shall be of ample size to permit the ready insertion and withdrawal of conductors without abrasion. All joints shall be cut square, reamed smooth and drawn up tight.

(2) Concealed conduits shall be run in as direct a manner as possible, and with as long a bend as possible. Exposed conduit shall be run parallel to or at right angles with the lines of the building. All bends shall be made with standard ells, conduit bent to a radius not less than shown in NEC, or screw jointed conduit fittings. All bends shall be free of dents or flattening. Not more than the equivalent of four quarter bends shall be used in any run between terminals and cabinets, or between outlets and junction or pull boxes.

(3) Conduits shall be continuous from outlet to outlet and from outlet to cabinets, junction or pull boxes, and shall enter and be secured at all boxes in such a manner that each system shall be electrically continuous throughout.

(4) A #14 galvanized iron or steel fish wire shall be left in all conduits in which the permanent wiring is not installed.

(5) Where conduits cross building joints, furnish and install O.Z. Electric Manufacturing Company expansion fittings for contraction, expansion and settlement.

(6) Open ends shall be capped with approved manufactured conduit seals as soon as installed and kept capped until conductors are ready to be installed.

(7) Conduit shall be securely fastened to all sheet metal outlets, junction and pull boxes with galvanized locknuts and bushings, care being observed to see that the full number of threads project through to permit the bushings to be drawn tight against the end of conduit, after which the locknut shall be made sufficiently tight to draw the bushings into firm electrical contact with the box.

(8) For all flush-mounted panels there shall be provided and installed 1¼" empty conduit up through wall and turned out above ceiling and one 1¼" empty conduit down to space below floor except where slab is on grade.

c. CONDUIT HANGERS AND SUPPORTS

(1) Conduit throughout the project shall be securely and rigidly supported to the building structure in a neat and workmanlike manner and wherever possible, parallel runs of horizontal conduit shall be grouped together on adjustable trapeze hangers. Support spacing shall not be more than eight feet.

(2) Exposed conduit shall be supported by one-hole malleable iron straps, two-hole straps, suitable beam clamps or split ring conduit hanger with support rod.

(3) Single conduit 1¼" and larger run concealed horizontally shall be supported by suitable beam clamps or split ring conduit hangers with support rod. Multiple runs of conduit shall be grouped together on trapeze hangers where possible. Vertical runs shall be supported by steel riser clamps.

(4) Conduit one inch and smaller run concealed above a ceiling may be supported directly to the building structure with strap hangers or No. 14 ga. galvanized wire provided the support spacing does not exceed four feet.

4. Outlet Boxes

a. GENERAL

(1) Before locating the outlet boxes check all of the architectural drawings for type of construction and to make sure that there is no conflict with other equipment. The outlet boxes shall be symmetrically located according to room layout and shall not interfere with other work or equipment. Also note any detail of the outlets shown on the drawings.

(2) Outlet boxes shall be made of galvanized sheet steel unless otherwise noted or required by the NEC and shall be of the proper code size for the required number of conductors. Outlet boxes shall be a minimum of 4 inches square unless specifically noted on the drawings with the exception that a box containing only two current-carrying conductors may be smaller. The outlet boxes shall be complete with the approved type of connectors and required accessories.

(3) The outlet boxes shall be complete with raised device covers as required to accept device installed. All outlet boxes must be securely fastened in position with the exposed edge of the raised device cover set flush with the finished surface. Approved factory-made knockout seals shall be installed where knockouts are not intact. Galvanized outlet boxes shall be manufactured by RACO, STEEL CITY, APPLETON or approved equal.

(4) Outlet boxes for exposed work shall be handy boxes with handy box covers unless otherwise noted.

(5) Outlet boxes located on the exterior in damp or wet locations or as otherwise noted shall be threaded cast aluminum device boxes such as CROUSE HINDS Type "FS" or "FD."

b. RECEPTACLE OUTLET BOXES: Wall receptacles shall be mounted approximately 18" above the finished floor (AFF) unless otherwise noted. When the receptacle is mounted in a masonry wall the bottom of the outlet box shall be in line with the bottom of a masonry unit. Receptacles for electric water coolers shall be installed behind the coolers in accordance with manufacturer's recommendations. All receptacle outlet boxes shall be equipped with grounding lead which shall be connected to grounding terminal of the device.

c. SWITCH OUTLET BOXES: Wall switches shall be mounted approximately 54 inches above the finished floor (AFF) unless otherwise noted. When the switch is mounted in a masonry wall the bottom of the outlet box shall be in line with the bottom of a masonry unit. Where more than two switches are located, the switches shall be mounted in a gang outlet box with gang cover. Dimmer switches shall be individually mounted unless otherwise noted. Switches with pilot lights, switches with overload motor protection and other special switches that will not conveniently fit under gang wall plates may be individually mounted.

d. LIGHTING FIXTURE OUTLET BOXES: The lighting fixture outlet boxes shall be furnished with the necessary accessories to install the fixture. The supports must be such as not to depend on the outlet box supporting the fixture. The supports for the lighting fixture shall be independent of the ceiling system. All ceiling outlet boxes shall be equipped with raised circular cover plates with their edges set flush with surface of finished ceiling.

5. Pull Boxes

a. Pull boxes shall be installed at all necessary points, whether indicated on the drawings or not, to prevent injury to the insulation or other damage that might result from pulling resistance for other reasons necessary to proper installation. Pull box locations shall be approved by the Architect prior to installation. Minimum dimensions shall be not less than NEC requirements and shall be increased if necessary for practical reasons or where required to fit a job condition.

b. All pull boxes shall be constructed of galvanized sheet steel, code gauge, except that no less than 12 gauge shall be used for any box.

c. Where boxes are used in connection with exposed conduit, plain covers attached to the box with a suitable number of countersunk flat head machine screws may be used.

d. Where so indicated, certain pull boxes shall be provided with barriers. These pull boxes shall have a single cover plate, and the barriers shall be of the same gauge as the pull box.

e. Each circuit in pull box shall be marked with a tag guide denoting panels to which it connects.

f. Exposed pull boxes will not be permitted in the public spaces.

6. Wireways or Wire Troughs

a. Wireways shall be used where indicated on the drawings and for mounting groups of switches and/or starters. Wireways shall be the standard manufactured product of a company regularly producing wireway and shall not be a local shop-assembled unit. Wireway shall be of the hinged cover type, Underwriters' Listed, and of sizes indicated or as required by NEC. Finish shall be medium light gray enamel over rust inhibitor. Wireways shall be of the raintight construction where required. Wireways shall be General Electric Type HS or approved equal.

7. Conductor Material and Workmanship

a. GENERAL

(1) The Electric Contractor shall provide and install a complete wiring system as shown on the drawing or specifications herein. All conductors used in the wiring

system shall be soft drawn copper wire having a conductivity of not less than 98% of that of pure copper, with 600-volt rating, unless otherwise noted. Wire shall be as manufactured by General Cable, Triangle or approved equal.

(2) The wire shall be delivered to the site in their original unbroken packages, plainly marked or tagged as follows: (a) Underwriters' Labels, (b) Size, kind and insulation of the wire, (c) Name of manufacturing company and the trade name of the wire.

b. CONDUCTOR WORKMANSHIP

(1) Install conductors in all raceways as required, unless otherwise noted, in a neat and workmanlike manner. Telephone conduits and empty conduits, as noted, shall have a No. 14 ga. galvanized pull wire left in place for future use.

(2) Conductors shall be color-coded in accordance with the National Electric Code. Mains, feeders, and subfeeders shall be tagged in all pull, junction and outlet boxes and in the gutter of panels with approved code type wire markers.

(3) No lubricant other than powdered soapstone or approved pulling compound may be used to pull conductors.

(4) At least eight (8) inches of slack wire shall be left in every outlet box whether it be in use or left for future use.

(5) All conductors and connections shall test free of grounds, shorts and opens before the job is turned over to the Owner.

8. Lugs, Taps and Splices

a. Joints on branch circuits shall occur only where such circuits divide and shall consist of one through circuit to which shall be spliced the branch from the circuit. In no case shall joints in branch circuits be left for the fixture hanger to make. No splices shall be made in conductor except at outlet boxes, junction boxes, or splice boxes.

(1) All joints or splices for No. 10 AWG or smaller shall be made with UL approved wire nuts or compression-type connectors.

(2) All joints or splices for No. 8 AWG or larger shall be made with a mechanical compression connector. After the conductors have been made mechanically and electrically secure, the entire joint or splice shall be covered with Scotch No. 33 tape or approved equal to make the insulation of the joint or splice equal to the insulation of the conductors. The connector shall be UL approved.

9. Access Doors

a. The Electrical Contractor shall furnish to the lather the access doors as shown on the drawings or required for access to junction boxes, etc. The doors shall be 12″ square, unless otherwise noted, hinged metal door with metal frames. Door and frame shall be not lighter than 16 gauge sheet steel. The access door shall be of the flush type with screwdriver latching device. The frame shall be constructed so that it can be secured to building material as required. The access doors shall be Milcor or equal. Access door and location shall meet the approval of the Architect.

10. Fuses

a. Fuses manufactured by Buss or Shawmut shall be furnished and installed as required. Motor protection fuses shall be dual element.

11. Cutting and Patching

a. On new work the Electrical Contractor shall furnish sketches to the General Contractor showing the locations and sizes of all openings and chases, and furnish and locate all sleeves and inserts required for the installation of the electrical work before the walls, floors and roof are built. The Electrical Contractor shall be responsible for the cost of cutting and patching where any electrical items were not installed or where incorrectly sized or located. The Contractor shall do all drilling required for the installation of his hangers.

b. On alterations and additions to existing projects, the Electrical Contractor shall be responsible for the cost of all cutting and patching, unless otherwise noted.

c. No structural members shall be cut without the approval of the Architect, and all such cutting shall be done in a manner directed by him. All patching shall be performed in a neat and workmanlike manner acceptable to the Architect.

12. Excavation and Backfilling

a. The Electrical Contractor shall be responsible for excavation, backfill, tamping, shoring, bracing, pumping, street cuts, repairing of finished surface and all protection for safety of persons and property as required for installing a complete electrical system. All excavation and backfill shall conform to the Architectural Section of the specifications.

b. Excavation shall be made in a manner to provide a uniform bearing for conduit. Where rock is encountered, excavate 3″ below conduit grade and fill with gravel to grade.

c. After required test and inspections, backfill the ditch and tamp. The first foot above the conduit shall be hand-backfilled with rock-free clean earth. The backfill in the ditches on the exterior and interior of the building shall be tamped to 90%. The Electrical Contractor shall be responsible for any ditches that go down.

13. Equipment and Installation Workmanship

a. All equipment and material shall be new and shall bear the manufacturer's name and trade name. The equipment and material shall be essentially the standard product of a manufacturer regularly engaged in the production of the required type of equipment and shall be the manufacturer's latest approved design.

b. The Electrical Contractor shall receive and properly store the equipment and material pertaining to the electrical work. The equipment shall be tightly covered and protected against dirt, water, chemical or mechanical injury and theft. The manufacturer's directions shall be followed completely in the delivery, storage, protection and installation of all equipment and materials.

c. The Electrical Contractor shall provide and install all items necessary for the complete installation of the equipment as recommended or as required by the manufacturer of the equipment or required by code without additional cost to the owner, regardless whether the items are shown on the plans or covered in the Specifications.

d. It shall be the responsibility of the Electrical Contractor to clean the electrical equipment, make necessary adjustments and place the equipment into operation before turning equipment over to Owner. Any paint that was scratched during construction shall be "touched up" with factory color paint to the satisfaction of the Architect. Any items that were damaged during construction shall be replaced.

14. Concrete Pads, Supports and Encasement

a. The Electrical Contractor shall be responsible for all concrete pads, supports, piers, bases, foundations and encasement required for the electrical equipment and conduit. The concrete pads for the electrical equipment shall be six (6) inches larger all around than the base of the equipment and a minimum of 4 inches thick unless specifically indicated otherwise.

15. Waterproofing

a. The Electrical Contractor shall provide all flashing, caulking and sleeves required where his items pass through the outside walls or roof. The waterproofing of the openings shall be made absolutely watertight. The method of installation shall conform to the requirements of Division 7—Moisture Control and/or meet the approval of the Architect.

Section 16C—Service Entrance System

1. Portions of the sections of the Documents designated by the letters "A," "B," "C" and "DIVISION ONE—GENERAL REQUIREMENTS" apply to this Division.
2. Consult Index to be certain that set of Documents and Specifications is complete. Report omissions or discrepancies to the Architect.

3. Service Entrance

a. GENERAL: The Electrical Contractor shall provide and install a complete service entrance system as shown on the drawings or as required for a complete system. All material and workmanship shall conform with Section 16B of the specifications, National Electric Code and the electric code. The electric service entrance shall conform to the requirements and regulations of the electric utility serving the project.

b. ELECTRIC UTILITY CHARGE: The Electrical Contractor shall make all arrangements with the electric utility and pay all charges made by the electric utility for permanent electric service to the project. In the event that the electric utility's charges are not available at the time the project is bid, the Electrical Contractor shall qualify his bid to notify the Owner that such charges are not included.

c. METERING: The Electrical Contractor shall provide and install raceway, install current transformer cabinet and/or meter trim for metering facilities, as required by the electric utility serving the project. The electric utility will provide the meter installation including meter, current transformers and connections.

d. GROUNDING: The Electrical Contractor shall properly ground the electrical system as required by the National Electrical Code.

e. CONDUIT: The conduit used for service entrance shall be galvanized rigid steel conduit unless otherwise noted on drawings.

f. CONDUCTORS: Conductors for the service entrance shall be copper type RHW or THW rated at 75° C unless otherwise noted. The conductors indicated on the drawings are based on aluminum.

Section 16D—Electrical Distribution System

1. Portions of the sections of the Documents designated by the letters "A," "B," and "C" and "DIVISION ONE—GENERAL REQUIREMENTS" apply to this Division.
2. Consult Index to be certain that set of Documents and Specifications is complete. Report omissions or discrepancies to the Architect.

3. Feeders and Branch Circuits

a. GENERAL: The Electrical Contractor shall provide and install a complete electrical distribution system as shown on the drawings or as required for a complete system. All materials and workmanship shall conform with Section 16B of the Specifications, National Electrical Code and the local electric code.

b. CONDUIT MATERIALS

(1) *Rigid Conduit (Heavy Wall)*: Rigid conduit shall be galvanized rigid steel conduit with a minimum size of ¾ inch unless otherwise stated. Rigid steel conduit shall be installed for the following services and locations: service entrance, underground in contact with earth, in concrete slab, panel feeders, exterior of building walls, motor feeders over 10 HP, electrical

equipment feeders over 10 KW, "wet" locations, and as required by the National Electric Code and local codes.

(2) *Electrical Metallic Tubing (EMT)*: Electrical metallic tubing shall be galvanized steel with a minimum size of ¾ inch. Electrical metallic tubing shall be used in all locations not otherwise specified for rigid or flexible conduit and where not in violation of the National Electric Code.

(3) *Flexible Metal Conduit*: Flexible metal conduit shall be galvanized steel. Flexible metal conduit located in wet locations shall be the Liquid-Tight type. Flexible metal conduit may be used in place of EMT where completely accessible, such as above removable acoustical tile ceilings and for exposed work in unfinished spaces.

A short piece of flexible metal conduit shall be used for the connection to all motors and vibrating equipment, connection between recessed light fixtures and junction box, and as otherwise noted, provided the use meets the requirements of the National Electric Code and local codes. The flexible metal conduit shall be the type approved for continuous grounding.

c. CONDUCTOR MATERIAL

(1) The conductor material shall be as follows, unless otherwise noted:

(a) *Feeders*: Shall be type RHW or THW rated at 75°C.

(b) *Branch Circuits*: Shall be Type THW rated at 75°C, except branch circuits with conductor sizes of No. 10 and smaller in dry locations may be type TW rated at 60°C.

(c) *Special Locations*: Conductors in special locations such as range hoods, lighting fixtures, etc., shall be as required by the National Electrical Code, local code or as otherwise stated.

(2) No Conductor shall be smaller than No. 12 wire, except for the control wiring and as stated in other sections of the Specifications or on the drawings. Wiring to switches shall not be considered as control wiring.

(3) Conductors indicated on the drawings are based on copper. Panel, motor and electrical equipment feeders with a size of No. 1/0 and larger may be aluminum, providing the size of the conductor is increased to have the same or more current carrying capacity as the copper conductors. Also, the conduit sizes shall be increased accordingly.

(4) All conductors with the size of No. 8 or larger shall be stranded.

(5) All lighting and receptacle branch circuits in excess of 100 linear feet shall be increased one size to prevent excessive voltage drop.

4. Safety Switches (FSS) (NFSS)

a. GENERAL: Furnish and install safety switches as indicated on the drawings or as required. All safety switches shall be *NEMA General Duty Type* and Underwriters' Laboratories Listed. The switches shall be Fused Safety Switches (FSS) or Nonfused Safety Switches (NFSS) as shown on the drawings or required.

b. SWITCHES: Switches shall have a quick-make and quick-break operating handle and mechanism which shall be an integral part of the box. Padlocking provisions shall be provided for padlocking in the "OFF" position with at least three padlocks. Switches shall be horsepower rated for 250 volts AC or DC as required. Lugs shall be UL listed for copper and aluminum cable.

c. ENCLOSURES: Switches shall be furnished in NEMA 1 general-purpose enclosures with knockouts unless otherwise noted or required. Switches located on the exterior of the building or in "wet" locations shall have NEMA 3R enclosures (WP).

d. INSTALLATION: The safety switches shall be securely mounted in accordance with the NEC. The Contractor shall provide all mounting materials. Install fuses in the FSS. The fuses shall be dual element on motor circuits.

e. MANUFACTURER: Square "D", General Electric, Cutler-Hammer, Westinghouse, or ITE.

5. Panel Boards—Circuit Breaker

a. GENERAL: Furnish and install circuit-breaker panel boards as indicated in the panelboard schedule and where shown on the drawings. The panel board shall be dead front safety type equipped with molded case circuit breakers and shall be the type as listed in the panelboard schedule: Service entrance panel boards shall include a full capacity box bonding strap and approved for service entrance. The acceptable manufacturers of the panel boards are ITE, General Electric, Cutler-Hammer, and Westinghouse, provided that they are fully equal to the type listed on the drawings. The panel board shall be listed by Underwriters' Laboratories and bear the UL Label

b. CIRCUIT BREAKERS: Provide molded case circuit breakers of frame, trip rating and interrupting capacity as shown on the schedule. Also, provide the number of spaces for future circuit breakers as shown in the schedule. The circuit breakers shall be quick-make, quick-break, thermal-magnetic, trip indicating and have common trip on all multipole breakers with internal tie mechanism.

c. PANELBOARD BUS ASSEMBLY: Bus bar connections to the branch ircuit breakers shall be the "phase sequence" type. Single-phase 3-wire panelboard busing shall be such that any two adjacent single-pole breakers are connected to opposite polarities in such a manner that 2-pole breakers can be installed in any location. Three-phase 4-wire breakers are individually connected to each of the three different phases in such a manner that 2- or 3-pole breakers can be installed at any location. All current carrying parts of the bus assembly shall be plated. Mains ratings shall be as shown in the panelboard schedule on the plans. Provide solid neutral assembly (S/N) when required.

d. WIRING TERMINALS: Terminals for feeder conductors to the panelboard mains and neutral shall be suitable for the type of conductor specified. Terminals for branch circuit wiring, both breaker and neutral, shall be suitable for the type of conductor specified.

e. CABINETS AND FRONTS: The panelboard bus assembly shall be enclosed in a steel cabinet. The size of the wiring gutters and gauge of steel shall be in accordance with NEMA Standards. The box shall be fabricated from galvanized steel or equivalent rust resistant steel. Fronts shall include door and have flush, brushed stainless steel, spring-loaded door pulls. The flush lock shall not protrude beyond the front of the door. All panelboard locks shall be keyed alike. Fronts shall not be removable with door in the locked position.

f. DIRECTORY: On the inside of the door of each cabinet, provide a typewritten directory which will indicate the location of the equipment or outlets supplied by each circuit. The directory shall be mounted in a metal frame with a non-breakable transparent cover. The panelboard designation shall be typed on the directory card and panel designation stenciled in 1½" high letters on the inside of the door.

g. PANELBOARD INSTALLATION

(1) Before installing panel boards check all of the architectural drawings for possible conflict of space and adjust the location of the panel board to prevent such conflict with otter items.

(2) When the panel board is recessed into a wall serving an area with accessible ceiling space, provide and install an empty conduit system for future wiring. All 1¼" conduit shall be stubbed into the ceiling space above the panel board and under the panel board if such accessible ceiling space exists.

(3) The panel boards shall be mounted in accordance with Article 373 of the NEC. The Electrical Contractor shall furnish all material for mounting the panel boards.

6. Wiring Devices

a. GENERAL: The wiring devices specified below with Arrow Hart numbers may also be the equivalent wiring device as manufactured by Bryant Electric, Harvey Hubbell, or Pass & Seymour. All other items shall be as specified.

b. WALL SWITCHES: Where more than one flush wall switch is indicated in the same location, the switches shall be mounted in gangs undera common plate.

(1)	Single-Pole	AH#1991
(2)	Three-Way	AH#1993
(3)	Four-way	AH#1994
(4)	Switch with pilot light	AH#2999-R
(5)	Motor Switch—Surface	AH#6808
(6)	Motor Switch—Flush	AH#6808-F

c. RECEPTACLES

(1)	Duplex	AH#6739
(2)	Clock Outlet	AH#5708
(3)	Weatherproof	AH#5735-F

(4) Floor Receptacles—Steel City Series 600 floor box with bronze edge ring, floor plate P-60-1, bronze carpet plate and service fitting SFH-40-RG.

(5) Floor outlet for Telephone and Alarm—Steel City Series 600 floor box with bronze edge ring, floor plate P-60-1, bronze carpet plate and service fitting SFL-10.

d. WALL PLATE: Stainless steel wall plates with satin finish minimum .030 inch shall be provided for all outlets and switches.

Section 16E—Lighting Fixtures and Lamps

1. Portions of the sections of the Documents designated by the letters "A," "B," and "C" and "DIVISION ONE—GENERAL REQUIREMENTS" apply to this division.

2. Consult Index to be certain that set of Documents and Specifications is complete. Report omissions or discrepancies to the Architect.

3. Lighting Fixtures

a. GENERAL: The Electrical Contractor shall furnish, install and connect all lighting fixtures to the building wiring system unless otherwise noted.

b. FIXTURE TYPE: The fixture for each location is indicated by type letter. Refer to fixture schedule on the drawings for each type, manufacturer, catalog number and type of mounting.

c. FLUORESCENT BALLASTS: All fluorescent fixtures shall have ETL-CBM high power factor, quiet operating, Class "A" sound rated, thermally protected Class "P" cool-rated ballast with UL approval. Ballasts shall be as manufactured by General Electric, Advance, Jefferson, or approved equal. The ballasts shall be subject to a two (2) year manufacturer's guarantee. Guarantee shall be filed with the Owner.

d. SHOP DRAWINGS

(1) Shop drawings for lighting fixtures shall indicate each type together with manufacturer's name and catalog number, complete photometric data compiled by an independent testing laboratory and type of lamp(s) to be installed. No fixtures shall be delivered to the job until approved by the Architect.

(2) If the Electrical Contractor submits shop drawings on a fixture for approval other than those specified, he shall also submit a sample fixture when requested by the Architect. The sample fixture will be returned to the Electrical Contractor. The decision of the Architect shall be final.

e. COORDINATION: It shall be the responsibility of the Electrical Contractor to coordinate with the ceiling contractor and the General Contractor in order that the proper type fixture be furnished to match the ceiling suspension system being installed or building construction material.

4. Lamps

a. The Electrical Contractor shall furnish and install lamps in all fixtures as indicated on the drawings or as required. Fluorescent lamps shall be standard cool white and incandescent lamps shall be inside frosted unless otherwise noted on the drawings.

b. Lamps shall be manufactured by General Electric, Westinghouse or Sylvania.

Section 16F—Special Systems

1. Portions of the sections of the Documents designated by the letters "A," "B," and "C" and "DIVISION ONE—GENERAL REQUIREMENTS" apply to this division.
2. Consult Index to be certain that set of Documents and Specifications is complete. Report omissions or discrepancies to the Architect.

3. Telephone Raceway Systems

a. GENERAL: The Electrical Contractor shall provide and install empty raceway, outlet boxes, pull boxes and associated equipment required for a complete telephone system as indicated on the drawings and specified herein. All materials and workmanship shall conform with Section 16B of the Specifications. All wiring shall be installed by the local telephone company. The entire installation shall be in accordance with the requirements of the local telephone company.

b. RIGID CONDUIT (Heavy Wall): Rigid conduit shall be installed in the following locations: service entrance, underground in contact with earth, in concrete slab and "wet" locations.

c. ELECTRIC METALLIC CONDUIT (EMT): Electric metallic tubing shall be used in all locations not otherwise specified to be rigid conduit.

d. OUTLETS: Telephone wall outlets shall consist of a 4" two-gang outlet box, raised device cover and a telephone device plate of the same material as the receptacle device plates. The conduit shall extend from the outlet to the designated telephone space unless otherwise noted.

e. PULL WIRE: The Electrical Contractor shall install a No. 14 ga. galvanized pull wire in the raceway system for future use.

f. MOUNTING HEIGHTS: The wall outlets shall be mounted at approximately the following heights unless otherwise noted on the drawings or required by telephone company: Desk Phones—18" AFF, Wall Phone—58" AFF, Telephone Booth—7'6" AFF.

4. Emergency Lighting System

a. The Electrical Contractor shall provide and install a complete emergency lighting system as indicated on the drawings and specified herein. The system shall originate on the line side of the service entrance main switch, through overcurrent protective equipment to each exit light fixture and each fixture designated as being "emergency light." The switch shall be painted red. The Contractor shall be responsible for verification with local governing authorities of the proper letter and background colors of exit light fixtures before purchase of same. The entire installation shall be in accordance with the National Electric Code, the local electric code and the fire protection department having authority in the local jurisdiction.

QUESTIONS

Using the floor plan of the residences in Fig. 14-6 on page 197, answer the following questions:

1. How many duplex receptacles are shown on the drawing?
2. How many 240-volt receptacles are shown on the drawing? What appliances are these provided for?
3. How many fluorescent lighting fixtures are shown on the drawing?
4. How many wall-mounted incandescent lighting fixtures are shown?
5. How many single-pole wall switches are shown? How many 3-way switches?

Answer the following questions by filling in the missing blanks.

1. The National Electrical Code is the result of the best efforts of _____ _____, _____ of electrical equipment, _____ _____, _____ _____ and other concerned experts throughout the country.
2. The article listing definitions essential to the proper use of the NEC is number _____.
3. Mandatory rules of the NEC are characterized by the use of the word _____, whereas advisory rules are characterized by the use of the word _____.
4. Wiring Design and Protection is covered in Chapter _____ of the NEC.
5. Besides the NEC, many cities and towns have their own electrical requirements; these rules are known as _____ _____.
6. The electrical specifications for a building project are a _____ description of work and duties required of the _____, _____, and _____.
7. The paragraph in the specifications (in this book) that gives information concerning inspection and certificates of approval is number _____, subparagraph _____, and is in Section _____—General _____.
8. Conduit throughout the project shall be securely and rigidly supported to the _____.
9. Wall receptacles shall be mounted approximately _____ above the finished floor.
10. Motor-protection fuses shall be _____ element.

Chapter 15

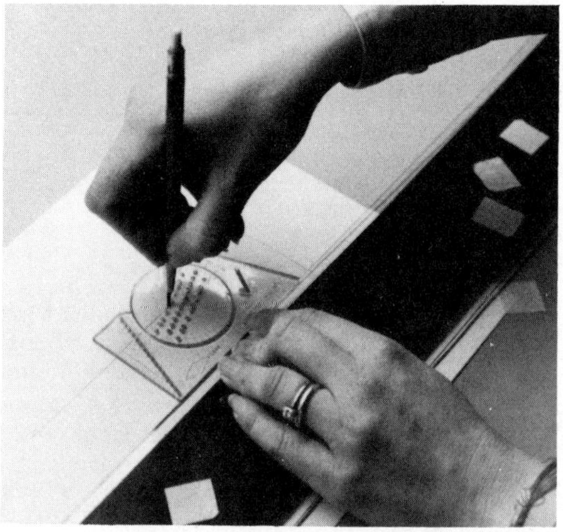

ENTERING THE DRAFTING PROFESSION

15.1 Drafting Opportunities

The opportunities in electrical and electronic drafting are almost unlimited, as nearly all large manufacturing plants along with architectural/engineering firms, electrical contractors, and utility companies have need for one or more draftsmen. However, in order to select the "right" job within this broad field, you must match your own interest, experience, and formal education to those required on a specific job. Still, it is not easy to choose a career if you do not know exactly what kinds of jobs are offered and for whom you can work. Therefore, it is important to take time to find out how the various organizations operate, the job titles, requirements, duties, employment possibilities, and in what part of the country these jobs are located.

ELECTRICAL DRAFTING IN INDUSTRY

As an industrial draftsman, you will be entering one of the most secure branches of the electrical drafting field. Benefits are usually good and include paid vacations, holidays, and insurance premiums; most have a generous retirement plan.

Typical assignments will include: industrial lighting layouts, power distribution diagrams, motor control diagrams, automation controls, and power house substation drawings. In nearly every case, the assignments will be laid out by experienced engineers in the form of rough sketches, notes, or verbal instruction.

Other duties may include the printing (blueprinting) of drawings, delivery of drawings, taking measurements or field notes, and assisting the engineering department in general.

Apprenticeship is essential before a person can assume full responsibilities as a senior draftsman. The apprenticeship period usually is from two to four years, depending upon the person's previous training and the availability of a senior position. The beginning draftsman may even find that his first job in industry involves many tasks besides actual drafting and affords only modest pay.

Draftsmen applying for a job in the larger industries usually are required to take an aptitude test and should have two years of training in drafting—either on the vocational level or actual experience. Advancement is normally slow, and the top position that one could hope for without further training or education is job captain or chief draftsman.

ELECTRICAL DRAFTING FOR CONTRACTING FIRMS

The position of electrical draftsman in an electrical contracting firm probably offers the most prestige and financial reward of any beginning drafting position—if you can qualify. For example, a Washington, D.C. electrical contracting firm requires the following qualifications for their electrical draftsmen:

JOB REQUIREMENTS

Electrical Draftsman

Education: B.S. Degree in Electrical Engineering from an accredited institution is preferred, but not mandatory. Additional training in air conditioning, refrigeration, illumination, and drafting desirable.

Special Knowledge: Familiarity with the National Electrical Code and ordinances pertaining to this geographical area.

Experience: Two years of experience in electrical design or drafting related to electrical installation.

Skills: Specific technical ability needed to properly design electrical systems, and to personally draft necessary plans and specifications.

Interaction: Willingness to communicate and consult with internal personnel and with outside contacts, including engineers, architects, and general contractors and their representatives.

Environmental: Adaptability to extended periods of work at the drawing board, and frequent field estimating assignments.

If you can qualify for this position, you can step into a "better-than-average" job with executive privileges. Your duties will include: to prepare engineering designs and specifications for electrical installations; to integrate installations of other building trades (plumbing and mechanical) into electrical installation plans and schedules; to draw electrical plans and write specifications; to recommend lighting fixtures and other electrical equipment for use in planned installations; and to make preliminary job estimates. In the smaller contracting firms, the position of electrical draftsman may be combined with that of estimator, in which case the draftsman is required to make detailed material take-offs and supply job pricing data (see Chapter 14).

The electrical draftsman who wants to choose this field for his career must carefully prepare himself before applying for the job. A job as an apprentice electrician between school vacations is extremely helpful. You should study the National Electrical Code—the Bible of the electrical industry—and learn basic electrical calculation. A course at a local community college in pre-engineering will add to your chances of landing the job. But above all, you must have practical experience and know the "talk of the trade."

UTILITY COMPANY DRAFTSMAN

Electrical draftsmen working for utility companies often perform duties similar to those in electrical contracting firms previously mentioned. However, it is often easier to obtain a position with a utility company than with a contracting firm, because the utility companies usually have trained engineers to supervise the draftsmen and qualifications are not as stringent. But the pay is much lower also, and advancement is about the same as for industrial draftsmen. One big advantage in working for utility companies is that most of them believe in the longevity rating of employees; that is, if you are with the firm long enough and apply yourself well, you are assured of advancement. In fact, the possibility of a beginning electrical draftman's eventually managing an entire office is not remote. Meter readers have advanced to this position.

Your first duties as an electrical draftsman for a utility company will be to assist the engineers in their many tasks; to make drawings of transmission lines, substations, switchgear, service-entrance routes to buildings, etc. Two years of drafting training in a vocational school and a good score on the company's aptitude test should be sufficient for the position.

DRAFTING FOR UNCLE SAM

There are over 90,000 engineers working on five continents for the U.S. Government. Their designs must be "translated" into exact drawings that can be understood by engineers, mechanics, and production men alike. This is where the electrical draftsman comes in.

Pay rates and scales are constantly being altered for several reasons. Cost of living oscillates over a period of years, and the pay rates must change accordingly. What the government calls a "general schedule" (GS) is merely a career niche. There are eighteen of these at present, ranging in pay from about $4000 for GS-1 to over $30,000 for GS-18 (1976 scale). Above these eighteen steps lie various appointive jobs, such as highly technical positions in the missile and nuclear ranks.

An electrical draftsman (depending upon the department he enters) should begin at a GS-5 level or even higher. He can advance to an Engineering Aide and possibly higher.

Although the beginning draftsman may make slightly less money than those in private firms, the benefits usually compensate for this; that is, vacation, sick leave, incentive bonuses and retirement pay.

For answers to your career questions, or to inquire about specific career opportunities in any U.S. Government department or agency, write or visit the nearest Interagency Board of the U.S. Civil Service Examiner's Office. All of them are listed below.

Southerland Building
806 Governors Drive S.W.
Huntsville, AL 35801

167 St. Louis St.
Mobile, AL 36602

Hill Building
632 6th Ave.
Anchorage, AK 99501

Balke Building
44 W. Adams St.
Phoenix, AZ 85003

923 W. Fourth St.
Little Rock, AR 72201

851 South Broadway
Los Angeles, CA 90014

425 Capitol Mall
Suite 125
Sacramento, CA 95814

380 W. Court St.
San Bernardino, CA 92401

1400 5th Ave.
Suite 100
San Diego, CA 92101

Box 36122
450 Golden Gate Ave.
San Francisco, CA 94102

Main Post Office Bldg.
1823 Stout St.
Room 112
Denver, CO 80201

Federal Bldg.
Room 716
450 Main St.
Hartford, CT 06103

U.S. Post Office & Courthouse
11th & King Sts.
Wilmington, DE 19801

123 South Court Ave.
Orlando, FL 32801

230 Peachtree St. N.W.
Atlanta, GA 30303

Federal Bldg.
451 College St.
Macon, GA 31201

Federal Building
Honolulu, HI 96813

Federal Building
Room 107
Boise, ID 83702

219 S. Dearborn St.
Chicago, IL 60604

36 S. Pennsylvania St.
Indianapolis, IN 46204

Old Federal Office Bldg.
Fifth & Court Sts.
Des Moines, IA 50309

Beacon Bldg.
114 South Main St.
Wichita, KS 67202

Heyburn Bldg.
721 S. 4th St.
Louisville, KY 40202

Federal Bldg. So.
600 South St.
New Orleans, LA 70130

Federal Bldg.
Augusta, ME 04330

Main Post Office Bldg.
Calvert & Fayette Sts.
Baltimore, MD 21202

Post Office & Courthouse Bldg.
Boston, MA 02109

144 West Lafayette St.
Detroit, MI 48226

Building 57
Fort Snelling, MN 55111

802 N. State St.
Jackson, MS 39201

Federal Bldg.
601 East 12th St.
Kansas City, MO 64106

Federal Bldg.
1520 Market St.
St. Louis, MO 63103

IBM Building
130 Neill Ave.
Helena, MT 59601

U.S. Court & Post Office Bldg.
215 N. 17th St.
Omaha, NE 68102

Federal Bldg.
300 Booth St.
Reno, NV 89502

Portsmouth Naval Shipyard
Building 59
Portsmouth, NH 03801

Post Office Bldg.
Federal Square
Newark, NJ 07102

Federal Bldg.
421 Gold Ave. S.W.
Albuquerque, NM 87101

News Building
220 East 42nd St.
New York, NY 10017

O'Donnell Bldg.
301 Erie Blvd.
Syracuse, NY 13202

415 W. Hillsborough St.
Raleigh, NC 27603

Manchester Bldg.
112 University Drive
Fargo, ND 58102

1909 E. 13th St.
Cleveland, OH 44114

Knott Bldg.
7 East Fourth St.
Dayton, OH 45402

American General Bldg.
210 N.W. 6th St.
Oklahoma City, OK 73102

319 S.W. Pine St.
Portland, OR 97204

128 N. Broad St.
Philadelphia, PA 19102

Federal Bldg.
1000 Liberty Ave.
Pittsburgh, PA 15222

Federal Office Bldg.
224 Meeting St.
Charleston, SC 29403

Dusek Bldg.
919 Main St.
Rapid City, SD 57701

Federal Office Bldg.
167 N. Main St.
Memphis, TN 38101

1114 Commerce St.
Dallas, TX 75202

El Paso National Bank Bldg.
411 N. Stanton St.
El Paso, TX 79901

702 Caroline St.
Houston, TX 77002

ELECTRICAL DRAFTING FOR CONSULTING ENGINEERING FIRMS

While possibly the least stable, a position as an electrical draftsman with a consulting engineering firm may offer the draftsman the greatest chance for personal achievement. It is also the easiest electrical drafting job to land—as well as to be dismissed from.

Most consulting engineering firms are small—from four to twenty-five employees—and are run by only one or two principals. They must keep a heavy workload, for obvious reasons; if additional work is available, they will take it. This is the time that they will employ practically any draftsman who is breathing. This is your chance to "get in."

In a consulting engineering firm, you will probably start your drafting career by tracing architectural floor plans so that the electrical system can be drawn in later. From this, you will probably prepare neat working drawings from designer's sketches, all the time being closely supervised by the engineer, designer, or the job captain over you.

If you apply yourself well, turn out extremely neat and accurate drawings, keep your errors and omissions to a minimum, and follow your supervisor's instructions, you will probably be one of the draftsmen who remains with the firm at the next layoff (when work gets slack).

For the two years to follow, you will continue to make working drawings from designer's sketches, and with most details of design laid out for you. At the end of this time, you should be making at least a moderate salary. If, during this two-year period, you study the National Electrical Code and lighting and other design techniques, you should be promoted to draftsman/designer. This means that you are beginning to design systems on your own.

For example, your designer/supervisor may have calculated the need for 20 lighting fixtures in a given area. Instead of sketching out the location and spacing of each fixture, he merely tells you to draw 20 fixtures in the area; you are supposed to be capable of placing the fixtures yourself. You should also be capable of circuiting the lighting fixtures and should know how many are allowed on a 20-ampere circuit.

After this period, you should receive pay raises annually or even semiannually until perhaps another two years have passed. You could then be promoted to "full" designer; you would design and still do some drafting, but you also might have one or two draftsmen working under you.

Another two years of faithful service, and you may be made an associate of the firm and share into the profits and perhaps be in the $15,000 to $20,000-a-year bracket.

After you have worked under a Professional Engineer for a period of ten or more years, most states will allow you to take the Professional Engineer's Exam—regardless of your education. If you can pass this exam, you may want to open your own consulting firm. Or if you prefer to stay with the original firm, you should now be in the $25,000-plus pay range.

DRAFTING SERVICE

From time to time, all of the organizations mentioned previously become extremely overloaded with work that must be turned out. If the firm has enough physical space, the firm may try to hire more draftsmen, but sometimes they are reluctant to do this if this extra burden will last only for a short time. Then what do they do? They either begin working their regular employees overtime or else "farm out" their work to someone else. This is where a drafting service comes into action.

Most firms will gladly pay a drafting service

twice the amount they pay their regular employees if they know the service is available when they need it, and, more important, they don't have to pay for it when they don't need it.

Such a service can be set up initially in your home and enlarged when you see fit. A very small investment is required, and if you are the only employee, your overhead should be next to nothing. You can expect to make from $8,000 to $15,000 a year in this business and be your own boss besides.

We have not gone into great detail with any of the fields previously mentioned. It is recommended that you choose a couple of the fields that you think would suit you best. Then visit a few firms who operate within these fields.

After visiting a dozen or so firms who hire electrical draftsmen, and after talking to the principals and perhaps some of the employees in these firms, you should be in a position to make up your mind as to the field you want to try.

One word of caution. Many people start with poorer positions than they need to because they do not know where to go, how to look, or how to present themselves. Many of these people do not even know what they are qualified to do. Do you?

SUMMARY

A technique in selecting a position as an electrical draftsman is to ask yourself, "What and how is best for me?" Therefore, the first concern is *choosing* the right job; the second, finding out the job qualifications; the third, preparing yourself for the job. The three items can save endless hours of writing letters, beating the pavement, and many fruitless interviews.

15.2 How To Apply for a Job in Electrical Drafting

In general, the basic "tools" for obtaining a job consist of letters asking for job requirements and application forms, a phone call, and a personal interview. Sometimes it is necessary to use all of them, while one will suffice in other instances. Regardless of which "tool" is used, it must be as effective as possible.

A letter to a prospective employer is perhaps one of the most effective approaches for obtaining work, and since this letter will probably be the very first contact that you make with the firm, make certain that the letter is prepared carefully; there is little chance of getting the job if you send out poor, uninformative letters.

There are many kinds of letters classified as job-getting letters. Some are called *lead letters*, because the sender learned of the position through an ad. Others may be written to various firms asking what might be available. However, all of the different types of letters have the same basic function: to secure an interview that may lead to a drafting position with the firm.

An example of a suitable letter for a job application might read as follows:

Return Address Here

Today's Date

Mr. Ronald D. Carter, Principal
Engineering Associates Ltd.
1000 Any Avenue
Anytown, USA.

Dear Mr. Carter:

For the past two years, I have devoted a strong, continuous effort to developing my knowledge and ability in the field of electrical drafting. At present, I am interested in obtaining work as an electrical draftsman with a consulting engineering firm. Naturally, my goal is firmly set on the position of draftsman/designer, but this will have to come in time.

I have had two years of formal training in electrical drafting at John Smith Vocational School under the supervision of Mr. Bob Jones, Instructor. I am now taking an advanced course by correspondence from CBC Correspondence School.

I will be in your area on Monday, June 1 and I will phone you on that date in the hope that your schedule may include an interview with me about a position with your firm.

Best regards,
/signed/
Your name

Although letter forms could change from year to year, a letter should look neat, be arranged in a logical sequence, and be consistent. Check all spelling of words and use correct grammar.

The personal interview comes next, and here is where you sink or swim. An interview is very much like a good letter—it takes careful planning beforehand. Find out what your prospective employer wants and then convince him that you have it.

First, you should learn specifically what job or job area is open, either through a letter, from a personal contact, or through an ad. Then be ready to have something definite to offer your prospective employer. Remember that labor is simple to come by, but effective services are exceedingly difficult to find.

Within a day or two after the interview, you should send the prospective employer a letter of thanks with an additional sales pitch in it. Or perhaps your former teacher could write you a letter of recommendation displaying interest in seeing that you get the job for which you were interviewed.

SUMMARY

Let us take a look at some of the important details to be kept in mind while preparing for an interview and then carrying it off in a professional manner.

Before the Interview

1. Know much more about the job than its title.
2. Be familiar with the qualifications and requirements for the position you are seeking. Also be familiar with the duties of the job.
3. Find information about the firm or agency.
    a. History and development.
    b. Organization and reorganization.
    c. Past personnel and their contributions and accomplishments.
    d. Serious problems that arose and how they were solved.
    e. Present problems and how you would suggest they be solved.
    f. Names of present administrative staff and something about them if possible.

During the Interview

1. Be pleasant; smile occasionally.
2. Be prepared to discuss some of the things that you have learned about the organization.
3. Both before and during the interview attempt to discover what your employer wants. Then help him to understand or realize that you have the attributes to fill his needs.
4. Demonstrate your knowledge of the agency's needs by asking leading or pertinent questions.
5. Continually point to your desire to provide a service by keeping out of the conversation the approach, "I want a job."
6. Provide as much direct proof of your accomplishments and abilities with good illustrations, letters from others, photographs, or anything else that will help.
7. Anticipate and be fully ready for confusing or demoralizing questions designed to put you on the spot, such as, "What do you expect to be paid?"
8. Be ready to turn shortcomings, poor records, or even physical handicaps to your advantage by *showing* you can do better than others because of these problems.
9. Leave the interview with an atmosphere of optimism.
10. Never fail to return a letter of thanks and, if appropriate, include an additional sales pitch.

15.3 How To Apply Yourself Once You Obtain the Job

Once you have obtained the drafting position, you cannot stop learning, or you are going to lose it. Continue your education by reading trade journals in your field; read at least one book on the subject of electricity monthly, and thumb through all the reference books that you can get your hands on. Strive to improve yourself daily.

Everyone likes to work around a pleasant person, so personality plays an important role in how you handle your job and get along with your boss and fellow employees. Do not be the kind of person who continually asks for extra favors, and never ask for a raise until you are certain that you deserve one. Many employers will recognize your progress and give you a raise in pay when the correct time comes.

Be careful in your work so as to avoid errors and omissions, and above all, be neat. Following the preceding suggestions will help you to keep the position and to earn top dollar for the work that you are performing.

15.4 What To Expect in the Future as an Electrical Draftsman

The future of electrical drafting can be somewhat predicted by comparing the use of electricity today with that of the past. The following figures should answer the question.

In 1940, the U.S. used approximately 124,270,000,000 kWh
In 1949, the U.S. used approximately 248,540,000,000 kWh
In 1959, the U.S. used approximately 626,740,000,000 kWh
In 1969, the U.S. used over 1,000,000,000,000 kWh

From the above figures, it can be seen that the United States is doubling its use of electricity approximately every ten years. What does all this mean to the electrical draftsman? It means more jobs—more opportunity in your chosen field if you qualify. If there is any increase in the use of electricity, there will be more designs of electrical systems, and these designs must be conveyed to the workman on the job so that the system will be installed correctly.

ASSIGNMENT

1. From the information you obtained from this chapter (or from other sources, for that matter) list your first and second choice of the type of firm you think you would like to work for. (1)____ (2)____.
2. What geographical location would you like to settle in when you begin working as an electrical or electronic draftsman?____
3. Obtain at least three names of firms for each of the two types listed in Question 1 located in or near the area you chose in Question 2. (*Hint*: Try the yellow pages in telephone directories, or write to the Chamber of Commerce in the area of your choice.)
4. Using the sample letter in this chapter as a guide, write a letter to each of the firms you obtained in Question 3. Be brief and to the point, stating your qualifications and when you will be available to begin work.

Appendix A

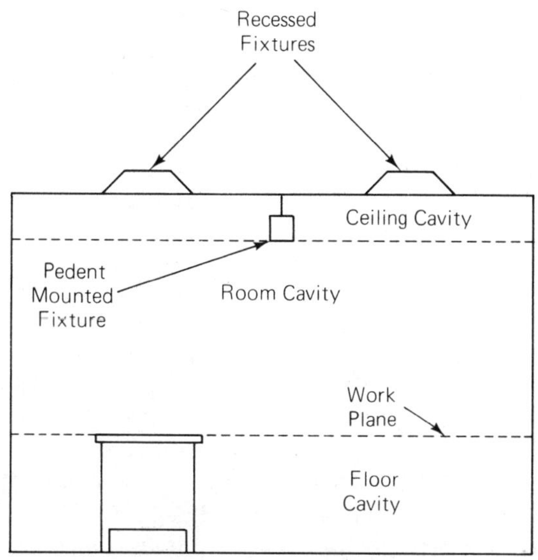

GLOSSARY

Accessible (as applied to wiring methods.) Capable of being removed or exposed without damaging the building structure or finish, or not permanently closed in by the structure or finish of the building.

Accessible (as applied to wiring methods). Capable of approach because not guarded by locked doors, elevation, or other effective means.

Aggregate. Inert material mixed with cement and water to produce concrete.

Ampacity. Current-carrying capacity expressed in amperes.

Amplification. Procedure of expanding the strength (current, power, or voltage) of a signal.

Amplifier. An instrument used to increase the signal voltage, current, or power. It is composed of a transistor or vacuum tube and an associated circuit called a *stage*. It may contain several stages in order to obtain a desired gain.

Amplitude. The maximum instantaneous value of an alternating voltage or current. It is measured in either the positive or negative direction.

Anode. Positive electrode or terminal; the plate of a vacuum tube. In some cases the most positive electrode.

Appliance. Utilization equipment, generally equipment other than industrial, normally built in standardized sizes or types and installed or connected as a unit to perform one or more functions, such as clothes washing, air conditioning, food mixing, deep frying, etc.

Appliance, fixed. An appliance that is fastened or otherwise secured at a specific location.

Appliance, portable. An appliance that is actually moved or can easily be moved from one place to another in normal use.

Appliance, stationary. An appliance that is not easily moved from one place to another in normal use.

Approved. Acceptable to the authority enforcing the Code.

Attachment plug (plug cap) (cap). A device that, upon insertion in a receptacle, establishes a connection between the conductors of the attached flexible cord and the conductors connected permanently to the receptacle.

Attenuation. The decrease in the strength of a signal.

Automatic. Self-acting, operating by its own mechanism when actuated by some impersonal influence, such as a change in current strength, pressure, temperature, or mechanical configuration.

Backfill. Loose earth placed outside foundation walls for filling and grading.

Base. One of the regions or terminals of a transistor.

Bearing plate. Steel plate placed under one end of a beam or truss for load distribution.

Bearing wall. Wall supporting a load other than its own weight.

Bench mark. Point of reference from which measurements are made.

Bias vacuum tube. Difference of potential between control grid and cathode. Transistor—difference of potential between base and emitter and base and collector. Magnetic amplifier—level of flux density in the magnetic amplifier core under no-signal condition.

Bonding jumper. A reliable conductor used to ensure the required electrical conductivity between metal parts required to be electrically connected.

Branch circuit. That portion of a wiring system extending beyond the final overcurrent device protecting the circuit.

Branch circuit, appliance. A circuit supplying energy to one or more outlets to which appliances are to be connected; such circuits have no permanently connected lighting fixtures that are not a part of an appliance.

Branch circuit, general-purpose. A branch circuit that supplies a number of outlets for lighting and appliances.

Branch circuit, individual. A branch circuit that supplies only one piece of utilization equipment.

Bridging. System of bracing between floor beams to distribute floor load.

Building. A structure that stands alone or that is cut off from adjoining structures by fire walls with all openings therein protected by approved fire doors.

Bus bar. The heavy copper or aluminum bar used on switchboards to carry current.

Cabinet. An enclosure designed for either surface or flush mounting and provided with a frame, mat, or trim in which swinging doors are hung.

Capacitor. An apparatus consisting of two conducting surfaces separated by an insulating material. It stores energy, blocks the flow of direct current, and permits the flow of alternating current to a degree depending on the capacitance and frequency.

Cathode. The electron-emitting electrode of an electron tube or semiconductor device.

Cavity wall. Wall built of solid masonry units arranged to provide air space within the wall.

Chase. Recess in inner face of masonry wall providing space for pipes and/or ducts.

Choke coil. A coil used to limit the flow of alternating current while permitting direct current to pass.

Circuit breaker. A device designed to open and close a circuit by nonautomatic means and to open the circuit automatically on a predetermined overload of current, without injury to itself when properly applied within its rating.

Coaxial cable. A cable consisting of two conductors concentric with and insulated from each other.

Cold cathode. A cathode that does not depend upon heat for electron emission.

Collector. The part of a transistor that collects electrons.

Column. Vertical load-carrying member of a structural frame.

Commutator. Device used on electric motors or generators to maintain a unidirectional current.

Concealed. Rendered inaccessible by the structure or finish of the building. Wires in concealed raceways are considered concealed, even though they may become accessible by withdrawing them.

Conductance. The ability of material to carry an electric current.

Conductor, bare. A conductor having no covering or insulation whatsoever.

Conductor, covered. A conductor having one or more layers of nonconducting materials that are not recognized as insulation under the Code.

Conductor, insulated. A conductor covered with material recognized as insulation.

Connector, pressure (solderless). A connector that establishes the connection between two or more conductors or between one or more conductors and a terminal by means of mechanical pressure and without the use of solder.

Continuous load. A load in which the maximum current is expected to continue for three hours or more.

Contour line. On a land map denoting elevations, a line connecting points with the same elevation.

Controller. A device, or group of devices, that serves to govern in some predetermined manner the electric power delivered to the apparatus to which it is connected.

Cooking unit, counter-mounted. An assembly of one or more domestic surface heating elements for cooking purposes, designed to be flush mounted in, or supported by, a counter and complete with internal wiring and inherent or separately mounted controls.

Crawl space. Shallow space between the first tier of beams and the ground (no basement).

Cryogen. Device becoming a superconductor at extremely cold temperatures.

Current. The rate of transfer of electricity.

Curtain wall. Nonbearing wall between piers or columns for the enclosure of the structure; not supported at each story.

Deflecting plate. The part of a certain type of electron tube which provides an electrical field to produce deflection of an electron beam.

Demand factor. In any system or part of a system, the ratio of the maximum demand of the system, or part of the system, to the total connected load of the system, or part of the system under consideration.

Detection. The process of obtaining the separation of the modulation component from the received signal.

Dielectric. An insulator or a term referring to the insulation between the plates of a capacitor.

Diode. A device having two electrodes, the cathode and the plate or anode. Used as rectifiers and detectors.

Disconnecting means. A device, a group of devices, or other means whereby the conductors of a circuit can be disconnected from their source of supply.

Double-strength glass. Sheet glass that is one-eighth inch thick (single-strength glass is one-tenth inch thick.)

Dry wall. Interior wall construction consisting of plaster boards, wood paneling, or plywood nailed directly to the studs without application of plaster.

Duty, continuous. A requirement of service that demands operation at a substantially constant load for an indefinitely long time.

Duty, intermittent. A requirement of service that demands operation for alternate intervals of (1) load and no load, (2) load and rest, or (3) load, no load, and rest.

Duty, periodic. A type of intermittent duty in which the load conditions regularly recur.

Duty, short-time. A requirement of service that demands operations at loads and for intervals of time, both of which may be subject to wide variation.

Electrode. A conducting element used to emit, collect, or control electrons and ions.

Electron. A particle of matter netatively charged.

Electron emission. The release of electrons from the surface of a material into surrounding space, due to heat, light, high voltage, or other causes.

Elevation. Drawing showing the projection of a building on a vertical plane.

Emitter. The part of a transistor that emits electrons.

Enclosed. Surrounded by a case that will prevent anyone from accidentally touching live parts.

Epitaxial. A very significant thin-film type of deposition for making certain devices in microcircuits involving a realignment of molecules.

Equipment. A general term including material, fittings, devices, appliances, fixtures, apparatus, and the like used as part of, or in connection with, an electrical installation.

Eyelet. Something used on printed-circuit boards to make reliable connections from one side of the board to the other.

Expansion bolt. Bolt with a casing arranged to wedge the bolt into a masonry wall to provide an anchorage.

Expansion joint. Joint between two adjoining concrete members arranged to permit expansion and contraction with changes in temperature.

Exposed (as applied to live parts). Live parts that a person could inadvertently touch or approach nearer than a safe distance. This term is applied to parts not suitably guarded, isolated, or insulated.

Exposed (as applied to wiring method). Not concealed.

Externally operable. Capable of being operated without exposing the operator to contact with live parts.

Facade. Main front of a building.

Farad. The basic unit of capacity.

Feedback. The process of transferring energy from the output circuit of a device back to its input.

Feeder. The conductors between the service equipment, or the generator switchboard of an isolated plant, and the branch-circuit overcurrent device.

Filament. A wire, heated electrically, which emits electrons or heats a cathode emitting electrons.

Filter. A combination of circuit elements designed to pass a definite range of frequencies, reducing all others.

Fire stop. Incombustible filler material used to block interior draft spaces.

Fitting. An accessory such as a locknut, bushing, or other part of a wiring system that is intended primarily to perform a mechanical rather than an electrical function.

Flashing. Strips of sheet metal bent into an angle between the roof and wall to make a watertight joint.

Footing. Structural unit used to distribute loads to the bearing materials.

Frequency. The number of complete cycles an alternating electric current, a sound wave, or a vibrating object undergoes per second.

Frostline. Deepest level below grade to which frost penetrates in a geographic area.

Fuse. A protective device inserted in series with a circuit.

Gain. The ratio of output to input power, voltage, or current, respectively.

Garage. A building or portion of a building in which one or more self-propelled vehicles carrying volatile, flammable liquid for fuel or power are kept for use, sale, storage, rental, repair, etc.

Gate. A device that makes an electronic circuit operable for a short time.

Grid. An electrode having one or more openings for the passage of electrons or ions.

Grid leak. A resistor of high ohmic value connected between the control grid and the cathode in a grid-leak capacitor detector circuit, for automatic biasing.

Ground. A conducting connection, whether intentional or accidental, between an electrical circuit or piece of equipment, and earth or some other conducting body serving in place of the earth.

Grounded. Connected to earth or to some conducting body that serves in place of the earth.

Grounded conductor. A system or circuit conductor that is intentionally grounded.

Grounding conductor. A conductor used to connect equipment or the grounded circuit of a wiring system to a grounding electrode.

Henry. The basic unit of inductance.

I beam. Rolled steel beam or built-up beam of I section.

Impedance. The total opposition offered to the flow of an alternating current.

Incombustible material. Material that will not ignite or actively support combustion in a surrounding temperature of 1200°F during an exposure of five minutes; also material that will not melt when the temperature of the material is maintained at 900°F for a period of at least five minutes.

Inductance. The property of a circuit or two neighboring circuits that determines how much voltage will be induced in one circuit by a change of current in either circuit.

Inductor. A coil.

Integrated circuit. A circuit in which different types of devices, such as resistors, capacitors, and transistors, are made from a single piece of material and then connected to form a circuit.

Isolated. Not readily accessible to persons unless special means for access are used.

Jamb. Upright member forming the side of a door or window opening.

Lally column. Compression member consisting of a steel pipe filled with concrete under pressure.

Laminated wood. Wood built up of plies or laminations that have been joined either with glue or with mechanical fasteners. Usually, the plies are too thick to be classified as veneer, and the grain of all plies is parallel.

Lighting outlet. An outlet intended for the direct connection of a lampholder, lighting fixture, or pendant cord terminating in a lampholder.

Location, damp. A location subject to a moderate amount of moisture, such as some basements, some barns, some cold-storage warehouses, and the like.

Location, dry. A location not normally subject to dampness or wetness; a location classified as dry may be temporarily subject to dampness or wetness, as in the case of a building under construction.

Location, wet. A location subject to saturation with water or other liquids, such as locations exposed to weather, washrooms in garages, and similar locations. Installations that are located underground or in concrete slabs, or masonry in direct contact with the earth shall be considered wet locations.

Logic. The arrangement of circuitry designed to accomplish certain objectives.

Low-energy power circuit. A circuit that is not a remote-control or signal circuit but whose power supply is limited in accordance with the requirements of Class-2 remote-control circuits.

Modulation. The process of varying the amplitude, frequency, or the phase of a carrier wave.

Multioutlet assembly. A type of surface or flush raceway designed to hold conductors and attachment plug receptacles and assembled in the field or at the factory.

Nonautomatic. Used to describe an action requiring personal intervention for its control.

Nonbearing wall. Wall that carries no load other than its own weight.

Oscillator. A circuit that is designed to generate an audio or radio frequency.

Oscilloscope. A test instrument that shows visually on a screen the waveform of a varying current or voltage.

Outlet. In the wiring system, a point at which current is taken to supply utilization equipment.

Outline lighting. An arrangement of incandescent lamps or gaseous tubes to outline and call attention to certain features, such as the shape of a building or the decoration of a window.

Oven, wall-mounted. A domestic oven for cooking purposes designed for mounting into or onto a wall or other surface.

Panel board. A single panel or group of panel units designed for assembly in the form of a single panel; includes buses and may come with or without switches and/or automatic overcurrent protective devices for the control of light, heat, or power circuits of small individual as well as aggregate capacity. It is designed to be placed in a cabinet or cutout box placed in or against a wall or partition and accessible only from the front.

Pentode. An electron tube with five electrodes or elements.

Permalloy. An alloy of nickel and iron, having an unusually high magnetic permeability.

Pilaster. Flat square column attached to a wall and projecting about a fifth of its width from the face of the wall.

Plate. The principal anode in an electron tube to which the electron stream is attracted.

Plenum. Chamber or space forming a part of an air-conditioning system.

Potential. The difference in voltage between two points of a circuit. Frequently, one is assumed to be ground (zero potential).

Potentiometer. An instrument for measuring an unknown voltage or potential difference by balancing it, wholly or in part, by a known potential difference produced by the flow of known currents in a network of circuits of known electrical constants.

Power. The rate of doing work or expending energy.

Precast concrete. Concrete units (such as piles or vaults) cast away from the construction site and set in place.

Qualified person. One familiar with the construction and operation of the apparatus and the hazards involved.

Raceway. Any channel designed expressly for holding wire, cables, or bus bars and used solely for this purpose.

Rainproof. So constructed, protected, or treated as to prevent rain from interfering with successful operation of the apparatus.

Raintight. So constructed or protected that exposure to a beating rain will not result in the entrance of water.

Readily accessible. Capable of being reached quickly, for operation, renewal, or inspections, without requiring those to whom ready access is requisite to climb over or remove obstacles or resort to portable ladders, chairs, etc.

Receptacle (convenience outlet). A contact device installed at an outlet for the connection of an attachment plug.

Receptacle outlet. An outlet where one or more receptacles are installed.

Rectifiers. Devices used to change alternating current to unidirectional current.

Relay. An electromechanical switching device that can be used as a remote control.

Remote-control circuit. Any electrical circuit that controls any other circuit through a relay or equivalent device.

Resistance. The opposition that a device or material offers to the flow of current.

Resistor. A circuit element whose chief characteristic is resistance.

Resonance. In a circuit containing both inductance and capacitance, a condition in which the inductive reactance is equal to and cancels out the capacitance reactance.

Riser. Upright member of stair extending from tread to tread.

Roughing in. Installation of all concealed electrical wiring; includes all electrical work done before finishing.

Saturation. The condition existing in a circuit when an increase in the driving signal does not produce any further change in the resultant effect.

Sealed (hermetic-type) motor compressor. A mechanical compressor consisting of a compressor and a motor, both of which are enclosed in the same sealed housing, with no external shaft or shaft seals, the motor operating in the refrigerant atmosphere.

Semiconductor. A material midway between a conductor and an insulator.

Service. The conductors and equipment used for delivering energy from the electricity supply system to the wiring system of the premises served.

Service cable. The service conductors made up in the form of a cable.

Service conductors. The supply conductors that extend from the street main or transformers to the service equipment of the premises being supplied.

Service drop. The overhead service conductors from the last pole, or other aerial support, to and including the splices, if any, that connect to the service-entrance conductors at the building or other structure.

Service-entrance conductors, underground system. The service conductors between the terminals of the service equipment and the point of connection to the service lateral.

Service equipment. The necessary equipment, usually consisting of a circuit breaker, or switch and fuses, and their accessories, located near the point of entrance of supply conductors to a building and intended to constitute the main control and means of cutoff for the supply to that building.

Service lateral. The underground service conductors between the street main, including any risers at a pole or other structure or from transformers, and the first point of connection to the service-entrance conductors in a terminal box, meter, or other enclosure with adequate space, inside or outside the building wall. Where there is no terminal box, meter, or other enclosure with adequate space, the point of connection shall be considered to be the point of entrance of the service conductors into the building.

Service raceway. The rigid metal conduit, electrical metallic tubing, or other raceway that encloses the service-entrance conductors.

Setting (of circuit breaker). The value of the current at which the circuit breaker is set to trip.

Sheathing. First covering of boards or paneling nailed to the outside of the wood studs of a frame building.

Siding. Finishing material that is nailed to the sheathing of a wood frame building and that forms the exposed surface.

Signal circuit. Any electrical circuit supplying energy to an appliance that gives a recognizable signal.

Soffit. Underside of a stair, arch, or cornice.

Solenoid. An electromagnet having a movable iron core.

Soleplate. Horizontal bottom member of wood-stud partition.

Studs. Vertically set skeleton members of a partition or wall to which lath is nailed.

Switch, general-use. A switch intended for use in general distribution and branch circuits. It is rated in amperes and is capable of interrupting its rated voltage.

Switch, general-use snap. A form of general-use switch so constructed that it can be installed in flush device boxes or on outlet covers, or otherwise used in conjunction with wiring systems recognized by the Code.

Switch, ac general-use snap. A form of general-use snap switch suitable only for use on alternating-current circuits and for controlling the following:

1. Resistive and inductive loads (including electric discharge lamps) not exceeding the ampere rating at the voltage involved.
2. Tungsten-filament lamp loads not exceeding the ampere rating at 120 volts.
3. Motor loads not exceeding 80% of the ampere rating of the switches at the rated voltage.

Switch, ac-dc general-use snap. A form of general-use snap switch suitable for use on either direct- or alternating-current circuits and for controlling the following:

1. Resistive loads not exceeding the ampere rating at the voltage involved.
2. Inductive loads not exceeding one-half the ampere rating at the voltage involved, except that switches having a marked horsepower rating are suitable for controlling motors not exceeding the horsepower rating of the switch at the voltage involved.
3. Tungsten-filament lamp loads not exceeding the ampere rating at 125 volts, when marked with the letter T.

Switch, isolating. A switch intended for isolating an electrical circuit from the source of power. It has no interrupting rating and is intended to be operated only after the circuit has been opened by some other means.

Switch, motor-circuit. A switch, rated in horsepower, capable of interrupting the maximum operating overload current of a motor having the same horsepower rating as the switch at the rated voltage.

Switchboard. A large single panel, frame, or assembly of panels, having switches, overcurrent, and other protective devices, buses, and usually instruments, mounted on the face or back or both. Switchboards are generally accessible from the rear as well as from the front and are not intended to be installed in cabinets.

Synchronous. Simultaneous in action and in time (in phase).

Tachometer. An instrument for measuring revolutions per minute.

Thermal cutout. An overcurrent protective device containing a heater element in addition to, and affecting, a renewable fusible member that opens the circuit. It is not designed to interrupt short-circuit currents.

Thermally protected (as applied to motors). Refers to the words "Thermally Protected" appearing on the nameplate of a motor or motor-compressor and means that the motor is provided with a thermal protector.

Thermal protector (as applied to motors). A protective device that is assembled as an integral part of a motor or motor-compressor and that, when properly applied, protects the motor against dangerous overheating due to overload and failure to start.

Thyratron. A gas-filled triode tube that is used in electronic-control circutts.

Transducer. A device converting input into a different type of output.

Transformer. A device used to transfer energy from one circuit to another. It is composed of two or more coils linked by magnetic lines of force.

Triode. A three-electrode electron tube containing an anode, a cathode, and a control electrode.

Trusses. Framed structural pieces consisting of triangles in a single plane for supporting loads over spans.

Utilization equipment. Equipment that utilizes electric energy for mechanical, chemical, heating, lighting, or other similar useful purposes.

Ventilated. Provided with a means to permit circulation of air sufficient to remove an excess of heat fumes, or vapors.

Volt. The practical unit of voltage or electromotive force. One volt sends a current of one ampere through a resistance of one ohm.

Voltage (of a circuit). The greatest root-mean-square (effective difference of potential) between any two conductors of the circuit concerned.

Voltage to ground. In grounded circuits the voltage between the given conductor and that point or conductor of the circuit which is grounded; in ungrounded circuits, the greatest voltage between the given conductor and any other conductor of the circuit.

Watertight. So constructed that moisture will not enter the enclosing case or housing.

Watt. The practical unit of electrical power.

Weatherproof. So constructed or protected that exposure to the weather will not interfere with successful operation.

Web. Central portion of an I beam.

Appendix B

BARE AND INSULATED SOLID COPPER WIRE

	Section Area			Diam. Overall, In.		Bare Wire Breaking Strength Pounds		Weight, Pounds per 1000 Feet	
A.W.G. No.	Circular Mils	Square Inches	Square Milli-meters	Bare	Weather-proof (Min.)	Hard Drawn	Annealed	Bare	T.B.W.
0000	211600	.1662	107.0	.4600	.6163	8143	5320	641	767
000	167800	.1318	85.0	.4096	.5659	6722	4220	508	629
00	133100	.1045	67.4	.3648	.5211	5519	3340	403	502
0	105500	.08289	53.5	.3249	.4812	4517	2650	320	407
1	83690	.06573	42.4	.2893	.4456	3688	2100	253	316
2	66370	.05213	33.6	.2576	.3826	3003	1670	201	260
3	52640	.04134	26.7	.2294	.3544	2439	1325	159	199
4	41740	.03278	21.2	.2043	.3293	1970	1050	126	164
5	33100	.02600	16.8	.1819	.3069	1591	880	100	135
6	26250	.02062	13.3	.1620	.2870	1280	700	79	112
7	20870	.01635	10.6	.1443	.2693	1030	550	63	...
8	16510	.01297	8.37	.1285	.2535	826	440	50	75

USEFUL ELECTRICAL AND ELECTRONIC DATA

The following information should prove useful to all electrical and electronic draftsmen for use in their everyday work. No attempt has been made to fully explain the use of all the data, as it is assumed that the draftsman has gained a basic knowledge of electricity and electronics by other means, such as experience, books, etc. These data, therefore, are meant to be useful as reference only—not as a training aid.

METHOD OF COMPUTING WIRE SIZES

The proper size of wire required to carry a current any distance, with a given loss in volts, can be determined from the following table based on Ohm's law:

$$\frac{\text{Length of run in ft} \times \text{amp} \times 21.5}{\text{Volts lost}} = \text{circular mils}$$

In this formula the number of feet must be measured one way, not both sides of the circuit; volts lost should be taken as the drop allowed in volts. Circular mils show the size of wire to use.

Example: What size wire should be used on a 250-volt circuit where it is necessary to carry 150 amperes a distance of 300 feet to a center of distribution with a loss of 3 percent under full load?

$$\frac{300 \times 150 \times 21.5}{7.5} = 129{,}000 \text{ circular mils}$$

or number 00 AWG, which is the next size heavier.

For a faster method of determining the proper wire for a given load at a given distance, the following voltage drop tables may be used. They are used through the courtesy of Industry Committee on Interior Wiring Design. To use, first determine the load in amperes on the circuit, then find the table corresponding to the type of conductors used and the voltage on the circuit. In the left-hand column of the circuit, find the load in amperes of the circuit in question; if the exact figure is not in the column, go to the next *highest* amperage. Then read across to find the size wire required for the amount of load (in amperes). If the distance of the circuit run is fifty feet or greater, read across in the "Length of Run in Feet" column and find the correct size wire required for the distance of the run in question.

COPPER CONDUCTORS, 230-240 VOLTS, SINGLE PHASE, 1 PERCENT VOLTAGE DROP

Load in Amp	Minimum Allowable Size of Conductor: In Cable, Conduit, Earth: Types R, T, TW	Types RH, RHW, THW	Overhead in Air*: Bare & Covered Conductors	Length of Run in Feet (Compare size shown below with size shown to left of double line. Use the larger size.) 30	50	75	100	125	150	175	200	225	250	275	300	350	400	450	500	550	600
5	14	14	10	14	14	14	12	12	10	10	10	8	8	8	8	8	6	6	6	6	4
7	14	14	10	14	14	14	12	10	10	8	8	8	8	8	6	6	6	4	4	4	4
10	14	14	10	14	14	12	10	8	8	8	6	6	6	6	4	4	4	4	3	3	2
15	14	14	10	14	12	10	8	6	6	6	6	4	4	4	4	2	2	2	2	1	1
20	12	12	10	12	10	8	6	6	4	4	4	4	2	2	2	2	1	1	0	0	0
25	10	10	10	10	8	6	6	4	4	4	3	3	2	2	1	1	0	0	00	00	00
30	10	10	10	10	8	6	4	4	4	3	2	2	1	1	0	0	00	00	00	000	000
35	8	8	10	10	8	6	4	4	3	2	2	1	1	0	0	00	00	000	000	4/0	4/0
40	8	8	10	8	6	4	4	3	2	2	1	1	0	0	00	00	000	000	4/0	4/0	4/0
45	6	8	10	8	6	4	4	3	2	1	1	0	0	00	00	000	000	4/0	4/0	250	250
50	6	6	10	8	6	4	3	2	1	1	0	0	00	00	000	000	4/0	4/0	250	250	250
60	4	6	8	8	4	4	2	1	1	0	00	00	000	000	000	4/0	250	250	300	300	350
70	4	4	8	6	4	3	2	1	0	00	00	000	000	4/0	4/0	250	300	300	350	350	400
80	2	4	6	6	4	2	1	0	00	00	000	000	4/0	4/0	250	300	300	350	350	400	500
90	2	3	6	6	4	2	1	0	00	000	000	4/0	4/0	250	300	300	350	400	400	500	500
100	1	3	6	6	3	2	0	00	000	000	4/0	4/0	250	300	300	350	400	400	500	500	500
115	0	2	4	4	2	1	0	00	000	4/0	4/0	250	300	300	350	400	500	500	600	600	700
130	00	1	4	4	2	0	00	000	4/0	4/0	250	300	300	350	350	500	500	600	600	700	700
150	000	0	2	2	1	0	000	4/0	4/0	250	300	350	350	400	500	500	600	600	700	750	
175	4/0	00	2	2	1	00	000	4/0	250	300	350	350	400	500	500	600	700	700			
200	250	000	1	1	0	00	4/0	250	300	350	350	400	500	500	600	700	700				
225	300	4/0	0	0	0	000	4/0	300	350	350	400	500	500	600	700	700					
250	350	250	00	00	00	000	250	300	350	400	500	500	600	700	700						
275	400	300	00	00	000	4/0	250	350	400	500	500	600	700	700	750						
300	500	350	000	000	000	4/0	300	350	500	500	600	700	700	750							
325	600	400	4/0	4/0	4/0	250	300	400	500	600	600	700	750								
350	600	500	4/0	4/0	4/0	250	350	400	500	600	700	700									
375	700	500	250	250	250	250	350	400	500	600	700	750									
400	750	600	250	250	250	300	400	500	600	700	750										

COPPER CONDUCTORS, 230-240 VOLTS, SINGLE PHASE, 2 PERCENT VOLTAGE DROP

Load in Amp	Minimum Allowable Size of Conductor: In Cable, Conduit, Earth: Types R, T, TW	Types RH, RHW, THW	Overhead in Air*: Bare or Covered Conductors	Length of Run in Feet (Compare size shown below with size shown to left of double line. Use the larger size.): 50	75	100	125	150	175	200	225	250	275	300	350	400	450	500	550	600	650	700	750	800	900
5	14	14	10	14	14	14	14	14	12	12	12	12	10	10	10	10	8	8	8	8	8	6	6	6	6
7	14	14	10	14	14	14	14	12	12	12	10	10	10	10	8	8	8	8	6	6	6	6	6	6	4
10	14	14	10	14	14	12	12	10	10	10	10	8	8	8	8	6	6	6	6	6	4	4	4	4	4
15	14	14	10	14	12	12	10	10	8	8	8	8	6	6	6	4	4	4	4	4	2	2	2	2	2
20	12	12	10	12	12	10	8	8	8	6	6	6	6	4	4	4	4	2	2	2	2	2	2	1	1
25	10	10	10	10	10	8	8	6	6	6	6	4	4	4	4	3	2	2	2	2	1	1	1	0	0
30	10	10	10	10	10	8	6	6	6	6	4	4	4	4	3	2	2	2	1	1	0	0	0	0	00
35	8	8	10	10	8	8	6	6	4	4	4	4	3	3	2	2	1	1	1	0	0	00	00	00	00
40	8	8	10	10	8	6	6	4	4	4	4	3	3	2	2	1	1	0	0	0	00	00	00	000	000
45	6	8	10	10	8	6	6	4	4	4	3	3	2	2	1	1	0	0	0	00	00	000	000	000	4/0
50	6	6	10	8	8	6	4	4	4	3	3	2	2	2	1	0	0	00	00	00	000	000	000	4/0	4/0
60	4	6	8	8	6	4	4	4	3	2	2	1	1	1	0	00	00	00	000	000	4/0	4/0	4/0	4/0	250
70	4	4	8	8	6	4	4	3	2	2	1	1	0	0	00	00	00	000	000	4/0	4/0	4/0	250	250	300
80	2	4	6	6	6	4	3	2	2	1	1	0	0	00	00	000	000	4/0	4/0	4/0	250	250	300	300	350
90	2	3	6	6	4	4	3	2	1	1	0	0	00	00	000	000	4/0	4/0	4/0	250	300	300	300	300	350
100	1	3	6	6	4	3	2	2	1	0	0	00	00	000	000	4/0	4/0	250	250	300	300	300	350	350	400
115	0	2	4	6	4	2	2	1	0	00	00	00	000	000	4/0	250	250	250	300	350	350	350	400	400	500
130	00	1	4	4	3	2	1	0	0	00	000	000	000	4/0	4/0	250	300	300	350	350	400	400	500	500	500
150	000	0	2	4	2	1	0	0	00	00	000	4/0	4/0	4/0	250	300	300	350	350	400	500	500	500	600	600
175	4/0	00	2	2	2	1	0	00	000	000	4/0	4/0	250	300	300	350	350	400	500	500	500	600	600	600	700
200	250	000	1	1	1	0	00	00	000	4/0	4/0	250	300	300	350	400	400	500	500	600	600	600	700	700	750
225	300	4/0	0	0	0	0	00	000	4/0	4/0	250	300	300	350	400	500	500	500	600	600	700	700	750	750	
250	350	250	00	00	00	00	000	000	4/0	250	300	300	350	350	400	500	500	600	600	700	750	750			
275	400	300	00	00	00	00	000	4/0	250	300	300	350	350	400	500	500	600	600	700	750					
300	500	350	000	000	000	000	4/0	4/0	250	300	350	350	400	500	500	600	600	700	750						
325	600	400	4/0	4/0	4/0	4/0	4/0	250	300	300	350	400	500	500	500	600	700	700	750						
350	600	500	4/0	4/0	4/0	4/0	4/0	250	300	350	400	400	500	500	600	600	700	750							
375	700	500	250	250	250	250	250	300	350	350	400	500	500	500	600	700	750								
400	750	600	250	250	250	250	250	300	350	400	500	500	500	600	600	700									

—ALUMINUM CONDUCTORS, 230-240 VOLTS, SINGLE PHASE, 2 PERCENT VOLTAGE DROP

Load in Amp	Minimum Allowable Size of Conductor: In Cable, Conduit, Earth: Types R, T, TW	Types RH, RHW, THW	Overhead in Air*: Bare or Covered Conductors: Single	Triplex	Length of Run in Feet — Compare size shown below with size shown to left of double line. Use the larger size. 50	75	100	125	150	175	200	225	250	275	300	350	400	450	500	550	600	650	700	750	800	900
5	12	12	10		12	12	12	12	12	10	10	10	10	8	8	8	8	8	6	6	6	6	6	4	4	4
7	12	12	10		12	12	12	12	10	10	10	8	8	8	8	6	6	6	6	4	4	4	4	4	4	2
10	12	12	10		12	12	10	10	8	8	8	8	6	6	6	6	4	4	4	4	2	2	2	2	2	2
15	12	12	10		12	10	8	8	8	6	6	6	4	4	4	4	2	2	2	2	2	1	1	1	0	0
20	10	10	10		10	8	8	6	6	6	4	4	4	4	2	2	2	2	1	1	0	0	0	00	00	00
25	10	10	10		10	8	6	6	4	4	4	4	2	2	2	2	1	1	0	0	00	00	00	000	000	000
30	8	8	10		8	8	6	4	4	4	2	2	2	2	2	1	0	0	00	00	00	000	000	000	4/0	4/0
35	6	8	10		8	6	6	4	4	2	2	2	2	1	1	0	0	00	00	00	000	000	4/0	4/0	4/0	250
40	6	8	10		8	6	4	4	2	2	2	2	1	1	0	0	00	00	000	000	4/0	4/0	4/0	250	250	300
45	4	6	10		8	6	4	4	2	2	2	1	1	0	0	00	00	000	000	4/0	4/0	250	250	250	300	300
50	4	6	8		6	4	4	2	2	2	1	1	0	0	00	00	000	000	4/0	4/0	250	300	300	300	300	350
60	2	4	6	6	6	4	2	2	2	1	0	0	00	00	00	000	4/0	4/0	250	250	300	300	350	350	350	400
70	2	2(a)	6	4	6	4	2	2	1	0	0	00	00	000	000	4/0	250	250	300	300	350	350	400	400	500	500
80	1	2(a)	6	4	4	2	2	1	0	0	00	00	000	000	4/0	4/0	250	300	300	350	350	400	500	500	500	600
90	0	2(a)	4	2	4	2	2	1	0	00	00	000	000	4/0	4/0	250	300	300	350	400	400	500	500	500	600	600
100	0	1(a)	4	2	4	2	1	0	00	00	000	000	4/0	4/0	250	300	300	350	400	400	500	500	600	600	600	700
115	00	0(a)	2	1	2	2	0	0	00	000	000	4/0	4/0	250	250	300	350	400	500	500	600	600	600	700	700	800
130	000	00(a)	2	0	2	1	0	00	000	000	4/0	250	250	300	300	350	400	500	500	600	700	700	700	800	800	900
150	4/0	000(a)	1	00	1	1	00	000	000	4/0	250	250	300	300	350	400	500	500	600	600	700	800	800	900	900	1M
175	300	4/0(a)	0	000	0	0	00	000	4/0	250	300	300	350	400	400	500	600	600	700	700	800	900	900	1M	1M	
200	350	250	00	4/0	00	00	000	4/0	250	300	300	350	400	400	500	600	600	700	750	800	900	1M	1M			
225	400	300	000		000	000	000	4/0	250	300	350	400	500	500	500	600	700	750	900	900	1M					
250	500	350	000		000	000	4/0	250	300	350	400	500	500	500	600	700	750	900	1M	1M						
275	600	500	4/0		4/0	4/0	4/0	300	350	400	400	500	500	600	600	700	800	900	1M							
300	700	500	250		250	250	250	300	350	400	500	500	600	600	700	800	900	1M								
325	800	600	300		300	300	300	300	400	500	500	600	600	700	750	900	900									
350	900	700	300		300	300	300	350	400	500	600	600	700	750	800	900	1M									
375	1M	700	350		350	350	350	350	500	500	600	700	700	750	900	1M										
400		900	350		350	350	350	400	500	600	600	700	750	800	900											

—COPPER CONDUCTORS, 115-120 VOLTS, SINGLE PHASE, 2 PERCENT VOLTAGE DROP

Load in Amp	Minimum Allowable Size of Conductor: In Cable, Conduit, Earth: Types R,T,TW	Types-RH, RHW, THW	Overhead in Air*: Bare & Covered Conductors	Length of Run in Feet — Compare size shown below with size shown to left of double line. Use the larger size. 30	40	50	75	100	125	150	175	200	225	250	275	300	350	400	450	500	550	600
5	14	14	10	14	14	14	14	12	12	10	10	10	8	8	8	8	6	6	6	6	4	4
7	14	14	10	14	14	14	12	12	10	10	8	8	8	8	6	6	6	6	4	4	4	4
10	14	14	10	14	14	12	12	10	8	8	8	6	6	6	6	4	4	4	4	3	3	2
15	14	14	10	14	12	12	10	8	6	6	6	6	4	4	4	4	3	2	2	2	1	1
20	12	12	10	12	10	10	8	6	6	6	4	4	4	3	3	3	2	1	1	0	0	0
25	10	10	10	10	8	8	8	6	4	4	4	3	3	2	2	2	1	0	0	00	00	00
30	10	10	10	10	8	8	6	4	4	4	3	2	2	2	1	1	0	0	00	00	000	000
35	8	8	10	10	8	8	6	4	4	3	2	2	1	1	0	0	00	00	000	000	4/0	4/0
40	8	8	10	8	8	6	6	4	3	2	2	1	1	0	0	0	00	000	000	4/0	4/0	4/0
45	6	8	10	8	8	6	4	4	3	2	1	1	0	0	00	00	000	000	4/0	4/0	250	250
50	6	6	10	8	6	6	4	3	2	2	1	0	0	00	00	00	000	4/0	4/0	250	250	300
60	4	6	8	8	6	4	4	2	1	1	0	0	00	00	000	000	4/0	4/0	250	300	300	350
70	4	4	8	6	6	4	3	2	1	0	00	00	000	000	4/0	4/0	250	250	300	350	350	400
80	2	4	6	6	4	4	2	1	0	0	00	000	000	4/0	4/0	4/0	250	300	350	400	400	500
90	2	3	6	6	4	4	2	1	0	00	000	000	4/0	4/0	250	250	300	350	350	400	500	500
100	1	3	6	6	4	3	[illegible]	0	00	00	000	4/0	4/0	250	250	300	300	350	400	500	500	600
115	0	2	4	4	4	2	1	0	00	000	4/0	4/0	250	250	300	350	350	400	500	600	600	700
130	00	1	4	4	2	2	0	00	000	4/0	4/0	250	300	300	350	350	400	500	600	600	700	700
150	000	0	2	2	2	1	0	00	4/0	4/0	250	300	300	350	400	400	500	600	600	700	750	
175	4/0	00	2	2	2	1	00	000	4/0	250	300	350	350	400	500	500	600	700	700			
200	250	000	1	1	1	0	00	4/0	250	300	350	350	400	500	500	600	600	700				

—COPPER CONDUCTORS, SINGLE-PHASE MOTORS, 115-230 VOLT , 3 PERCENT VOLTAGE DROP

			Minimum Allowable Size of Conductor			Length of Run in Feet														
		Full	In Cable, Conduit, Earth		Overhead in Air*	Compare size shown below with size shown to left of double line. Use the larger size.														
Motor Hp	Rated Volts	Load Amp	Types R, T, TW	Types RH, RHW, THW	Bare & Covered Conductors	50	75	100	150	200	250	300	350	400	500	600	700	800	900	1000
1/6	115	4.4	14	14	10	14	14	14	12	12	10	10	8	8	8	6	6	6	6	4
1/4	115	5.8	14	14	10	14	14	14	12	10	10	8	8	8	6	6	6	4	4	4
1/3	115	7.2	14	14	10	14	14	12	12	10	8	8	8	6	6	6	4	4	4	3
1/2	115	9.8	14	14	10	14	12	12	10	8	8	6	6	6	4	4	4	3	2	2
3/4	115	13.8	12	12	10	12	12	10	8	6	6	6	4	4	4	3	2	1	1	0
1/6	230	2.2	14	14	10	14	14	14	14	14	14	14	14	12	12	10	10	10	10	8
1/4	230	2.9	14	14	10	14	14	14	14	14	14	14	14	12	12	10	10	10	8	8
1/3	230	3.6	14	14	10	14	14	14	14	14	14	12	12	12	10	10	10	8	8	8
1/2	230	4.9	14	14	10	14	14	14	14	14	12	12	12	12	10	10	8	8	8	8
3/4	230	6.9	14	14	10	14	14	14	14	12	12	12	12	10	10	8	8	8	6	6
1	230	8	14	14	10	14	14	14	14	12	12	10	10	10	8	8	8	6	6	6
1½	230	10	14	14	10	14	14	14	12	12	10	10	10	8	8	6	6	6	6	4
2	230	12	12	12	10	12	12	12	12	10	10	8	8	8	8	6	6	6	4	4
3	230	17	10	10	10	10	10	10	10	10	8	8	8	6	6	4	4	4	3	2
5	230	28	8	8	10	10	10	10	8	6	6	6	4	4	4	2	2	1	1	0
7½	230	40	6	6	10	10	10	8	6	6	6	4	3	3	2	1	1	0	0	00
10	230	50	4	6	8	8	8	8	6	4	4	3	2	2	1	0	0	00	00	000

—ALUMINUM CONDUCTORS, SINGLE-PHASE MOTORS, 115-230 VOLT , 2 PERCENT VOLTAGE DROP

			Minimum Allowable Size of Conductor			Length of Run in Feet														
		Full	In Cable, Conduit, Earth		Overhead in Air*	Compare size shown below with size shown to left of double line. Use the larger size.														
Motor Hp	Rated Volts	Load Amp	Types R, T, TW	Types RH, RHW, THW	Bare & Covered Conductors	50	75	100	150	200	250	300	350	400	500	600	700	800	900	1000
1/6	115	4.4	12	12	10	12	12	10	8	8	6	6	6	4	4	2	2	2	2	1
1/4	115	5.8	12	12	10	12	10	10	8	6	6	6	4	4	2	2	2	1	1	0
1/3	115	7.2	12	12	10	12	10	10	8	6	6	4	4	4	2	1	1	0	0	00
1/2	115	9.8	12	12	10	10	8	8	6	4	4	2	2	2	1	0	0	00	00	000
3/4	115	13.8	10	10	10	8	8	6	4	4	2	2	1	1	0	00	000	000	0000	0000
1/6	230	2.2	12	12	10	12	12	12	12	12	12	12	10	10	10	10	8	8	8	8
1/4	230	2.9	12	12	10	12	12	12	12	12	12	10	10	10	8	8	8	6	6	6
1/3	230	3.6	12	12	10	12	12	12	12	12	10	10	10	8	8	8	6	6	6	4
1/2	230	4.9	12	12	10	12	12	12	12	10	10	8	8	8	6	6	6	4	4	4
3/4	230	6.9	12	12	10	12	12	12	10	10	8	8	6	6	6	4	4	4	2	2
1	230	8	12	12	10	12	12	12	10	8	8	6	6	6	4	4	4	2	2	2
1½	230	10	12	12	10	12	12	10	8	8	6	6	6	4	4	2	2	2	2	1
2	230	12	12	12	10	12	10	10	8	6	6	6	4	4	2	2	2	1	1	0
3	230	17	10	10	10	10	10	8	6	6	4	4	2	2	2	1	0	0	00	00
5	230	28	6	8	8	8	8	6	4	2	2	2	1	0	00	00	000	0000	0000	0000
7½	230	40	4	6	8	8	6	4	2	2	1	0	0	00	000	0000	0000	250M	300M	300M
10	230	50	2	4	6	6	4	4	2	1	0	00	00	000	0000	250M	300M	300M	350M	400M

—ALUMINUM CONDUCTORS, THREE-PHASE MOTORS, 220 VOLTS, 2 PERCENT VOLTAGE DROP

		Minimum Allowable Size of Conductor					Length of Run in Feet											
		In Cable, Conduit, Earth		Overhead in Air*			Compare size shown below with size shown to left of double line. Use the larger size.											
				Bare & Covered Conductors														
Motor Hp	Full Load Amp	Types R,T,TW	Types RH, RHW,THW	Single	Triplex	Quadruplex	75	100	150	200	300	400	500	600	700	800	900	1000
1/2	2.0	12	12	10			12	12	12	12	12	12	10	10	10	8	8	8
3/4	2.8	12	12	10			12	12	12	12	12	10	10	8	8	8	6	6
1	3.5	12	12	10			12	12	12	12	10	10	8	8	8	6	6	6
1½	5.0	12	12	10			12	12	12	10	10	8	8	6	6	4	4	4
2	6.5	12	12	10			12	12	10	10	8	6	6	6	4	4	4	2
3	9	12	12	10			12	12	10	8	6	6	4	4	2	2	2	2
5	15	10	10	10			10	10	8	6	4	4	2	2	1	1	0	0
7½	22	8	8	10			8	8	6	4	2	2	1	0	0	00	00	000
10	27	6	8	10			8	6	4	4	2	1	0	00	00	000	000	4/0
15	40	4	6	8			6	4	2	2	1	00	000	000	4/0	250	250	300
20	52	2	4	6	6	6	6	4	2	1	0	000	000	4/0	250	300	350	350
25	64	1	2	4	4	4	4	2	2	0	00	4/0	250	300	350	350	300	500
30	78	0	1	4	4	2	4	2	1	00	000	250	300	350	400	500	500	600

—COPPER CONDUCTORS, THREE-PHASE MOTORS, 220 VOLTS, 3 PERCENT VOLTAGE DROP

		Minimum Allowable Size of Conductor			Length of Run in Feet										
		In Cable, Conduit, Earth		Overhead in Air*	Compare size shown below with size shown to left of double line. Use the larger size.										
Motor Hp	Full Load Amp	Types R,T,TW	Types RH, RHW, THW	Bare & Covered Conductors	100	150	200	300	400	500	600	700	800	900	1000
1/2	2.0	14	14	10	14	14	14	14	14	14	14	14	12	12	12
3/4	2.8	14	14	10	14	14	14	14	14	14	12	12	12	10	10
1	3.5	14	14	10	14	14	14	14	14	12	12	10	10	10	10
1½	5.0	14	14	10	14	14	14	12	12	10	10	10	8	8	8
2	6.5	14	14	10	14	14	14	12	10	10	8	8	8	6	6
3	9	14	14	10	14	14	12	10	10	8	8	6	6	6	6
5	15	12	12	10	12	12	10	8	6	6	6	4	4	4	3
7½	22	10	10	10	12	10	8	6	6	4	4	3	3	2	2
10	27	8	8	10	10	8	8	6	4	4	3	2	2	1	1
15	40	6	6	10	8	6	6	4	3	2	1	1	0	0	00
20	52	4	6	8	8	6	4	3	2	1	0	0	00	00	000
25	64	2	4	6	6	4	4	2	1	0	00	00	000	000	4/0
30	78	1	3	6	6	4	3	1	0	00	000	000	4/0	4/0	250

CURRENT-CARRYING CAPACITY OF INSULATED COPPER CONDUCTORS

Size AWG MCM	TYPES OF INSULATION				
	Types RUW (14-2), T, TW	Types RH RHW, RUH (14-2), THW, THWN, XHHW	Types V, MI	Types TA, TBS, SA AVB, SIS, FEP, FEPB, RHH, THHN, XHHW	Types AVA, AVL
14	15	15	25	25	30
12	20	20	30	30	35
10	30	30	40	40	45
8	40	45	50	50	60
6	55	65	70	70	80
4	70	85	90	90	105
3	80	100	105	105	120
2	95	115	120	120	135
1	110	130	140	140	160
0	125	150	155	155	190
00	145	175	185	185	215
000	165	200	210	210	245
0000	195	230	235	235	275
250	215	255	270	270	315
300	240	285	300	300	345
350	260	310	325	325	390
400	280	335	360	360	420
500	320	380	405	405	470
600	355	420	455	455	525
700	385	460	490	490	560
750	400	475	500	500	580
800	410	490	515	515	600
900	435	520	555	555	...
1000	455	545	585	585	680
1250	495	590	645	645	...
1500	520	625	700	700	785
1750	545	650	735	735	...
2000	560	665	775	775	840

CURRENT-CARRYING CAPACITY OF INSULATED ALUMINUM AND COPPER-CLAD ALUMINUM CONDUCTORS

Size AWG MCM	TYPES OF INSULATION				
	Types RUW (12-2), T, TW	Types RH, RHW, RUH (12-2), THW, THWN, XHHW	Types V, MI	Types TA, TBS, SA, AVB, SIS, RHH, THHN, XHHW	Types AVA, AVL
12	15	15	25	25	25
10	25	25	30	30	35
8	30	40	40	40	45
6	40	50	55	55	60
4	55	65	70	70	80
3	65	75	80	80	95
2	75	90	95	95	105
1	85	100	110	110	125
0	100	120	125	125	150
00	115	135	145	145	170
000	130	155	165	165	195
0000	155	180	185	185	215
250	170	205	215	215	250
300	190	230	240	240	275
350	210	250	260	260	310
400	225	270	290	290	335
500	260	310	330	330	380
600	285	340	370	370	425
700	310	375	395	395	455
750	320	385	405	405	470
800	330	395	415	415	485
900	355	425	455	455	...
1000	375	445	480	480	560
1250	405	485	530	530	...
1500	435	520	580	580	650
1750	455	545	615	615	...
2000	470	560	650	650	705

EQUATIONS FOR DETERMINING AMPERES, HORSEPOWER, KILOWATTS, AND KILOVOLT-AMPERES

TO FIND	ALTERNATING CURRENT SINGLE-PHASE	ALTERNATING CURRENT THREE PHASE
Amperes when horsepower is known	$\frac{hp \times 746}{E \times \% Eff. \times P\text{-}F}$	$\frac{hp \times 746}{1.73 \times E \times \% Eff. \times P\text{-}F}$
Amperes when kilowatts are known	$\frac{kW \times 1000}{E \times P\text{-}F}$	$\frac{kW \times 1000}{1.73 \times E \times P\text{-}F}$
Amperes when kilovolt-amperes are known	$\frac{kVA \times 1000}{E}$	$\frac{kVA \times 1000}{1.73 \times E}$
Kilowatts	$\frac{I \times E \times P\text{-}F}{1000}$	$\frac{I \times E \times 1.73 \times P\text{-}F}{1000}$
Kilovolt-amperes	$\frac{I \times E}{1000}$	$\frac{I \times E \times 1.73}{1000}$
Horsepower (output)	$\frac{I \times E \times \% Eff. \times P\text{-}F}{746}$	$\frac{I \times E \times 1.73 \times \% Eff. \times P\text{-}F}{746}$

BARE AND INSULATED SOLID COPPER WIRE

A.W.G. No.	Section Area: Circular Mils	Section Area: Square Inches	Section Area: Square Millimeters	Diam. Overall, In.: Bare	Diam. Overall, In.: Weatherproof (Min.)	Bare Wire Breaking Strength Pounds: Hard Drawn	Bare Wire Breaking Strength Pounds: Annealed	Weight, Pounds per 1000 Feet: Bare	Weight, Pounds per 1000 Feet: T.B.W.
0000	211600	.1662	107.0	.4600	.6163	8143	5320	641	767
000	167800	.1318	85.0	.4096	.5659	6722	4220	508	629
00	133100	.1045	67.4	.3648	.5211	5519	3340	403	502
0	105500	.08289	53.5	.3249	.4812	4517	2650	320	407
1	83690	.06573	42.4	.2893	.4456	3688	2100	253	316
2	66370	.05213	33.6	.2576	.3826	3003	1670	201	260
3	52640	.04134	26.7	.2294	.3544	2439	1325	159	199
4	41740	.03278	21.2	.2043	.3293	1970	1050	126	164
5	33100	.02600	16.8	.1819	.3069	1591	880	100	135
6	26250	.02062	13.3	.1620	.2870	1280	700	79	112
7	20870	.01635	10.6	.1443	.2693	1030	550	63	...
8	16510	.01297	8.37	.1285	.2535	826	440	50	75

BARE AND INSULATED STRANDED COPPER CABLE

A.W.G. No.	Section Area: Circular Mils	Section Area: Square Inches	Section Area: Square Milli-meters	No. of Wires in Strand	Diam. Overall, In.: Bare	Diam. Overall, In.: Weather-proof (Min.)	Bare Wire Breaking Strength, Pounds: Hard Drawn	Bare Wire Breaking Strength, Pounds: Soft (Min.)	Weight Pounds per 1000 Feet: Bare	Weight Pounds per 1000 Feet: T.B.W.
....	2000000	1.571	1014	91	1.630	1.880	87790	43830	6175	7008
....	1750000	1.374	887	91	1.526	1.776	77930	38350	5403	6193
....	1500000	1.178	760	61	1.411	1.661	65840	32870	4631	5380
....	1250000	.9817	633	61	1.288	1.538	55670	27390	3859	4508
....	1000000	.7854	507	61	1.152	1.402	45030	21910	3088	3674
....	950000	.7461	481	61	1.123	.....	42780	.....	2933	....
....	900000	.7069	456	61	1.094	1.313	40520	19720	2779	3332
....	850000	.6676	431	61	1.062	.....	38270	.....	2624	....
....	800000	.6283	405	61	1.031	1.250	36020	17530	2470	2992
....	750000	.5890	380	61	.998	1.217	34090	16430	2316	2822
....	700000	.5498	355	61	.964	1.183	31820	15340	2161	2650
....	650000	.5105	329	61	.929	.....	29770	.....	2007	....
....	600000	.4712	304	37	.893	1.112	27020	13150	1853	2235
....	550000	.4320	279	37	.853	.....	24760	.....	1698	....
....	500000	.3927	253	37	.813	1.001	22510	10960	1544	1894
....	450000	.3534	228	37	.772	.960	20450	9860	1389	1724
....	400000	.3142	203	19	.726	.914	17560	8765	1235	1553
....	350000	.2749	177	19	.679	.867	15590	7669	1081	1345
....	300000	.2356	152	19	.629	.817	13510	6574	926	1174
....	250000	.1963	127	19	.574	.762	11260	5478	772	985
0000	211600	.1662	107	*19 or 7	.528	.684	9617	4637	653	800
000	167800	.1318	85.0	*19 or 7	.470	.626	7366	3677	518	653
00	133100	.1045	67.4	7	.414	.570	5926	2916	411	522
0	105500	.08289	53.5	7	.368	.524	4752	2312	326	424
1	83690	.06573	42.4	7	.328	.484	3804	1834	258	328
2	66370	.05213	33.6	7	.292	.417	3045	1525	205	270
3	52640	.04134	26.7	7	.260	.385	2433	1209	163	206
4	41740	.03278	21.2	7	.232	.357	1938	959	129	170
5	33100	.02600	16.8	7	.206	.331	1542	761	102	140
6	26250	.02062	13.3	7	.184	.309	1228	603	81	115
8	16510	.01297	8.37	7	.146	.271	777	379	51	78

*A.W.G. Nos. 0000 and 000 cable are usually made of 7 strands when bare and 19 strands when insulated.

EQUIVALENTS OF WIRES

Showing Approximately Equal Capacities in the Use of Conductors of Different Sizes

Total Capacity C. M. or A.W.G. No.	Equivalent in No. of Smaller Conductors or Combinations: 2	3	4	5	6	Combinations	
1000000	....	....	....	0000	000	....	
800000	....	....	0000	000	00	....	
600000	....	0000	000	00	0	....	..
500000	....	000	00	0	1	....	
400000	0000	00	0	1	2	...	..
350000	000	....	1	2	...	0000 & 00	..
300000	....	0	2	....	3	000 & 00	0000 & 1
250000	00	1	....	3	4	000 & 1	0000 & 4
0000	0	2	3	4	5	00 & 1	000 & 3
000	1	3	4	5	6	0 & 2	00 & 5
00	2	4	5	6	...	1 & 3	0 & 6
0	3	5	6	....	8	2 & 4	....
1	4	6	...	8	...	3 & 5	....
2	5	....	8	....	...	4 & 6	....
3	6	8	....	10	...	....	....
4	....	....	10	....	...	6 & 8	....
5	8	10	....	12	...	....	....
6	....	....	12	....	14	8 & 10	....
8	....	....	14	....	16	10 & 12	....
10	....	....	16	....	18	12 & 14	....

TENSILE STRENGTH OF COPPER WIRE

A.W.G. No.	Breaking Wt., Lb.: Hard Drawn	Annealed
0000	9971	5650
000	7907	4480
00	6271	3553
0	4973	2818
1	3943	2234
2	3127	1772
3	2480	1405
4	1967	1114
5	1559	883
6	1237	700
7	980	555
8	778	440
9	617	349
10	489	277
11	388	219
12	307	174
13	244	138
14	193	109
15	153	87
16	133	69
17	97	55
18	77	43
19	61	34
20	48	27

DECIMAL EQUIVALENTS

Fraction Inches			Dec. In.
		1/64	.015625
	1/32	....	.03125
		3/64	.046875
1/16	...	....	.0625
		5/64	.078125
	3/32	....	.09375
		7/64	.109375
1/8	...	....	.125
		9/64	.140625
	5/32	....	.15625
		11/64	.171875
3/16	...	....	.1875
		13/64	.203125
	7/32	....	.21875
		15/64	.234375
1/4	...	....	.25

Fraction Inches			Dec. In.
		17/64	.265625
	9/32	....	.28125
		19/64	.296875
5/16	....	....	.3125
		21/64	.328125
	11/32	....	.34375
		23/64	.359375
3/8	....	....	.375
		25/64	.390625
	13/32	....	.40625
		27/64	.421895
7/16	....	....	.4375
		29/64	.453125
	15/32	....	.46875
		31/64	.484375
1/2	....	....	.5

Fraction Inches			Dec. In.
		33/64	.515625
	17/32	....	.53125
		35/64	.546875
9/16	....	....	.5625
		37/64	.578125
	19/32	....	.59375
		39/64	.609375
5/8	....	....	.625
		41/64	.640625
	21/32	....	.65625
		43/64	.671875
11/16	....	....	.6875
		45/64	.703125
	23/32	....	.71875
		47/64	.734375
3/4	....	....	.75

Fraction Inches			Dec. In.
		49/64	.765625
	25/32	....	.78125
	....	51/64	.796875
13/16	....	....	.8125
		53/64	.828125
	27/32	....	.84375
		55/64	.859375
7/8	....	....	.875
		57/64	.890625
	29/32	....	.90625
		59/64	.921875
15/16	....	....	.9375
		61/64	.953125
	31/32	....	.96875
		63/64	.984375
1	....	....	1.0

COMPARATIVE WEIGHTS AND RESISTANCES OF SOLID ALUMINUM AND COPPER WIRE

A.W.G. No.	Diameter Mils	Area: Circular Mils	Area: Square Inches	Weight Pounds per 1000 Feet: Aluminum Hard Drawn	Weight Pounds per 1000 Feet: Copper Annealed	Resistance Ohms per 1000 Feet at 20° C. (68° F.): Aluminum Hard Drawn	Resistance Ohms per 1000 Feet at 20° C. (68° F.): Copper Annealed	Diameter Bare Milli-meters
0000	460.0	211,600	.1662	195.	641.	.0804	.0490	11.68
000	409.6	167,800	.1318	154.	508.	.101	.0618	10.40
00	364.8	133,100	.1045	122.	403.	.128	.0779	9.266
0	324.9	105,500	.08289	97.1	320.	.161	.0983	8.252
1	289.3	83,690	.06573	77.0	253.	.203	.124	7.348
2	257.6	66,370	.05213	61.1	201.	.256	.156	6.544
3	229.4	52,640	.04134	48.4	159.	.323	.197	5.827
4	204.3	41,740	.03278	38.4	126.	.408	.249	5.189
5	181.9	33,100	.02600	30.4	100.	.514	.313	4.621
6	162.0	26,250	.02062	24.2	79.5	.648	.395	4.115
7	144.3	20,820	.01635	19.2	63.0	.817	.498	3.665
8	128.5	16,510	.01297	15.2	50.0	1.03	.628	3.264
9	114.4	13,090	.01028	12.1	39.6	1.30	.792	2.906
10	101.9	10,380	.008155	9.55	31.4	1.64	.999	2.588
11	90.74	8,234	.006467	7.57	24.9	2.07	1.26	2.305
12	80.81	6,530	.005129	6.00	19.8	2.61	1.59	2.053
13	71.96	5,178	.004067	4.76	15.7	3.29	2.00	1.828
14	64.08	4,107	.003225	3.78	12.4	4.14	2.53	1.628
15	57.07	3,257	.002558	2.99	9.9	5.22	3.18	1.450
16	50.82	2,583	.002028	2.38	7.8	6.59	4.02	1.291
17	45.26	2,048	.001609	1.88	6.2	8.31	5.06	1.150
18	40.30	1,624	.001276	1.50	4.9	10.50	6.39	1.024

Based on hard drawn aluminum and standard annealed copper. For weight or resistance per mile multiply values per 1000 feet by 5.28. Weight and resistance of actual wire may vary from calculated quantities in table.

CONVERSION TABLE

Diameter of a circle ×3.1416 = Circumference.
Radius of a circle ×6.283185 = Circumference.
Square of the radius of a circle ×3.1416 = Area.
Square of the diameter of a circle ×0.7854 = Area.
Square of the circumference of a circle ×0.07958 = Area.
Half the circumference of a circle ×half its diameter = Area.
Circumference of a circle ×0.159155 = Radius.
Square root of the area of a circle ×0.56419 = Radius.
Circumference of a circle ×0.31831 = Diameter.
Square root of the area of a circle ×1.12838 = Diameter.
Diameter of a circle ×0.86 = Side of an inscribed equilateral triangle
Diameter of a circle ×0.7071 = Side of an inscribed square.
Circumference of a circle ×0.225 = Side of an inscribed square.
Circumference of a circle ×0.282 = Side of an equal square.
Diameter of a circle ×0.8862 = Side of an equal square.
Base of a triangle ×one-half the altitude = Area.
Multiplying both diameter and .7854 together = Area of an ellipse.
Surface of a sphere ×one-sixth of its diameter = Solidity.
Circumference of a sphere ×its diameter = Surface.
Square of the diameter of a sphere ×3.1416 = Surface.
Square of the circumference of a sphere ×0.3183 = Surface.
Cube of the diameter of a sphere ×0.5236 = Solidity.
Cube of the circumference of a sphere ×0.016887 = Solidity.
Radius of a sphere ×1.1547 = Side of an inscribed cube.
Square root of (one-third of the square of) the diameter of a sphere = Side of an inscribed cube.
Area of its base ×one-third of its altitude = Solidity of a cone or pyramid, whether round, square or triangular.
Area of one of its sides ×6 = Surface of a cube.
Altitude of trapezoid ×one-half the sum of its parallel sides = Area.

Appendix C

INTRODUCTION

The metric system is an international language of measurement. Its symbols are identical in all languages. Just as the English language is governed by rules of spelling, punctuation, and pronunciation, so is the language of measurement. Uniformity of usage facilitates comprehension, and minimizes the chance of decimal point errors.

This pamphlet covers the editorial rules of the metric system, and is designed to serve as a convenient reference guide. It uses the modern form of the metric system, called "The International System of Units" (abbreviated SI). A limited number of the units outside SI that are acceptable for use with SI are also included.

Except for the use of American instead of British spelling, this Guide is consistent with the International Standard ISO 1000 and the National Bureau of Standards Special Publication 330, which is the approved English-language translation of the International Bureau of Weights and Measures publication "Le Système International d'Unités". However, greater standardization of punctuation, and a selection of preferred prefixes, are recommended in this Guide, and the pronunciation of unit names is given.

THE METRIC SYSTEM

RULES FOR WRITING METRIC QUANTITIES

1. CAPITALS

Units. Unit names, including prefixes, are not capitalized except at the beginning of a sentence or in titles. Note that in degree Celsius the unit "degree" is lowercase. (In the term "degree Celsius", "degree" is considered to be the unit name, modified by the adjective "Celsius", which is always capitalized. The "degree centigrade" is now obsolete.)

Symbols: The short forms for metric units are called symbols. They are lowercase except that the first letter is uppercase when the name of the unit is derived from the name of a person.

Examples:	Unit Name	Unit Symbol
	meter*	m
	liter*	l
	gram	g
	newton	N
	pascal	Pa

Printed unit symbols should have upright letters, because sloping letters are reserved for quantity symbols, such as *m* for mass and *l* for length.

Prefix Symbols: All prefix names, their symbols, and pronunciation are listed on page 12. Notice that the top five are uppercase and all the rest lowercase.

The importance of following the precise use of uppercase and lowercase letters is shown by the following examples:

G for giga; g for gram
K for kelvin; k for kilo
M for mega; m for milli
N for newton; n for nano
T for tera; t for metric ton

2. PLURALS AND FRACTIONS

a. **Names** of units are plural when appropriate.

Examples: 1 meter, 100 meters, 0 degrees Celsius

*The spellings "meter" and "liter" are preferred by ANMC but "metre" and "litre" are recognized to be in widespread use.

Values less than 1 take the singular form of the unit name.

Examples: 0.5 kilogram or 1/2 kilogram

While decimal notation (0.5, 0.35, 6.87) is generally preferred, the most simple fractions are acceptable, such as those where the denominator is 2, 3, 4, or 5.

b. **Symbols** of units are the same in singular and plural.

Examples: 1 m, 100 m

3. PERIODS

A period is NOT used after a symbol, except at the end of a sentence.

Examples: A current of 15 mA is found . . .
The field measured 350 x 125 m.

4. THE DECIMAL MARKER

A dot on the line is used as the decimal marker. In numbers less than one, a zero should be written before the decimal point (to prevent the possibility that a faint decimal point will be overlooked).

Example: The oral expression "point seven five" is written 0.75

5. GROUPING OF NUMBERS

a. Separate digits into groups of three, counting from the decimal marker. A comma should not be used. Instead, a space is left to avoid confusion, because many countries use a comma for the decimal marker.

b. In a four-digit number, the space is not required unless the four-digit number is in a column with numbers of five digits or more.

Examples:		
	For 4,720,525	write 4 720 525
	For 0.52875	write 0.528 75
	For 6,875	write 6875 or 6 875
	For 0.6875	write 0.6875 or 0.687 5

6. SPACING

a. In symbols or names for units having prefixes, no space is left between letters making up the symbol or the name.

Examples: kA, kiloampere; mg, milligram

b. When a symbol follows a number to which it refers, a space must be left between the number and the symbol, except when the symbol (such as °) appears in the superscript position. The symbol for degree Celsius may be written either with or without a space before the degree symbol.

Examples: 455 kHz, 22 mg, 20 mm, 10^6 N, 30°, 20°C or 20 °C

c. When a quantity is used in an adjectival sense, a hyphen should be used between the number and the symbol (except ° and °C).

Examples: It is a 35-mm film; but, the film width is 35 mm. I bought a 6-kg turkey; but, the turkey weighs 6 kg.

d. Leave a space on each side of signs for multiplication, division, addition, and subtraction—except within a compound symbol.

Examples: 4 m x 3 m (not 4 mx3 m); kg/m³; N·m

7. SQUARES, CUBES, ETC.

When writing symbols for units such as square meter or cubic centimeter, write the symbol for the unit, followed by the superscript ² or ³, respectively.

Examples: For 14 square meters, write 14 m²
For 26 cubic centimeters, write 26 cm³

8. COMPOUND UNITS

a. For a unit **name** (not a symbol) derived as a **quotient**—for example, kilometers per hour—it is preferable not to use a slash (/) as a substitute for "per" except where space is limited and a symbol might not be understood. Avoid other mixtures of words and symbols.

Examples: Use meter per second; not meter/second, meter/s, or m/second

Use only one "per" in any combination of units.

Example: meter per second squared, not meter per second per second.

b. For a unit **symbol** derived as a **quotient**—for example, km/h—do not write k.p.h. or kph because these are understood only in the English language, whereas km/h is used in all languages. The symbol km/h can also be written by means of a negative exponent—for example, $km \cdot h^{-1}$.

Do not use more than one slash (/) in any combination of symbols unless parentheses are used to avoid ambiguity.

Examples: m/s², not m/s/s; W/(m·K), not W/m/K

c. For a unit **name** derived as a **product**, a hyphen is recommended (or a space is permissible) but never a "product dot" (a period raised to a centered position)—for example, write newton-meter or newton meter, not newton·meter.

d. For a unit **symbol** derived as a **product**, use a product dot—for example, N·m.

Do not use the product dot as a multiplier symbol for calculations.

Example: Use 6.2 x 5, not 6.2·5

e. Do not mix nonmetric units with metric units, except units for time, plane angle, or rotation.

Example: Use kg/m³, not kg/ft³

f. To eliminate the problem of what units and multiples to use, a quantity that constitutes a ratio of two like quantities should be expressed as a fraction (either common or decimal) or as a percentage.

Examples: The slope is 1/100 or 0.01 or 1%, not 10 mm/m or 10 m/km.

9. PREFIX USAGE

a. While hecto, deka, deci, and centi are SI prefixes, their use should generally be avoided except for the SI unit-multiples for area and volume and nontechnical use of centimeter, as for body and clothing measurements.

b. In broad fields of use, or in a table of values for the same quantity, or in a discussion of such values within a given context, a common unit-multiple should be used even when some of the numerical values may require up to five or six digits before the decimal point.

Examples:

mm	for mechanical engineering drawings
kPa	for fluid pressure
MPa	for stress
kg/m³	for density

c. When convenient, choose prefixes resulting in numerical values between 0.1 and 1000, but only if this can be done without violating Rule 9a or b.

Quantity (1)	Common Units	Symbol	Acceptable Equivalent	Symbol
plane angle	degree (2)	°		
length	kilometer	km		
	meter (3)	m		
	centimeter	cm		
	millimeter	mm		
	micrometer	μm		
area	square kilometer	km^2		
	square hectometer	hm^2	hectare	ha
	square meter	m^2		
	square centimeter	cm^2		
	square millimeter	mm^2		
volume	cubic meter	m^3		
	cubic decimeter	dm^3	liter (3,4,5)	l
	cubic centimeter	cm^3	milliliter (4)	ml
velocity	meter per second	m/s		
	kilometer per hour	km/h		
acceleration	meter per second squared	m/s^2		
frequency	megahertz	MHz		
	kilohertz	kHz		
	hertz	Hz		
rotational frequency	revolution per second	r/s		
	revolution per minute	r/min		
mass	megagram	Mg	metric ton	t
	kilogram	kg		
	gram	g		
	milligram	mg		
density	kilogram per cubic meter	kg/m^3	gram per liter	g/l
force	kilonewton	kN		
	newton	N		
moment of force (6)	newton-meter	N•m		
pressure	kilopascal (7)	kPa		
stress	megapascal (8)	MPa		
energy, work, or quantity of heat	megajoule	MJ		
	kilojoule	kJ		
	joule	J		
	kilowatt-hour (9)	kW•h	kilowatthour	kWh
power or heat flow rate	kilowatt	kW		
	watt	W		
temperature	kelvin	K		
	degree Celsius	°C		
electric current	ampere	A		
quantity of electricity	coulomb	C		
	ampere-hour (10)	A•h		Ah
electromotive force	volt	V		
electric resistance	ohm	Ω		
luminous intensity	candela	cd		
luminous flux	lumen	lm		
illuminance	lux	lx		
sound level	decibel	dB		

NOTES

(1) Listed in same sequence as ISO 1000.

(2) For efficiency in calculations, plane angles should be expressed with decimal subdivisions rather than minutes (′) and seconds (″).

The SI unit for plane angle is radian (symbol rad), defined as "the plane angle between two radii of a circle which cut off on the circumference an arc equal in length to the radius".

(3) For information regarding spelling, see Preface. The spellings "meter" and "liter" are preferred by ANMC but "metre" and "litre" are recognized to be in widespread use.

(4) To be used only for fluids (both gases and liquids) and for dry ingredients in recipes.

(5) Do not use any prefix with "liter" except "milli".

(6) Torque or bending moment.

(7) See Rule 10.

(8) Except for very weak materials.

(9) To be abandoned eventually. 1 kW•h = 3.6 MJ

(10) 1 A•h = 3600 C

d. To avoid errors in calculations, prefixes may be replaced with powers of ten—for example, 1 MJ = 10^6 J.

e. Generally avoid use of prefixes in a denominator, except kilogram—for example kJ/m³, not J/dm³—and kJ/kg, not J/g.

f. Do not use a mixture of prefixes, unless the difference in size is extreme.

Examples: Use 40 mm wide and 1500 mm long, not 40 mm wide and 1.5 m long; but 1500 meters of 2-mm diameter wire.

g. Do not use multiple prefixes.

Examples: Use 13.58 m, not 13 m 580 mm
Use nm, not mμm
Use milligram, not microkilogram

h. Do not use a prefix without a unit—for example, use kilogram, not kilo.

10. UNITS FOR PRESSURE

Kilopascal (kPa) is the only unit recommended for fluid pressure for all fields of use except high vacuum measurements of absolute pressure for which Pa, mPa, etc. may be more convenient.

Do not use bar (10^5 Pa) or millibar (10^2 Pa) because they are not SI units, and are accepted internationally **only for a limited time in special fields** because of existing usage. They are also objectionable because they introduce too many different units, requiring frequent conversions to the preferred SI unit kPa (10^3 Pa), with consequent chance for decimal point errors. (1 bar = 100 kPa)

Gage pressure is absolute pressure minus ambient pressure (usually atmospheric pressure). It is positive or negative (called vacuum) according to whether the pressure is higher or lower, respectively, than the ambient pressure. Absolute pressure is specified either by (1) using the identifying phrase "absolute pressure", or (2) adding the word "absolute" after the unit symbol, separating the two by a comma or a space. Do not add (to the unit symbol) either "g" for gage or "a" for absolute.

11. SPELLING OF VOWEL PAIRS

There are three cases where the final vowel in a prefix is omitted—namely, megohm, kilohm, and hectare. In all other cases, both vowels are retained and both are pronounced. No space or hyphen should be used.

PRONUNCIATION OF METRIC TERMS

The pronunciation of most of the unit names is well-known and uniformly described in American dictionaries, but four have been pronounced in various ways. The following pronunciations are recommended:

candela — Put the accent on the second syllable and pronounce it like *dell*.

joule —Pronounce it to rhyme with *pool*.

pascal — The preferred pronunciation rhymes with *rascal*. An acceptable second choice puts the accent on the second syllable.

siemens — Pronounce it like *seamen's*.

For pronunciation of unit prefixes, see table on page 12, especially footnote 1.

WEIGHT, MASS, AND FORCE

Considerable confusion exists in the use of the term weight as a quantity to mean either force or mass. In commercial and everyday use, the term weight nearly always means mass; thus, when one speaks of a person's weight, the quantity referred to is mass. This nontechnical use of the term weight in everyday life will probably persist. On the other hand, in physics, weight has usually meant the force of gravity. American dictionaries define weight as either the heaviness or the mass of an object.

To avoid the ambiguity of this dual use of the term weight, it is better to avoid its use in technical practice except under circumstances in which its meaning is completely clear. Instead, the terms mass and force of gravity (or gravity force) should be used, together with the appropriate units *kilogram* (kg) and *newton* (N), respectively.

A floor load rating or a capacity rating of a vehicle (or other load-supporting machine) is intended to define the mass that can be supported safely. Hence the rating should be expressed in kilograms rather than newtons. Similarly, the area density of floor loading should be expressed in kilograms per square meter rather than pascals (newtons per square meter). In engineering calculations involving structures, vehicles, or machines on the surface of the earth, the mass in kilograms is multiplied by 9.8 to obtain the approximate force of gravity in newtons. (The force of gravity acting on a mass of 1 kg varies from about 9.77 N to 9.83 N in various parts of the world.)

TYPEWRITING RECOMMENDATIONS

SUPERSCRIPTS

The question of how superscripts should be typed on a machine with a conventional keyboard is raised by Rule 7.

With an ordinary keyboard, figures and the minus sign can be raised to the superscript position by rolling the platen half a space before typing the figure. When this is done, the figure may tend to run into the text in the line above. This interference can be avoided by using care. In printing, interference is avoided by making superscripts of smaller type than the body of the text, but this procedure would be available with a typewriter only by modifying the keyboard.

SPECIAL CHARACTERS

For technical work it is useful to have a number of Greek letters available on the typewriter. If all metric symbols for units are to be properly typed, a key with the Greek lowercase μ (pronounced "mew" like a cat, not "moo" like a cow) is necessary, since this is the symbol for "micro", meaning one millionth. This can be approximated on a conventional machine by using a lowercase u, and adding the tail by hand (µ). While this is not very elegant, it is unmistakable, and those planning typed work can choose whether to use this procedure, or use the transfer characters known as press-ons, or spell the unit name out in full.

For units of electricity, the Greek uppercase omega (Ω) for ohm will also be useful, but when it is not available, the word "ohm" can be spelled out.

It should be noted that, except for the more extensive use of the Greek μ for "micro", and Ω for ohm, the change to metric measurements will cause no more difficulty than has already existed in typing of material containing quantitative information.

THE LETTER "EL" FOR LITER

On most typewriters there is no difference at all between the lowercase "el" (the recognized symbol for liter) and the figure "one". Accordingly, **it is preferable to spell the word in full**—for example, 24 liters/100 km (fuel consumption). However there is no problem with ml (milliliter).

In a situation where you cannot afford the space for the full word liter, and you do not want to use a lowercase symbol "l" because it looks like a numeral "1", a capital "L" (although not yet recognized in any American National Standard or International Standard) is preferable to a script ℓ, because it does not preempt a typewriter character.

A better symbol than "l" is badly needed, and is now being considered by national and international standards bodies.

TYPEWRITER MODIFICATION

Where frequently used, the following symbols should be included on typewriters:

superscripts ^{2} and ^{3} for squared and cubed; Greek μ for micro; ° for degree; · for a product dot, for symbols derived as a product; Greek Ω for ohm.

The positions in which these are placed on the keyboard will depend partly on the design of the keyboard, and partly on the frequency with which the new characters are likely to be needed. A special type-ball is available for some typewriters, which contains all the superscripts, μ, Ω, and many other characters used in technical reports. Some typewriters have replaceable character keys.

LONGHAND

To assure legibility of the symbols m, n, and μ, it is recommended that these three symbols be written to resemble printing. For example, write nm —not *nm*. The symbol μ should have a long distinct tail, μ .

SHORTHAND

Stenographers will find that the SI symbols are generally quicker to write than the shorthand forms for the unit names.

SI UNIT PREFIXES

Multiplication Factor	Prefix	Symbol	Pronunciation (USA) (1)	Meaning (in USA)	In Other Countries
1 000 000 000 000 000 000 = 10^{18}	exa (2)	E	ex' a (a as in about)	One quintillion times (3)	trillion
1 000 000 000 000 000 = 10^{15}	peta (2)	P	as in petal	One quadrillion times (3)	thousand billion
1 000 000 000 000 = 10^{12}	tera	T	as in terrace	One trillion times (3)	billion
1 000 000 000 = 10^{9}	giga	G	jig' a (a as in about)	One billion times (3)	milliard
1 000 000 = 10^{6}	mega	M	as in megaphone	One million times	
1 000 = 10^{3}	kilo	k	as in kilowatt	One thousand times	
100 = 10^{2}	hecto	h (4)	heck' toe	One hundred times	
10 = 10	deka	da (4)	deck' a (a as in about)	Ten times	
0.1 = 10^{-1}	deci	d (4)	as in decimal	One tenth of	
0.01 = 10^{-2}	centi	c (4)	as in sentiment	One hundredth of	
0.001 = 10^{-3}	milli	m	as in military	One thousandth of	
0.000 001 = 10^{-6}	micro	μ (5)	as in microphone	One millionth of	
0.000 000 001 = 10^{-9}	nano	n	nan' oh (an as in ant)	One billionth of (3)	milliardth
0.000 000 000 001 = 10^{-12}	pico	p	peek' oh	One trillionth of (3)	billionth
0.000 000 000 000 001 = 10^{-15}	femto	f	fem' toe (fem as in feminine)	One quadrillionth of (3)	thousand billionth
0.000 000 000 000 000 001 = 10^{-18}	atto	a	as in anatomy	One quintillionth of (3)	trillionth

(1) The first syllable of every prefix is accented to assure that the prefix will retain its identity. Therefore, the preferred pronunciation of kilometer places the accent on the first syllable, not the second.

(2) Approved by the 15th General Conference of Weights and Measures (CGPM), May-June 1975.

(3) These terms should be avoided in technical writing because the denominations above one million are different in most other countries, as indicated in the last column.

(4) While hecto, deka, deci, and centi are SI prefixes, their use should generally be avoided except for the SI unit-multiples for area and volume and nontechnical use of centimeter, as for body and clothing measurement. The prefix hecto should be avoided also because the longhand symbol h may be confused with k.

(5) Although Rule 1 prescribes upright type, the sloping form is sometimes tolerated in the USA for the Greek letter μ because of the scarcity of the upright style.

INDEX

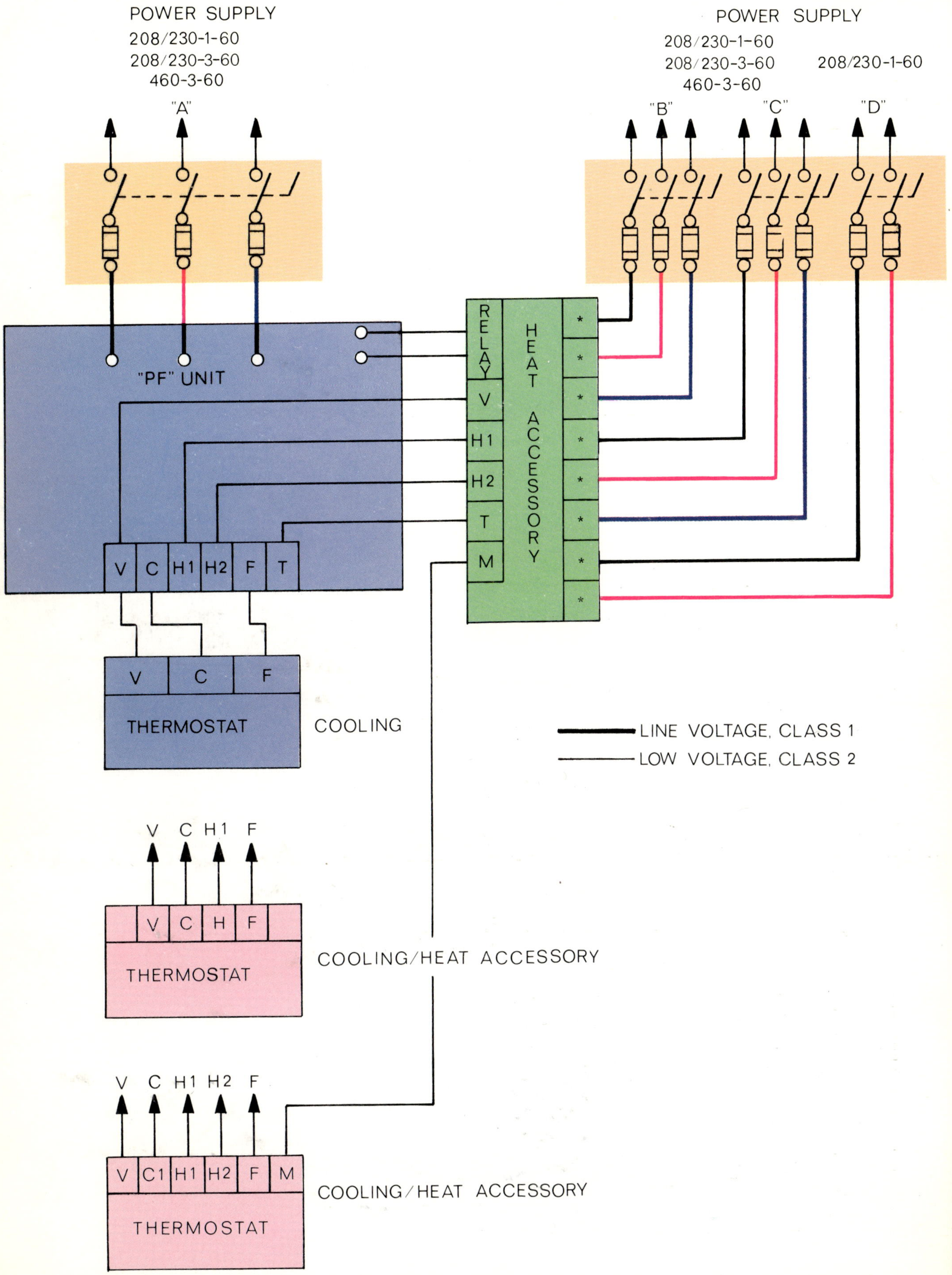
POWER SUPPLY
208/230-1-60
208/230-3-60
460-3-60
"A"
POWER SUPPLY
208/230-1-60
208/230-3-60
460-3-60
208/230-1-60
"B"
"C"
"D"
"PF" UNIT
RELAY
V
H1
H2
T
M
HEAT ACCESSORY
V C H1 H2 F T
V C F
THERMOSTAT
COOLING
LINE VOLTAGE, CLASS 1
LOW VOLTAGE, CLASS 2
V C H1 F
V C H F
THERMOSTAT
COOLING/HEAT ACCESSORY
V C H1 H2 F
V C1 H1 H2 F M
THERMOSTAT
COOLING/HEAT ACCESSORY